Y0-BSD-714

Modern Techniques in Applied Molecular Spectroscopy

TECHNIQUES IN ANALYTICAL CHEMISTRY SERIES

BAKER · CAPILLARY ELECTROPHORESIS

CUNNINGHAM · INTRODUCTION TO BIOANALYTICAL SENSORS

LACOURSE · PULSED ELECTROCHEMICAL DETECTORS IN HPLC

MCNAIR AND MILLER · BASIC GAS CHROMATOGRAPHY

MIRABELLA · MODERN TECHNIQUES IN APPLIED MOLECULAR SPECTROSCOPY

STEWART and EBEL · CHEMICAL MEASUREMENTS IN BIOLOGICAL SYSTEMS

TAYLOR · SUPERCRITICAL FLUID EXTRACTION

Modern Techniques in Applied Molecular Spectroscopy

Edited by

FRANCIS M. MIRABELLA
Equistar Chemicals, LP

A Wiley-Interscience Publication
JOHN WILEY & SONS, INC.
New York • Chichester • Weinheim • Brisbane • Singapore • Toronto

This book is printed on acid-free paper. ∞

Published simultaneously in Canada.

Library of Congress Cataloging-in-Publication Data:

Modern techniques in applied molecular spectroscopy / edited by
Francis M. Mirabella.
p. cm. -- (Techniques in analytical chemistry series)
"A Wiley-Interscience publication."
Includes bibliographical references and index.
ISBN 0-471-12359-5 (cloth : alk. paper)
1. Molecular spectroscopy. I. Mirabella, Francis M. II. Series.
QD96.M65M63 1998
543'.0858--dc21

Printed in the United States of America.

10 9 8 7 6 5 4 3 2 1

Contents

Preface

"To the making of books there is no end, and much devotion to them is wearisome to the flesh" Ecclesiastes 12:12.

King Solomon wisely observed the proliferation of books and the potential burden of assimilating the contained information. So why should we want a book on the techniques used in molecular spectroscopy? The rapid improvements made during the 70's, 80's and 90's in the instrumentation employed to obtain molecular spectra has been truly remarkable. The revolutionary improvements in the instrument hardware, per se, along with the advent of personal computer operating systems with graphical interface-based software, and at ever decreasing prices, has placed this type of instrumentation into the hands of a much wider user group with a much wider agenda of applications. This has primarily resulted in a shift of application of these devices from research instruments to routine tools. It has further had the effect of widening the user base from a highly trained cadre of scientists and their technicians, familiar with the chemistry and physics involved with the operation of the instrumentation and the interpretation of the data obtained, to a much larger group who are primarily interested in the application of the instrumentation to their activities. This latter group may be limited in time, background and interest, as to the instrument operation and data interpretation, and may rather require reliable control of their process, rapid resolution of a problem, or the like.

Another aspect of these developments is that instrument operation has been made so convenient that it is typically not a significant problematic element to many, or probably most, of the users. The problematic element, instead, is the presentation of the user's sample or system to the instrument in order to analyze it. What is meant by, the "presentation of the sample or system", is the particular technique used to interact the instrument's energy source with the material to be analyzed, so that it can absorb or emit energy that can be detected by the instrument's detector. Further, the data obtained in the analysis may be required to be treated and interpreted differently, according to the technique used. For these reasons, a knowledge or source of knowledge is required in order to intelligently choose a technique for a particular application and to correctly interpret the analytical data obtained.

This book contains a series of chapters each comprised of a comprehensive discussion of one of the techniques of modern molecular spectroscopy. It is hoped that this book will be a useful source of information for those interested in applying modern molecular spectroscopy techniques.

FRANCIS M. MIRABELLA

Contributors

S. J. Bajic
Ames Laboratory
Iowa State University
Ames, IA 50011

Jonathan P. Blitz
Department of Chemistry
Eastern Illinois University
Charleston, IL 61920

Chris W. Brown
University of Rhode Island
Kingston, RI 02881

Marilyn D. Duerst
Department of Chemistry
University of Wisconsin
River Falls, WI 54022

Richard W. Duerst
Abbott Laboratories
Abbott Park, IL

F. S. Franke
University of New Mexico
Alberquerque, NM 87131

R. W. Jones
Ames Laboratory
Iowa State University
Ames, IA 50011

Jack E. Katon
Miami University
Oxford, OH 45056

Brian D. Lamp
Ames Laboratory
Iowa State University
Ames, IA 50011

Robert J. Lipert
Ames Laboratory
Iowa State University
Ames, IA 50011

J. F. McClelland
Ames Laboratory
Iowa State University
Ames, IA 50011

Francis M. Mirabella
Equistar Chemicals, LP
Morris, IL 60450

T. M. Niemczyk
University of New Mexico
Albequerque, NM 87131

Marc D. Porter
Ames Laboratory
Iowa State University
Ames, IA 50011

L. M. Seaverson
Ames Laboratory
Iowa State University
Ames, IA 50011

Andre J. Sommer
Department of Chemistry
Miami University
Oxford, OH 45056

William L. Stebbings
3M Co.
St. Paul, MN 55144

S. Zhang
University of New Mexico
Alberquerque, NM 87131

Modern Techniques in Applied Molecular Spectroscopy

1 Introduction

Francis M. Mirabella

The past several decades have been remarkable due to the great ease of use of the instrumentation employed for the analysis of a wide variety of systems by molecular spectroscopy. In a recent survey [1] of U.S. industry it was found that the majority of molecular spectroscopy is done in chemical companies (organic, inorganic, and petrochemicals), followed by pharmaceuticals, plastics, polymers and rubbers, metals, foods and beverages, paints, dyes, colorants and inks, medical, scientific and photographic, household and personal care products, wood, paper and pulp, automotive, electronics, semiconductors, contract testing, cement, and office products. This vast array of industrial applications, along with similar applications in academic research, present a particularly diverse collection of specific problems in need of solution in order to characterize the systems involved.

It was mentioned above that the primary molecular spectroscopy instrumentation—the spectrometers—has become extremely easy to operate. This is especially due to user-friendly software that makes operation

Modern Techniques in Applied Molecular Spectroscopy, Edited by Francis M. Mirabella.
Techniques in Analytical Chemistry Series.
ISBN 0-471-12359-5

simple and intuitive, and also to well-designed hardware that is easy to operate and service. The attendant benefits are many and obvious; speed, accuracy, precision, convenience, unique capabilities, unique information, and so on. However, there are attendant detriments. These detriments primarily revolve around the use of the instrumentation by untrained or undertrained operators, which occurs because the instrumentation is so easy to operate.

Anyone familiar with graphical-interface software can immediately begin operating the instrumentation with little or no training. The software is so intuitive that it can usually be used immediately, and the instrumentation so well-designed for convenience that samples can be readily mounted into the equipment. This may cause the appearance that training is largely unnecessary. The data can then be generated easily and will be most likely in a form already modified by the software. All these conveniences may lead to serious errors in the interpretation and application of the data, and this may be done unknowlingly by the operator.

Some readers who have considerable experience in working with the particular systems discussed above may consider this assessment to be condescending. However, the point is that the instrumentation and computer software have undergone such rapid evolutionary modifications that there remain hidden many aspects of their mode of operation, especially to those not trained in molecular spectroscopy. Without background training in molecular spectroscopy, the knowledge required to interpret and apply the data, which can be so easily generated with the current instrumentation, will be lacking. This can lead to the generation of problems, rather than their solution, due to the availability and modest cost of the instrumentation. This current situation has been discussed by a panel of experts in the field in a recent article [2]. This book addresses issues that are twofold: first, general issues involved in applying the techniques of molecular spectroscopy to a wide variety of systems, and second, issues relative to the application of current instrumentation to specific systems in order to extract specialized information about these systems.

To avoid these detrimental aspects of the application of molecular spectroscopy, it would be beneficial, in fact required, to acquire the background knowledge necessary to guide the intelligent interpretation and application of the data generated. This might be imparted by a knowledgeable spectroscopist or by reading the many fine publications available. An excellent introduction to the origin and character of molecular spectra is available in the book by Colthup, Daly, and Wiberley [3]. A basic introduction to the measurement and applications of molecular spectra is available in the book by George and McIntyre [4]. There are certainly many other excellent sources of this background knowledge that can be found in a technical library.

Although molecular spectroscopy is used to solve problems in a majority of the cases noted above, the information required to characterize this array of systems cannot be obtained from just one type of measurement. The wide variety of systems encountered in industrial and academic applications of molecular spectroscopy require different techniques to extract the appropriate information for each type of system.

These techniques are made possible by a variety of accessories that are typically fit into the spectrometer in order to satisfy particular analytical requirements. These requirements determine the type of technique needed to extract the desired characterization data for each type of system. One basic requirement is the analysis of many systems without modification, namely without modification by preparation of any kind. Another common requirement is the analysis of specific parts of a system. For example, the bulk or surface of a system may be desired to be analyzed separately. Some systems require microscopic analysis. In some cases the spectrometer and system to be analyzed may have to be remote from one another.

The fundamental areas of investigation of molecular spectroscopy, namely the origin and character of spectra, have been relatively less actively pursued in recent decades. In the recent past there was a major period of optimization of spectrometers, which has resulted in the highly user-friendly lines of spectrometers available in the 1990s. Since then the major focus has been on the development, refinement, automation, and routinization of specialized techniques. These developments are realized in the current lines of accessories and associated software available for emplacement into the spectrometers to practice these specialized techniques.

Another advantage of the modern spectrometers available as the year 2000 approaches is the enormous improvement made by these instruments for performing specialized experimental techniques. In the not too distant past some techniques were extremely difficult or impossible to practice because of instrument limitations. The limitations included low energy throughput, low detector sensitivity, high detector noise level, poor instrument stability, slow computational speed of computers, and low mass storage capability of computers, for example. These limitations inhibited spectroscopy on very small masses or areal sizes and very weak absorbers. The sequential co-adding of spectra was limited by instrument stability. The processing of data for purposes, such as imaging, was limited by low computational speed and mass storage limitations. Currently many of these applications have become routine due to the revolutionary improvements in spectrometers. This is especially true of Fourier transform infrared and Raman spectrometers which have virtually eliminated many of these limitations. The effect of improved instrument performance on the development and applications of the techniques of molecular spectroscopy are discussed in some detail for each technique in each chapter of this book.

The primary purpose of this book is to describe these spectroscopy techniques in detail and to explain how to use the accessories in order to collect the specialized data afforded by the techniques. Each chapter is therefore dedicated to one of the important techniques used to perform molecular spectroscopy on a wide variety of system types. Each chapter includes the important area of data handling and analysis. The molecular spectroscopy techniques described are in the vast majority drawn from cases in which the mid-infrared region of the spectrum was applied to vibrational spectroscopy. This is simply due to the fact that the overwhelming majority of published work in molecular spectroscopy was done using the mid-IR. Additional spectral regions and spectroscopic techniques are discussed in a minority of cases, including UV-visible, near- and far-infrared spectroscopy, and Raman spectroscopy. Other spectroscopies that may be classified as molecular spectroscopy, such as nuclear magnetic resonance spectroscopy and mass spectrometry, are not discussed in this book.

1.1 TRANSMISSION SPECTROSCOPY

The fascinating history of the development of spectroscopy is traced in Chapter 2. The observations of ancient investigators is given as a basis for the modern development of spectroscopy, as an analytical tool, which has occurred over the last two hundred years. Various modes of interaction of radiation with matter are discussed, since they form the basis of the different spectroscopic techniques employed in modern molecular spectroscopy.

The discussion is then focused on mid-infrared absorption spectroscopy. A summary of the theory of the interaction of radiation with molecules is given, and this is used to describe the characteristics of the spectra.

The development of infrared spectrometers is discussed in some detail. The capabilities and limitations of modern dispersive and Fourier transform infrared spectrometers are considered, along with the associated equipment for obtaining spectra. Sample-handling techniques are examined in detail with many useful suggestions for improving the spectra obtained. Specific operational parameters of modern instrumentation are discussed, and practical suggestions are given for setting the instrument parameters for various desired applications.

The important subject of the interpretation of spectra and the qualitative identification of molecules by spectral searching techniques are discussed. Methods and theory of quantitative analysis are considered. Several case studies are described for the qualitative and quantitative analysis by transmission infrared spectroscopy of the chemical composition and molecular orientation of several systems.

1.2 SPECULAR REFLECTANCE SPECTROSCOPY

Next to transmission, specular (or external) reflectance spectroscopy is probably the oldest technique of molecular spectroscopy to be widely applied. Specular reflectance involves the application of molecular spectroscopy to systems with "mirrorlike" finishes. The angle of incidence is equal to the angle of reflection for such systems. These systems have imperfections that are small in comparison to the wavelength of the radiation employed. Specular reflectance spectroscopy is primarily used to characterize thin films at surfaces and interfaces. The experimental requirements for these analyses are often very demanding. The quantity of material is usually exceedingly small, so modern high-sensitivity systems, such as FTIR spectrometers, have made this technique far more versatile and straightforward to apply.

Chapter 3 gives a thorough, step-by-step explanation of these requirements and a description of the available instrumentation to satisfy the requirements. A detailed explanation of the theory of specular reflection is given. Experimental studies of actual systems are given, covering both highly reflecting and weakly reflecting substrates. Practical examples are described, demonstrating the application of the technique to specific systems and explaining how the data can be obtained and interpreted. The use of the technique for compositional and molecular orientation characterizations is described. Applications to monomolecular organic films, electrochemical studies at interfaces, and inorganic thin films are given.

1.3 ATTENUATED TOTAL REFLECTANCE SPECTROSCOPY

The analysis of near-surface layers has become extremely popular in applications to many types of systems. These involve surface chemical composition, physical structure, and molecular orientation characterization in applications such as chemical and physical modification of surfaces, adhesion, printing, and surface crystallinity, compatibility of agents such as blood with materials, monitoring of reactions, and so on. Applications of attenuated total reflection (ATR) spectroscopy for these types of analyses have increased, and continue to increase, at a tremendous rate.

The enormous popularity of ATR is mainly because it is so easy to perform, especially with the modern equipment now available. ATR is done by reflecting the incident radiation inside a high refractive index crystal while pressing the sample against the crystal surface. It is much less demanding than specular reflectance. The wide acceptance of ATR has been aided by the ease of use of the new ATR accessories, the great improvement in its sensitivity due to the use of modern spectrometers, and the improved quantitative reproducibility of the technique due to the use of more soundly based methods. It is useful for characterizing near-surface phenomena, but it is not truly surface sensitive. The layers observed in ATR spectroscopy are on the order of a few tenths to a few micrometers in thickness.

Chapter 4 describes in detail the accessories used to perform ATR spectroscopy, practical descriptions of how the experiments are done, the character of the data obtained, and the theory used to interpret the data. Many tables are presented that will be of use to the practitioner. The widespread use of horizontally mounted and microsampling accessories, which have become very popular due to their ease of use, is described.

Specific examples of the application of ATR are given, including chemical modification of polymer surfaces, concentration depth-profiling of polymer surfaces, quantitative determination of diffusion coefficients, and quantitative determination of chemical composition and molecular orientation of polymer surfaces.

1.4 DIFFUSE REFLECTANCE SPECTROSCOPY

For systems that are in the form of finely ground powders, the diffuse reflectance technique is especially useful. The reflection of incident radiation from powders will of course tend to be diffusely scattered in all directions. The devices used to collect diffuse reflectance spectra are discussed in Chapter 5, along with recommendations of the advantages and disadvantages of various types of accessories. This is a demanding technique, and the difficulties are described in detail, along with solutions to typical problems. The signal intensity obtained in diffuse reflectance tends to be very small and the strategies to overcome this limitation are considered in detail.

The theoretical basis of diffuse reflectance is considered in detail, and this basis is used to explain the correct interpretation of data obtained from this technique. Examples of diffuse reflectance analyses in the near-infrared spectral region are drawn from the biomedical and pharmaceutical industries. Applications in the mid-IR region for diffuse reflectance as a chromatographic detector, for surface science studies of particles, and for *insitu* and *exsitu* catalyst studies are discussed.

Although this technique is not easy to perform, it has undergone explosive growth, especially due to the availability of FTIR spectrometers with their high-energy throughput and sensitivity which have eased the complications of performing it.

1.5 PHOTOACOUSTIC SPECTROSCOPY

Photoacoustic spectroscopy (PAS) has been successfully applied only during recent decades. It became a technique of choice and value for a variety of applications in molecular spectroscopy during the 1980s, due to improvements in the PAS accessories and also to the spectrometers. Its major virtues include these features: nondestructive, noncontact, applicable to microsamples and macrosamples, insensitive to surface morphology, spectral range from UV to far-IR and others.

The mechanisms that permit the collection of molecular spectra by the PAS technique are described in detail in Chapter 6. The ramifications of these mechanisms are likewise described, especially the depth-profiling information available by this technique.

The PAS technique is quite complex and difficult to perform. The relatively complex instrumentation is described in detail. Each of the significant parameters are considered, and the effects of these parameters are examined in order to guide the correct interpretation of PAS data. Many useful tables and figures provide specific information for successfully applying PAS to a variety of sample types.

Special attention is given to step-scan spectrometric techniques in conjunction with PAS as a means for varying the sampling depth. Useful tables and plots are provided to aid in applying these specialized techniques.

Applications of PAS are given for qualitative and quantitative applications in plastics recycling, paper products, nuclear waste sludges, lime chemistry, polymer additive analysis and modification of polymers, and more.

1.6 INFRARED MICROSPECTROSCOPY

The application of molecular spectroscopy to specimens of microscopic dimensions was a rather obvious objective. This was especially true for the case of infrared spectroscopy. However, for a long period of time efforts to achieve this objective were largely unsuccessful from a practical standpoint. Chapter 7 describes the interesting historical development of the instrumentation that has made possible the collection of high-quality infrared spectral data on microscopic specimens. The further development of associated optical microscopy instrumentation to also permit the collection of high-quality optical images of the specimen on the same instrument used to collect the infrared spectrum is described, along with the principles of the optical microscope and the optics used to record IR spectra on microscopic specimens and the stringent requirements of the spectrometer used with the IR microscopy accessories.

The effects of the sample, which becomes part of the optical system, are described with reference to spectral quality and spatial resolution. The important issue of the diffraction of the infrared radiation by the sample or the aperture edges used in the microscope is considered, and recommendations are given for operating under the most advantageous conditions. The effects of sample refractive index and geometry and the various methods of obtaining and handling the spectral data are fully discussed.

The maximum spatial resolution in the infrared is on the order of the wavelength of light employed, so it is typically on the order of 2.5 to 15 μm across the spectral range in the mid-IR. This sets the limits of the diameter of the sample that can be analyzed spectroscopically, depending on the spectral region that is of interest.

1.7 RAMAN MICROSPECTROSCOPY

Chapter 8 gives a brief review of the Raman effect for producing molecular spectra by scattering of radiation. This instrumentation used to combine Raman spectroscopy and optical microscopy is considered in some detail, and the advantages and limitations of this combination are pointed out. The important contribution of Fourier transform Raman spectroscopy toward revolutionizing Raman microspectroscopy is discussed along with the optics involved in optimizing the Raman spectrum. Useful tables are presented that will aid the user in understanding and choosing the best optical arrangement.

Sampling requirements for solid, liquid, and gaseous samples are discussed with suggestions for optimizing the handling of each type of sample. The spatial resolution achievable is explained on the basis of the relevant optical considerations in order to help the practitioner to choose the best conditions for each sample type. Again, the important influence of the sample being part of the instrument optics, as in infrared microspectroscopy, is considered. The Raman microspectroscopy technique permits spatial resolution of about one order of magnitude better than infrared microspectroscopy, typically down to a diameter of about 1 μm, due to the use of the laser source which has much shorter wavelength than the infrared source.

The important causes, effects, and strategies to overcome sample heating problems are discussed at length. Similarly the causes, effects, and strategies to overcome problems associated with sample fluorescence and polarization effects are covered. Suggestions are given on all these important considerations that will be useful to the practitioner.

1.8 FIBER OPTICS TECHNIQUES

The use of fiber optics is becoming increasingly common in molecular spectroscopy applications. Optical fibers are convenient in applications where the environment to be monitored is desired to be left remote from the spectrometer for reasons such as hostility, inaccessibility, or multiplicity of the environment.

Chapter 9 gives a review of the developments in optical fiber applications in molecular spectroscopy. The optics are covered, along with the materials, construction, wavelength applications, spectrometer interfacing, and sampling configurations used in optical fiber applications. The roles of transmission, reflection-absorption, evanescent wave, and graded-index lenses in interacting the radiation carried in optical fibers with the sample are discussed. The advantages and disadvantages of each type of optical probe are noted for the various spectral regions.

Applications of optical fibers are drawn from examples involving pharmaceutical, water, natural gas, and polymer monitoring.

1.9 EMISSION SPECTROSCOPY

Emission spectroscopy is probably the most underutilized and underappreciated technique of modern molecular spectroscopy. Chapter 10 describes the many important applications of emission spectroscopy to rapid, nondestructive analyses. The chapter focuses on infrared emission spectroscopy.

The pertinent theory is given covering blackbody radiation, real-sample (graybody) radiation and infrared emission in thin films. Common problems related to emission spectra are stray light and band distortions. Their origins are considered, and practical examples are provided that minimize or correct for these effects. The examples given demonstrate that emission spectra can be obtained with the sample at higher or lower temperature with respect to the detector.

Most emission spectra are obtained by heating the sample with respect to the detector. Since many sample types can tolerate only limited heating, signal intensity is often very low. The stringent requirements for detecting these low-intensity signals to yield useful emission spectra were not practically fulfilled until the commercial availability of FTIR. The spectrometer requirements are described, including the accessories used for sample heating and data-handling techniques.

Examples of emission spectroscopy are drawn from applications involving process monitoring. Among the applications are gas phase studies, small molecules on surfaces, characterization of solids, atmospheric studies, planetary observations, and organic and inorganic thin films.

REFERENCES

1. M. MacRae, *Spectrosc,* **11**(2), 52 (1996).
2. M. MacRae, *Spectrosc,* **10**(8), 52 (1995).
3. N. B. Colthup, L. H. Daly, and S. E. Wiberley, *Introduction to Infrared and Raman Spectroscopy*, Academic Press, New York, 1975.
4. W. O. George and P. S. McIntyre, *Infrared Spectroscopy.* Wiley, New York, 1987.

2 Transmission Infrared Spectroscopy

Richard W. Duerst, Marilyn D. Duerst, and William L. Stebbings

2.1 HISTORY

Discovery of Infrared Radiation

The first clue that visible light possesses fascinating properties is likely to have been the ancients simply observing the rainbow. Although many explanations for these properties were conceived over thousands of years after these first observations, it was not until Sir Isaac Newton that the correct theory of the visible spectrum was developed through experiments with prisms. His first publication on light appeared in 1672 [1, p. 73]. In 1785 an American astronomer, David Rittenhouse, designed the first diffraction grating, a device made for the separation of light into its component colors [2, p. 4]. Unfortunately, he never used it for scientific investigations.

Modern Techniques in Applied Molecular Spectroscopy, Edited by Francis M. Mirabella.
Techniques in Analytical Chemistry Series.
ISBN 0-471-12359-5

In 1821, Frauenhofer independently invented a grating of sufficient quality to be able to study the solar spectrum. Over the last two centuries sophisticated instruments have been devised using prisms, gratings, and interferometers to separate, examine and utilize light for extensive scientific investigation.

The nature of light has been the subject of speculation for millennia. The idea that light consists of particles, either sent out from the eye or emitted by the object viewed, can be traced back to the Greeks. In the seventeenth century the Dutch physicist Huygens worked out a detailed theory that light consists of rapid vibrations (waves) in a medium. His perhaps more famous contemporary, Sir Isaac Newton, maintained, on the other hand, the "corpuscular theory" of light. His ideas were mainly based on the existence of shadows and the fact that light cannot penetrate solid objects. Ironically, in order to explain some properties of light, Newton had to suppose that light particles "stir up Vibrations in what they act upon," thus endowing his "light corpuscles" with some wave properties [1, p. 74]. The phenomena of light polarization and interference were studied intensely in the nineteenth century by Hooke, Young, and Fresnel, thus strengthening the wave theory of light. Fresnel, Green, MacCullagh, Cauchy, and Stokes all contributed to the development of the wave theory. Maxwell finally showed that light is an electromagnetic wave, and the old view that a medium or "aether" was needed to transport light waves finally fell by the wayside [1, pp. 99–100].

The discovery of the infrared portion of the electromagnetic spectrum in 1800 is attributed to Sir William Herschel. While investigating the heating effect of the various wavelengths in a dispersed spectrum, using sensitive thermometers as sensors, he found that heating did not cease at the red end of the visible spectrum, but rather the greatest heating effect occurred just beyond red light, which we know as the "infrared" (literally, "lower than red"). Herschel showed that this invisible radiation follows the laws of reflection and refraction, just as visible light does, through investigations of radiations emitted from fire and of the transmitting properties of glasses and liquids [3, p. 1].

In 1840, his son, Sir John Herschel, was the first to discover absorption bands when he soaked a piece of black paper in alcohol and placed the paper in the infrared region of a dispersed solar spectrum. The alcohol evaporated more rapidly from some regions of the paper. Those observations are now attributed to the presence of water vapor and carbon dioxide in earth's atmosphere, which selectively absorb some parts of the solar infrared spectrum [3, p. 1]. However, as Thomas [4] points out, John Herschel's main interest was astronomy, so he did not investigate the phenomenon further [5].

Between 1830 and 1880 the development of more sensitive detectors and expansion of understanding the theory promoted rapid advances in infrared physics. Seebeck discovered the thermoelectric effect in 1826, and Nobili invented a radiation thermocouple in 1830, a device consisting of two different metals, often antimony and bismuth. The first thermopile was created in 1833 by Melloni, and was the only detector available until Langley's invention of a bolometer in 1880. (The Greek work *βολοη* means "beam of light.")

In 1882, Abney and Festing [6] obtained infrared spectra for over 50 compounds and correlated absorption bands with the presence of certain functional groups in the molecules. Within a decade Julius [7] had obtained spectra of 20 organic compounds and noted that methyl groups absorb at characteristic wavelengths. By the turn of the century interest in the technique was rapidly growing. W. W. Coblentz was probably the most important early twentieth century contributor to the development of the infrared spectroscopy technique, having published spectral information on hundreds of organic and inorganic compounds [4].

Evolution of Instrumentation for Transmission Infrared Spectroscopy

Early twentieth-century spectoscopists had to face significant challenges, since they had to design, construct, and calibrate their own instruments. It often took several hours to obtain a single spectrum, which quite limited the applicability of the technique. Thus the technique remained rather dormant for a number of decades until just after World War II when it enjoyed a dramatic resurgence [4]. It was in the 1940s that a company founded by Richard S. Perkin and Charles W. Elmer, in collaboration with Cyanamid Corporation, built a prototype for commercially available dispersive infrared spectrometers. At about the same time, the Beckman Company, in collaboration with Shell Development Company, developed their Model IR-1 instrument. Why then? During World War II the U. S. government was interested in producing synthetic rubber by polymerizing butadiene, but the process required the analysis of C_4 hydrocarbon isomers. To that end it supported the efforts of those two industrial research laboratories to design an instrument for this type of analysis. The wartime development of the phase-sensitive detector and low-noise amplifier, together with more recent photoconductive detectors and bolometers, brought infrared spectroscopy to the status of a routine tool in astronomy, physics, atmospheric science, and chemistry.

The classical infrared spectrometers utilized either a prism or diffraction grating to selectively choose wavelengths of incident radiation to be impinged upon the sample. Since the early 1970s the Fourier transform infrared (FTIR) spectrometer has gained in popularity. Although the FTIR may seem to be a recent development, the foundations of the technique were surprisingly laid over a century ago. In 1891 and 1892, A. A. Michelson [8],

[9] published descriptions of an interferometer which he had constructed that created a mixture of constructive and destructive electromagnetic waves. In 1892 Lord Rayleigh realized that the interferogram was mathematically related to the spectrum by the Fourier transformation (FT). Rubens and Wood recorded the first true interferogram in 1911.

The FT concept did not become a practical technique for decades because of the overwhelming number of computations required. In 1949 Peter Fellgett first transformed an interferogram into its corresponding infrared spectrum. Computers in the 1950s were very slow and cumbersome, rarely being dedicated to a single instrument as they are today. Processing of the tape of information could take hours or days. It was not until the development of fast digital computers that the required transformation computations became practical and the FTIR technique became economical [10].

By 1965, Cooley and Tukey [11] had developed an algorithm that sped up the Fourier transform computation. Forman [12] recognized its application to interferometry, and soon a few commercial instruments appeared on the market for far-IR work. By 1969 a low-resolution mid-IR interferometer with a dedicated computer became available, but its price was much higher than the typical dispersive instrument. Today's instruments with dedicated computers with memories large enough to accommodate data-processing programs are now able to transform interferograms into spectra in a few seconds or faster, with less than one wave-number resolution. Inexpensive, powerful software has been developed for a wide variety of applications [13], and the tunable laser as a new infrared source has vastly improved the resolution obtainable by infrared spectroscopy.

Niche Applications

Over four thousand books as well as tens of thousands of articles in professional journals are in print on the subject of infrared radiation and infrared spectroscopy, indicating the extremely wide applicability, utility, and interest in this analytical technique.

Many of the niche applications for infrared spectroscopy are covered in more depth in other chapters in this book. Very important, for example, is the ATR-IR technique for obtaining spectra of samples pressed onto the surface of an infrared-transparent crystal such as Ge or ZnSe, developed about three decades ago by Fahrenfort and significantly advanced by Crawford.

The FTIR for remote measurements and environmental emissions in the field has been a high-priority recent advancement [13]. Infrared microspectroscopy is an important niche application for defect analysis. Depth-profile analysis of layered products such as tapes [14] is an important application, and utilization of infrared techniques has become important in some biological studies [15]. Near-field and 2-D photoacoustic spectroscopy are other developing fields of infrared spectroscopy. The application of infrared spectroscopy to on-line and near-line process measurements continues to grow in interest and practicality. As an example, a near-IR analyzer for process analysis at petroleum refineries has been available for a number of years [15].

2.2 PRINCIPLES AND THEORY

Molecular Interactions with Electromagnetic Radiation

Light waves are characterized by frequency (ν), wavelength (λ), amplitude, polarization, phase and direction of propagation [16] (see the Glossary at the end of this chapter for definitions, description, and appropriate units). Matter, on the other hand, can be classified into two main categories: classical dielectrics, which have localized, bound electrons that are able to interact with photons, and conductors, which have quite different properties due to the presence of free electrons [17, pp. 33-11 to 33-13] [18, p. 622], [19, p. 38].

If infrared light, or any type of electromagnetic radiation, could simply pass through any substance and be totally unchanged and unaffected, it would be of little theoretical interest or analytical value. But this is not the case. Photons of light striking matter may be *absorbed* by the molecules or atoms and changed into thermal energy. Or, the incident energy may excite "oscillating dipoles" [17, pp. 32-1] in that matter, which reemit energy manifested as *transmission, refraction, or reflection.* These three phenomena determine the direction in which the reemitted radiation emerges. Longer delays in the reemission result in phenomena such as fluorescence and phosphorescence. If the matter is discontinuous with respect to the refractive index, diffraction, scattering or dispersion may occur [20, p. 7] [21]. Photons may also interact with each other, resulting in interference.

Absorption

Absorption refers to matter's converting the incident radiation into increased molecular motion (including increased electronic energy). When atoms or molecules absorb light, the incoming energy excites a quantized structure to a higher-energy level. The type of excitation depends on the wavelength of the photon; infrared photons are able to excite molecular vibrations and rotations. Since the wavelengths of infrared absorption bands are characteristic of specific types of molecular vibrations and rotations, infrared spectroscopy finds its primary utility in identification of organic, organometallic and inorganic compounds, which will be dealt with in some detail below.

Transmission

When electromagnetic waves pass through matter without changing direction (are transmitted), it may appear that nothing whatsoever has happened to that radiation. However, since transmission of radiation through matter actually involves numerous absorptions and reemissions, an apparent decrease in velocity through the entire sample is observed.

Refraction

Refraction is really a variant of transmission; it refers to the passing of light through matter with a change in direction, which occurs when the light impinges at nonnormal angles to the surface. The direction of the beam is

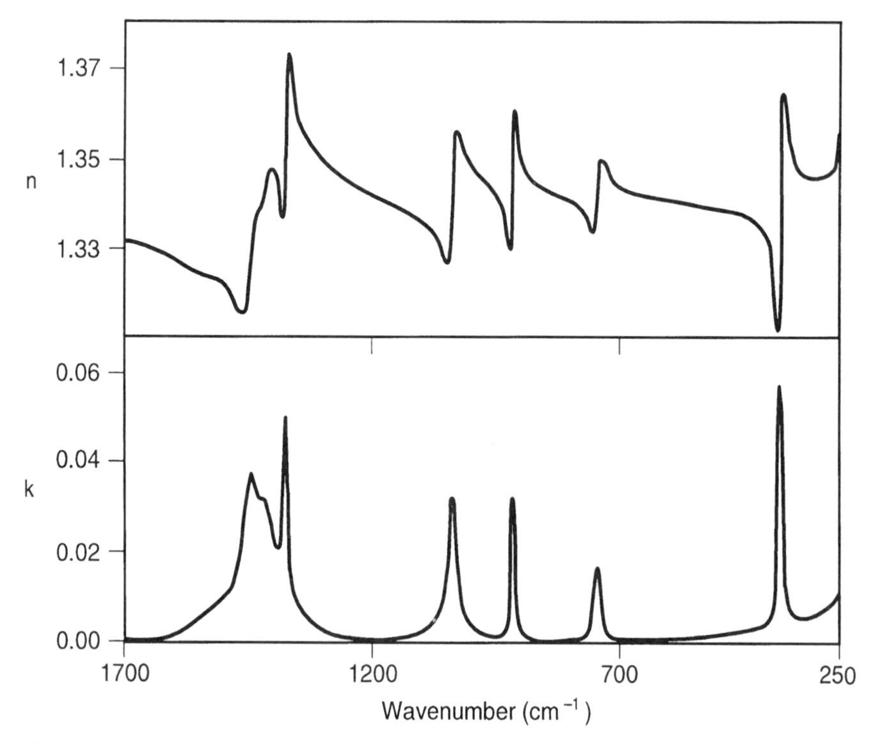

Fig. 2.1 Dispersion and spectral curves as a function of wavenumber. For the example of acetonitrile, the refractive index curve (*upper curve*) can be compared to the absorption curve (*lower curve*), over the range from 250 to 1700 cm^{-1}.

changed according to the law of refraction (Snell's law, see [22, pp. 12f]). The amount of refraction depends on the wavelength of the radiation, the differences in velocity of the photon in the "incident" medium and in the sample, the polarization of the light, and the angle of incidence [17, p. 14] [23].

Since the speed of light in matter depends on the type of matter, a material may be characterized by its refractive indexes for different wavelengths (the refractive index is defined as the ratio of the velocity of a photon of a particular energy in a vacuum to its velocity in that matter). A well-known example of how the refractive index varies for different wavelengths is "white" light ($\sim 350 < \lambda < \sim 700$ nm) which, if allowed to enter a glass prism at an angle other than the normal will emerge from the prism as a "rainbow" (a visible spectrum). This phenomenon of spatial separation of polychromatic radiation into its components by use of prisms is known as *dispersion* [18].

A graph of the relationship between refractive index and wavelength is called a dispersion curve (see Fig. 2.1 for an example using acetonitrile, shown together with its spectral curve for the same wavenumber range). Note that dispersion curves and absorption spectra offer the same information via the Kramers-Kronig relationship. The refractive index and absorption are known as *conjugate pairs* and together serve as an example of the common relationship between an *elastic* response and an *inelastic* response of an oscillating system [25, pp. 7–11] [26, p. 33] [27, pp. 108f].

Reflection

Reflection is a change in the direction of the impinging beam back into the medium from which it came. The angle follows the law of reflection (the angle of incidence equals the angle of reflection) with respect to the direction of the beam. The intensity follows Fresnel's law for dielectrics and Drude's formula for conductors [28, p. 399] [29, p. 502]. Some amount of phase change usually accompanies reflection [30, p. 830]. Reflection is most pronounced on macro-sized, homogeneous surfaces that are extremely smooth, such as water surfaces and metallic surfaces, especially silvered or aluminized mirrors. In this case, where any minute irregularities of the surface are small compared with the wavelength of the incident light, the reflection is called *specular*. At the opposite extreme, if the surface of the solid is rough (the magnitude of imperfections or "bumps" being of the order of magnitude of λ of the incident radiation), the incident light is scattered in all directions. Much of the scientific literature describes such a surface as a *Lambertian radiator*; and categorizes the phenomenon as *diffuse reflection* [31, p. 22] [18, p. 182] [32, p. 42], which involves diffraction (see below) and interference.

Diffraction

Diffraction refers to the spreading out (a directional change) of a beam of photons when it passes by a lateral boundary between regions of matter with different refractive indexes. Diffraction phenomena are most easily observed when light passes between tiny particles or through tiny holes or slits that have dimensions comparable to the magnitude of the wavelength of the photon [33]. The boundary particles act as new sources for the waves that propagate away from the aperture in all directions [34]. Diffraction may be observed for both monochromatic and polychromatic radiation. An example of the latter is the rainbow observed from a laser disc, which is due both to diffraction and to constructive and destructive interference.

Scattering

Scattering is primarily a diffraction phenomenon. As the name implies, scattering is a process by which incident energy is redistributed into many angular directions of propagation [35] [36]. The amount of scattering is determined by the size of the particles, the ratio of the index of refraction of the particles to that of the medium, the wavelength, angle of incidence, and polarization. Everyday examples where scattering is observed include the apparent color of clouds, minute dust particles seen in a light beam coming through a window (the Tyndall effect), and eyecolor due to scattering of light from small, widely separated particles in the iris.

On the topic of *scattering*, it is convenient to consider three cases based on the ratio of the diameter (d) of the particles of matter to the wavelength of the incident radiation (λ). The spatial intensity distribution of scattered radiation varies with particle diameter, and it is complicated due to constructive and destructive interference in turn due to different optical path lengths and/or changes in phase.

In the first case, for extremely small particles (when the diameter of the particles is much smaller than the wavelength ($d < 0.1\lambda$)), the phenomenon is defined as *Rayleigh scattering*, which arises from fluctuations of the refractive index. The spatial energy distribution is shown in Fig. 2.2*a* [37] [38]. Rayleigh scattering accounts for the blueness of the sky. Raman effects are also included in this category [see 3, p. 281, for a more extensive introduction]. A mathematical treatment of this topic can be found in Kruse, McLaughlin and McQuistan [32, pp. 107f], Chantry [3, pp. 280f], Feynman [17], and Partington [29, p. 247].

The second case, when the particle size is within an order of magnitude of the incident wavelength (when $0.1\lambda < d < 10\lambda$), is called *Mie scattering* [20, p. 205] [39]. An example of the spatial energy distribution is shown in Fig. 2.2*b*. Partington [29, p. 249] describes the distribution as having a

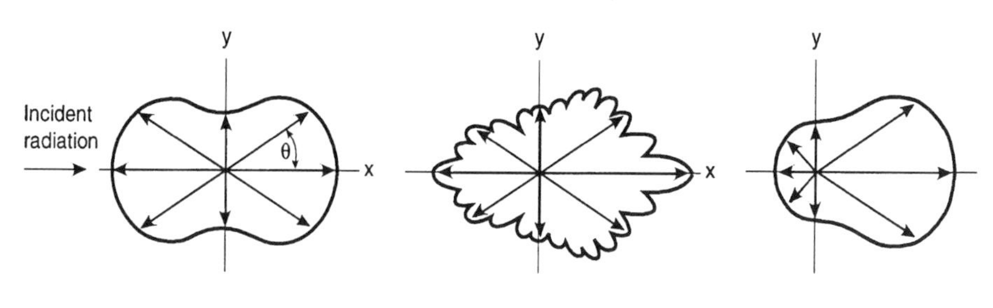

Fig. 2.2 Spatial energy distribution for three sizes of particles. Shown on a logarithmic scale, three sizes of particles experience different scattering patterns [38, pp. 25–26] [37]. (*a*) Small particle scattering, (*b*) Mie scattering, and (*c*) large particle scattering.

rosette shape. To the spectroscopist, the importance of this dependence of scattering effects on the size of the particles present may be related to the patience one shows in grinding the sample into pellets for analysis. Particles in the range from 1 to 20 μm (when $d = 0.4\lambda$ [28, p. 415]) are to be avoided, since they exhibit the greatest scattering and have the most detrimental effect on the appearance of the spectrum. On the other hand, Mie scattering theory is often used for the analysis of atmospheric transmission IR experiments [15].

The third case, when the particles are much larger than the wavelength ($d > 10\lambda$), is sometimes referred to as *nonselective scattering*. Figure 2.2c shows that in this case a higher percentage of radiation is being scattered in the forward direction and a lower percentage scattered backward. This phenomenon may be primarily reflectance, but it could have significant amounts of radiation involved in diffraction at the edges of the particles, or even refraction. Generally, there is little or no wavelength dependence [3, pp. 280–281] [20, p. 206].

Molecular Vibration-Rotation Theory

Molecular Energies

Let's examine the absorption process in greater depth. The total internal energy E of any molecule is a sum of its electronic energy E_e, its vibrational energy E_v, its rotational energy E_r, and its translational energy E_t. Depending on the physical state of the molecule, the amount of each contribution can vary tremendously. For example, for a small, gaseous molecule the primary contributor to that molecule's internal energy is its energy of motion E_t, whereas for a molecule in a solid crystal, E_t is a minor contributor.

When a molecule absorbs electromagnetic radiation, there is an increase in its internal energy. According to quantum theory only radiation of specific narrow ranges of energies can be absorbed by a molecule, raising it to an excited state. This increase in energy is equal to the difference in energy between the ground and the excited state (whether electronic, vibrational, rotational, or translational). This energy difference is proportional to the frequency v and is equal to hv, according to Planck's equation ($h = 6.63 \times 10^{-34}\,\mathrm{J \cdot s}$), for the interacting photon of radiation.

Infrared radiation has sufficient energy to cause transitions among the vibrational, translational, and rotational energy levels of a molecule. For a molecule to change its vibrational energy, a changing dipole moment must be present as the molecule moves through its equilibrium position, whereas for a molecule to change its rotational energy, a permanent dipole moment must exist in that molecule. Translational transitions are too close together to be analyzed by the currently available technology. Thus infrared spectroscopy is mainly a study of transitions within vibrational and rotational energy levels due to the absorption of infrared radiation.

Allowed motions of atoms within these vibrational levels (*normal vibrational modes*) are described as simple harmonic oscillations, for which the atoms all reach the extremes of the displacement at the same time (i.e., they vibrate at the same frequency and phase) [40]. Note that these motions are also constrained by the Eckart (Sayvetz) conditions which ensure that no distortion will constitute either a translation or rotation of the molecule [41, p. 17] [42, pp. 22–29].

The positions of absorption peaks in an infrared spectrum are determined by the forces between the atoms involved in the vibrational motion; in other words, by the "strength" of the vibration and the mass, and are quantitatively defined by force constants via

$$F = 4\pi^2 v^2 u \tag{2.1}$$

with u related to the masses of the atoms involved [43, p. 11] [25, p. 4] [27, p. 4]. See [44, p. 77] for some examples of force constants.

Peak intensity γ is related to the magnitude of the dipole moment change $\Delta\mu$ [45, p. 238] via

$$\gamma \simeq \left(\frac{\Delta\mu}{\Delta\xi}\right)^2. \tag{2.2}$$

Thus the intensity is actually proportional to the square of the dipole moment change ($\Delta\mu$) with respect to a change in a physical dimension ($\Delta\xi$) of the molecule that is a mixture of bond distances and bond angle changes.

Peak intensities may be quantitatively specified in terms of an absorption cross-section, which is given in units of cm^2/molecule or per mole [46] [47] [48] [49] [50] [51, p. 57] [52].

Degrees of Freedom

The distributions of energies of molecules due to vibrational, rotational and translational motions are commonly referred to as degrees of freedom. When a molecule is irradiated in a spectrometer with a whole range of infrared frequencies, the molecule is capable of absorbing energy at certain specific frequencies related to these degrees of freedom [43, p. 12].

For linear or nonlinear molecules containing N atoms there are $3N$ degrees of freedom. For nonlinear molecules there are three each for translation and rotation and the remaining ($3N$-6) for vibration [53, pp. 2–3] [41, p. 17]; that is, there are $3N$-6 normal vibrational modes. For a three-atom nonlinear molecule such as water, for example, Fig. 2.3 illustrates the nine degrees of freedom. For linear molecules there are three degrees of translational freedom but only two degrees of rotational freedom (the molecule, for all practical purposes, is unable to access the energy levels related to rotation around the bond axis) and thus exhibits $3N$-5 vibrational degrees of freedom.

Modes of vibration are classified as stretching modes or deformation modes. Stretching modes are described as changes in bond lengths;

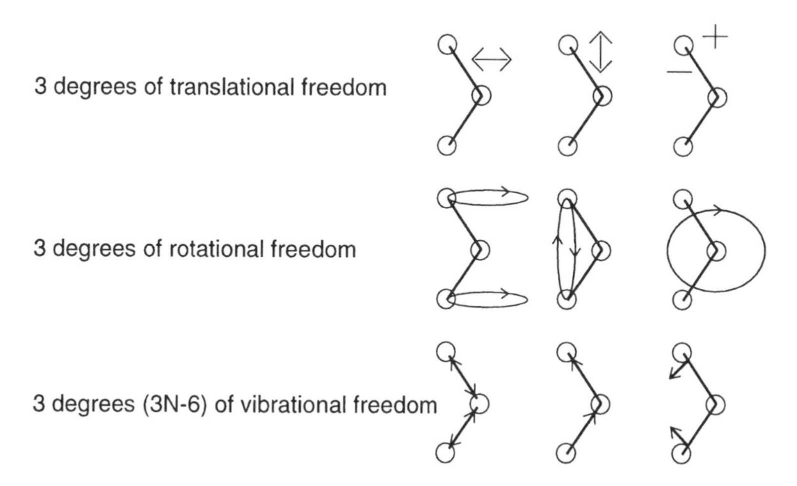

Fig. 2.3 Degrees of freedom for a 3-atom nonlinear molecule. Nine degrees of freedom, three each for translational, rotational, and vibrational degrees of freedom, are illustrated for a molecule such as water.

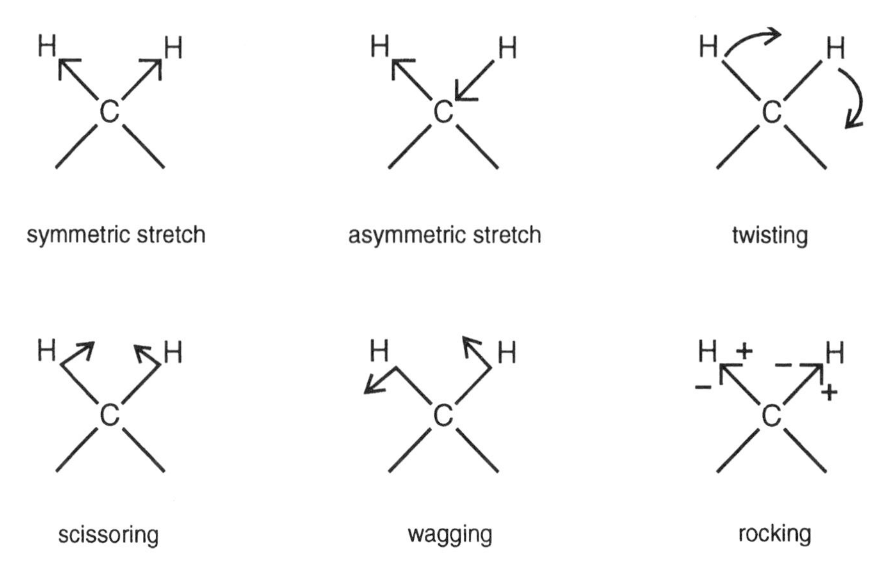

Fig. 2.4 Stretching and deformation modes for the $-CH_2-$ moiety. Six types of stretching and deformation can occur in three-dimensional space for the methylene moiety in molecules.

stretching may be symmetrical or asymmetrical. Deformation modes are envisioned as changes in bond angles. For nonlinear molecules, N-1 stretching modes exist and 2N-5 deformation modes exist. For a linear molecule, N-1 stretching modes exist and 2N-4 deformation modes exist. For the three-atom, nonlinear molecule in Fig. 2.3, one of the vibrations is symmetric stretching, one is asymmetric stretching, and the third is a deformation (bending).

For molecules with other functional groups attached to a central atom, a wider variety of deformation modes can exist because the geometry of the entire molecule must be considered. Deformation modes may be classified as in-plane (called scissoring in some cases), out-of-plane, wagging, torsional (twisting), and rocking [54, p. 93] [42, pp. 83f] [55, pp. 76f]. For example, the H–C–H bond moiety in a larger molecule can undergo deformation changes described as scissoring, wagging, twisting, and rocking [53, p. 3] [51], as shown in Fig. 2.4. Each functional group typically will exhibit vibrational energy transitions from absorption of infrared radiation in a specific region of the spectrum. This fact is the basis for qualitative analysis, namely, structural determinations, by infrared spectroscopy, which will be discussed below in our section on interpretation.

The effects of the rotational degrees of freedom become apparent for gaseous molecules and, to a lesser degree, for molecules in the liquid state, as will be discussed in more detail later. This means that infrared spectra of gaseous molecules provide information on both vibrational and rotational transitions, since both contribute significantly to the spectral characteristics.

Role of Symmetry and Group Theory

Symmetry plays a major role in the appearance of a spectrum. Early crystallographers realized the importance of symmetry in crystals, and spectroscopists later realized that the symmetry of a molecule plays an important role in the exclusion and the allowance of some types of vibrations [43, p. 109] as well as in how they respond to polarized light. Thus a basic knowledge of group theory is desirable for the practicing spectroscopist in order to determine the infrared-active modes of vibration and, in some cases, ultimately to calculate their expected frequencies.

A knowledge of the symmetry of the modes aids in the assignment of bands [54, pp. 90ff] [56] [57] [58] and in the calculation of the vibrational frequencies for each normal mode of vibration (*normal coordinate analysis*) from structural parameters (bond distances and angles), atomic masses, and force constants that may be obtained empirically or from quantum mechanics [59] [60]; see [59, p. 175] for stretching force constants and [59, p. 176] for bending force constants.

To apply group theory to a molecule, the hypothesized equilibrium geometry of the molecule is drawn with each of the atoms having arrows pointing in the cartesian x, y, and z directions. Symmetry operations (using symmetry elements such as E, C_n, S_n, σ and i) are performed on the molecule. The molecular *point group* is determined (for water this is C_{2v}), and the character table is established from mathematical manipulations of matrices summarizing the symmetry operations on the arrows, with $+1$ representing unchanged and -1 representing reversed. For molecules with

TABLE 2.1 Designations for Classes of Vibrations

Designation	Properties
A	Symmetrical to the main symmetry axis of the molecule
B	Antisymmetrical to the main symmetry axis of the molecule
E	Doubly degenerate; the representation of the normal vibration is 2-dimensional
F	Triply degenerate; the representation of the normal vibration is 3-dimensional
g or u (as indexes)	Symmetrical or antisymmetrical to a symmetry center
1 or 2 (as indexes)	Symmetrical or antisymmetrical to a rotation axis (C_n) or inversion axis (S_n), which is not at the same time the main symmetry axis.
$'$ or $''$	Symmetrical or antisymmetrical to a symmetry plane

more than two equivalent atoms (they exhibit higher symmetry), a more complex treatment is required [61, p. 173] [54, p. 90] [62, pp. 255f].

The types of symmetry for molecular vibrations are called the group theory classes (or *symmetry species*) with symbols such as A_1, A_2, B_1, and so on [61, p. 159f] [42, p. 102f] [54, p. 90]. See Table 2.1 for what these symbols represent. Any vibrational mode may be identified as infrared-active if any of the translational components T_x, T_y, and/or T_z are listed for a group theory class.

Although force constants used in normal coordinate analyses are usually taken from empirical tables, recent studies have involved the calculation of the force constants from first principles [63, pp. 267f] [64]. Currently the density functional/Hartree-Fock hybrid using the Gaussian 94 program is the most successful. The inaccuracies in the calculated vibrational frequencies, about 4% of the expected values, are attributed to inaccuracies in the force matrix.

A good introduction to normal coordinate analysis, together with personal computer algorithms, is available [65] [51] [42] as are test data for normal coordinate calculations [66] [67].

Other Types of Transitions

In addition to the fundamental bands involving "normal" transitions between rotational-vibrational levels, as populated by the Boltzman distribution in the ground and first vibrational levels, other types of transitions are present in the infrared spectrum, including overtones, combination, Fermi resonance [43, p. 31], [68] [69] [70, p. 378] [71, p. 141], and hot bands [43].

Overtones are near multiples of the fundamental energy in one vibrational mode. Combination bands occur at frequencies that are the sum or difference of two or more fundamental frequencies [72, p. 4]. Fermi resonance, observed as bands with considerably enhanced intensity, results when two energy levels are close to one another so that the two states "mix" [73, p. 116], the two levels being either a combination band and a fundamental or an overtone and a fundamental, both with the same symmetry species resulting in the two energy levels spreading apart. Hot bands originate from nonground state vibrational energy levels [43, p. 27].

Peak Shape

Besides being characterized by its peak position and intensity, an absorption is characterized by its peak shape. Although absorption of one photon by one molecule ought to result in an infinitesimally narrow line, actual spectra reflect the statistics of interactions, since, in reality, a large number of photons are typically impinged on a large number of molecules during the

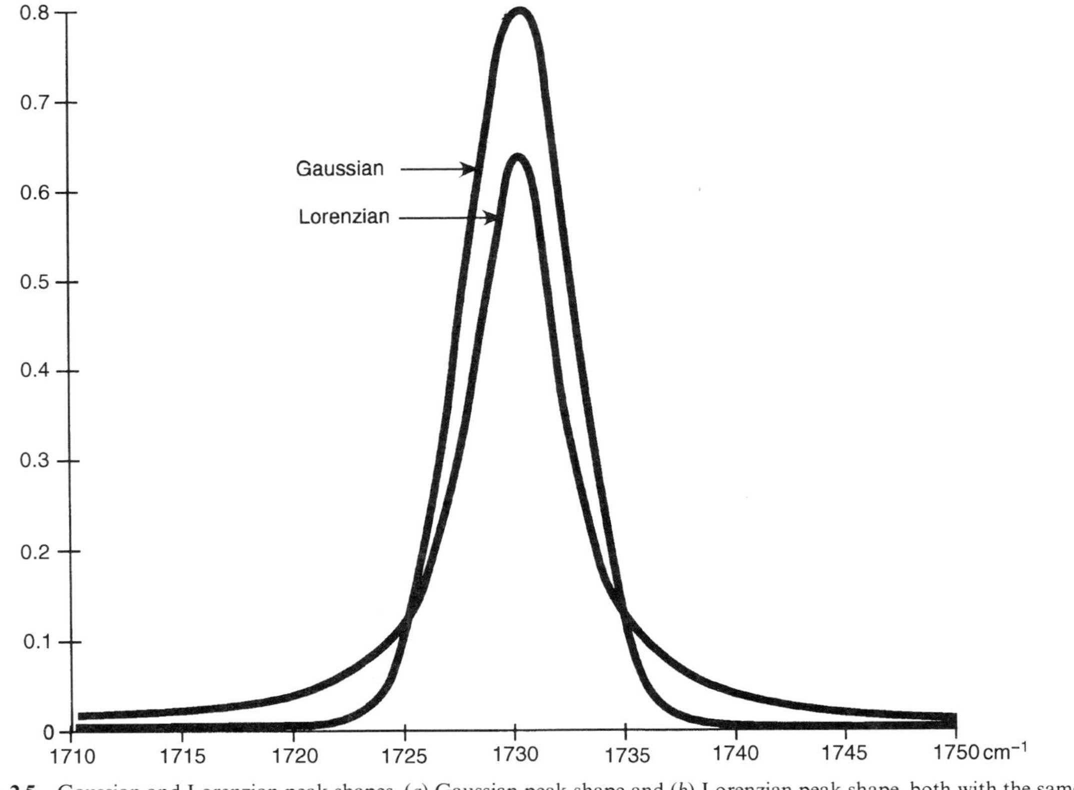

Fig. 2.5 Gaussian and Lorenzian peak shapes. (*a*) Gaussian peak shape and (*b*) Lorenzian peak shape, both with the same area under the curve.

course of obtaining a spectrum. The line broadening is primarily due to varying lifetimes of the energy states and/or environmental differences [74, p. 548] [75, p. 24] [76, p. 20].

"Peaks" of finite widths having Gaussian, Lorenzian, or other shapes are observed (see Fig. 2.5) [77, pp. 9–10] [78] [3, p. 136]. Lorentzian shapes are obtained if the molecules are able to move around or exchange their positions in physical space. Thus all the molecules in the sample are able to affect all parts of the entire peak shape profile, and the line broadening is classified as *homogeneous*. Collision broadening in vapor phase infrared spectra or high-resolution NMR of liquids are examples.

Gaussian peak shapes are obtained if selected groups of molecules contribute to only narrow regions of the line shape profile (classified as *heterogeneous*). A rigid system such as a crystal exhibits Gaussian peak shapes because the molecules are interacting with the same nearest neighbors and not rearranging to establish new nearest neighbors. Gaussian peak shapes also result from certain groups of molecules in a gas having a specific velocity, namely the Doppler effect [79].

The *line width* (usually defined as the width at half height), is determined by at least four factors: radiation damping, the Doppler effect, collision (pressure) broadening [80] [3] [81] [32, p. 69] [82, p. 336] and intermolecular interactions. In gases at low pressure, the Doppler effect is greater than the effect of collision broadening. For example, for a gaseous molecule with a molar mass of 100 at 20°C, with an absorption at $1000\,cm^{-1}$, Doppler broadening contributes $0.0012\,cm^{-1}$, in contrast to collision broadening, which contributes $0.0001\,cm^{-1}$/torr pressure [83]. For liquids, the magnitude of intermolecular interactions is much greater than the other effects [82, p. 336] [27, p. 4] [78] [84] [85] [76, p. 20] [86, p. 108] [83] [80].

An important result of molecular interaction in the gas phase is the effect of these interactions on the intensity of the absorptions. Quantitative analysis is at times extremely difficult unless a second gas, called a *foreign* gas, is introduced into the sample chamber. The presence of this foreign gas results in a pronounced increase in the apparent intensity of the absorptions of the gas under investigation. This is attributed to shortening of the lifetime of the vibrationally excited state [87].

Band Envelopes and Band Shapes

We must be careful to distinguish the terms "peak" and "band." The band envelope (colloquially referred to simply as a "band") refers to any portion of a spectrum that consists of multiple related absorption peaks [70, p. 388]. The band envelope may consist of two branches with "symmetric" shapes, with or without a central "peak"; see Figs. 2.6*a* and 2.6*b*, designated as

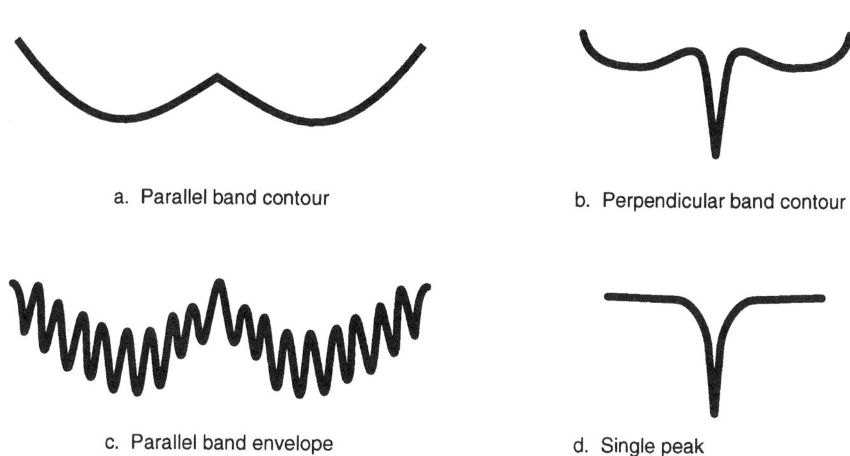

Fig. 2.6 Band shapes and contours. (*a*) and (*c*) show the parallel band contour and envelope, respectively, whereas (*b*) shows a perpendicular band contour and (*d*) a single peak.

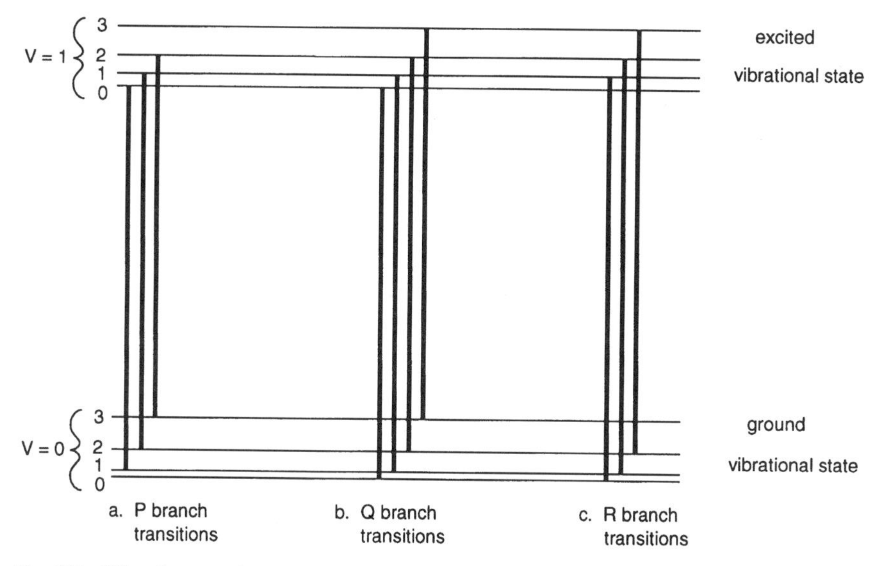

Fig. 2.7 Vibration-rotation transitions. The ground vibrational state ($v = 0$) and the first excited vibrational state ($v = 1$) together with its ground and three excited rotational states (0, 1, 2, and 3, respectively) are illustrated representationally. P branch transitions (*a*) are those from an excited rotational state in $v = 0$ to one lower rotational state of an excited vibrational state in $v = 1$. Q branch transitions (*b*) are between identical rotational states. R branch transitions (*c*) are those from a given rotational state of $v = 0$ to one higher excited rotational state in $v = 1$.

parallel and *perpendicular*, respectively, or a large number of lines or peaks whose outer contour maps out the band envelope (see Fig. 2.6*c*), or simply a single peak (see Fig. 2.6*d*). Vapor phase spectra of small molecules typically give the multiple lines shown in Fig. 2.6*c*. Condensed phase spectra typically show just the outer band contours of Fig. 2.6*a*, *b* or *d* [80].

The lower energy portion of the "parallel" bands shown in Fig. 2.6 are designated as P branches; they are a result of vibrational transitions that go from one rotational state of the ground vibrational state to a lower numbered rotational state of the next excited vibrational level (see Fig. 2.7*a*). The higher-energy portions of such bands are collectively referred to as the R branches; they reflect vibrational transitions that go from one rotational state of the ground vibrational state to a higher numbered rotational state of the next excited vibrational state (see Fig. 2.7*c*). Thus, such bands are referred to as *vibration-rotational* bands. These two "wings" [43, p. 48] result when the direction of molecular rotation changes the orientation of the vibrationally caused dipole moment oscillation.

The center portion of a band, called the Q branch, is a purely vibrational transition because it exists when excitation occurs between identical rotational states of the ground and next excited vibrational level (see Fig. 2.7*b*). It does not involve rotational transitions at all, since rotation in that particular axis causes no orientation change in the oscillating dipole moment. Thus, the so-called perpendicular band (Fig. 2.6*b*) is a combination of vibration-rotational band wings and a purely vibrational center peak.

Gas Phase Spectra

We have mentioned above how infrared spectra of gases are often different from the spectra obtained from other phases. Since the spectra of gases show mainly vibration-rotational bands, it is most advantageous to study the gas phase infrared spectrum of a particular molecule of interest, if possible, in order to determine its conformation (geometry). The rotational fine structure of the IR spectra of many small molecules in the gas phase has been analyzed, and structural parameters such as bond angles and interatomic distances have been obtained [49].

The observed transitions are dictated by the selection rules, which are classified according to the type of rotor and the direction of the changing dipole moment with respect to the vibration. The types of rotators and their classifications with examples are listed in Table 2.2 [43, p. 44] [88, pp. 38f]. For further discussion of the mathematical bases for allowed energy transitions (selection rules), see [61, p. 113] [43, p. 44]. Table 2.3 gives a summary of the differences in selection rules for molecules with either permanent dipole moment components or with dipole moment changes along one or more of the three principal moments of inertia directions [82, p. 94] [70, p. 468]. We must realize that the transitions involving rotational energy levels

TABLE 2.2 Types of Rotators

Type	Moments of Inertia	Examples
Linear	$I_a = 0, I_b = I_c$	HF
Spherical top	$I_a = I_b = I_c$	CH_4
Symmetric top		
Oblate	$I_a = I_b \neq I_c$	BF_3, C_6H_6. $CHCl_3$
Prolate	$I_a \neq I_b = I_c$	CH_3Cl
Asymmetric top	$I_a \neq I_b \neq I_c$	H_2O, C_5H_5N

TABLE 2.3 Selection Rules for Asymmetric Tops; *e* = even, *o* = odd

Axes Parallel to Dipole Moment	Allowed Transitions
a (least)	$ee \leftrightarrow eo$
	$oo \leftrightarrow oe$
b (intermediate)	$ee \leftrightarrow oo$
	$eo \leftrightarrow oe$
c (greatest)	$ee \leftrightarrow oe$
	$oo \leftrightarrow eo$

may not involve the three rotational constants independently but that the energy level positions may involve all three rotational constants simultaneously. This is quite unlike the first approximations involving vibrations which can be separately treated for each vibrational mode as an independent harmonic oscillator [43, pp. 46–54].

Bear in mind that although the rotational spectrum intensity is determined by the direction of the permanent dipole moment along each principal axis, the vibrational-rotational spectrum intensity involves the changing dipole moment magnitude along each principal axis. The intensity of the rotation-vibrational band (for a gaseous molecule) will be related to the magnitude of the dipole moment change along a specified rotational axis of the molecule [70]. An examination of the rotational-vibrational spectrum of water nicely portrays these interrelationships, and it is presented under examples in a later section [3, p. 455].

Liquids, Solids, and Polymers

The spectra of liquids usually do not contain any rotational fine structure because molecular collisions hinder free rotation. In the crystalline state the interactions are well defined, making bands appear sharper. Interestingly enough, the infrared spectra of giant molecules (polymers) are generally relatively simple because (1) many of the normal vibrations have almost the same frequency (the molecules consist of repeating simple units), and thus appear as one band in the spectrum, and (2) only a few of the normal vibrations are infrared-active [89, p. 4] [43, p. 16] [90, p. 180] [91].

Of particular interest in solids is that the changing dipole moments have a specific orientation that may be determined by polarization studies depending on the sample. An experimentally determined quantity called the "dichroic ratio" of a particular mode of vibration, which is equal to the ratio of the peak heights of the band with the electric vector parallel and then perpendicular to a chosen reference frame, is employed in these studies (see later example in this chapter) [43, p. 96].

2.3 INSTRUMENTATION

Two main types of infrared spectrometers are in active use, the dispersive instrument and the Fourier transform infrared spectrometer. A less common instrument uses filters to effect a scan across the entire infrared range of wavelengths. Also tunable lasers and free electron laser systems are available.

Components of a typical dispersive instrument are illustrated in Fig. 2.8; see [92, p. 2]. Components of a typical FTIR instrument are illustrated in Fig. 2.9; see [93, p. 88]. Ciurczak [94] has an excellent comparative summary of various spectrometer components, including tables of operational and financial parameters for spectrometer sources, monochromators, and detectors. A spectrometer consists of a source, its optics, an associated wavelength separation device, a detector, electronics, and, if applicable, software. We will now address each component separately.

Infrared Sources

The ideal infrared source is one that emits infrared radiation at controllable variable wavelengths, at controllable, sufficient, and variable intensity, and at controllable variable bandwidths over the entire infrared range [95, pp. 20f] at a reasonable price. Currently no sources exhibiting such ideal characteristics exist. Therefore, the currently accepted practical infrared source is one that *approximates* a "perfect" blackbody* because it represents the maximum energy that can be received from a heated source. All other heated "bodies" emit less energy at any specified temperature. Since

*A "perfect" blackbody would emit radiation according to Planck's radiation law, [96] [97] [98, pp. 141f] [99, pp. 103–104] [74, pp. 171f].

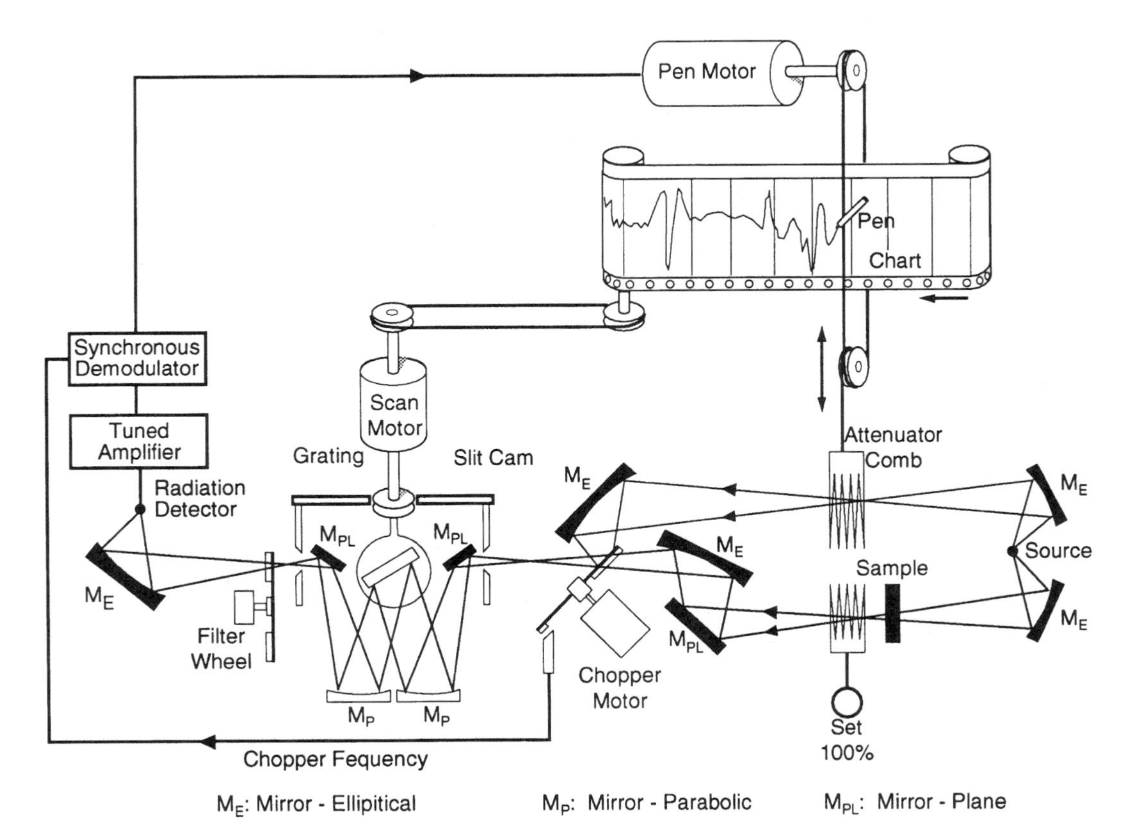

Fig. 2.8 Components of a typical dispersive infrared spectrometer.

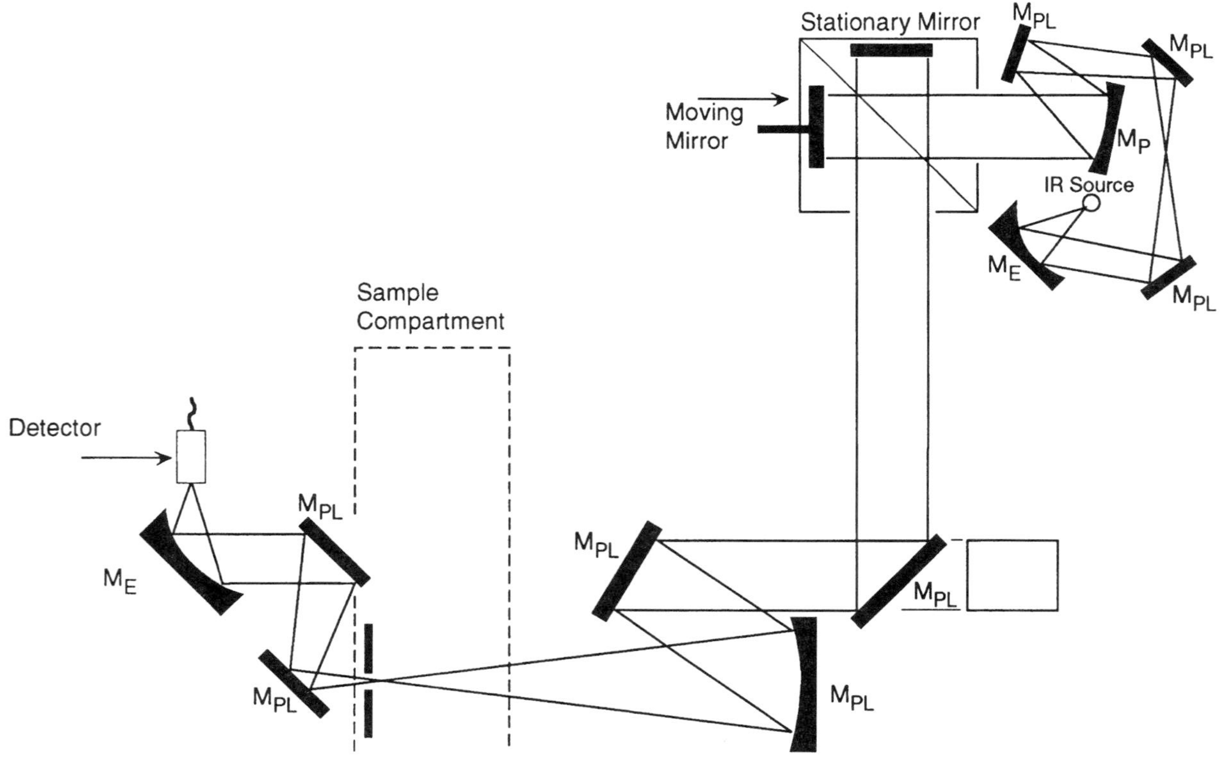

Fig. 2.9 Components of a typical fourier transform infrared spectrometer.

"perfect" blackbodies do not exist either, we must use materials that only closely resemble a blackbody, and correctly configure then with a hole [100, p. 55] so as to trap incoming radiation to avoid reflections. Most IR sources are heated to at least 1450 K so as to have the energy maximum near the short wavelength limit of the spectrum [43, p. 76] [101, pp. 326f].

In commercial IR spectrometers, the source is usually a solid material heated to incandescence by an electric current. A rugged, inexpensive source is a piece of nichrome wire wound around ceramic. Another infrared source is a rod (about 6 by 55 mm) of carborundum (SiC) heated to 1750 K by an electric current. This so-called Globar (or glowbar) source which has been around for many years is still extensively used [100, p. 63] [94, p. 12]. Disadvantages include the fact that a completely enclosed, water-cooling casing is required and that a significant amount of power (200 W) is needed. Some fluctuation in the radiation intensity can occur because of the fluctuation of the contact resistance between the rod and the hold-down clamps [100, p. 64], but an advantage is that the glowbar source can be used down to 200 cm^{-1}.

An early infrared source still in common use is the Nernst glower [92, p. 4] [100, pp. 62–63] [102, p. 24] [99, pp. 104f]), which is composed of refractory oxides of rare earths, mainly zirconium, often containing yttrium, thorium, or cerium [103, p. 181]. The Nernst glower is much smaller than the SiC source, less than 2 mm in diameter. It uses much less current than the glowbar, can be used in air, and has an operating temperature of between 1700 and 1900 K. A disadvantage is its brittleness and fragility, so it must be mounted in flexible holders. Radiation emitted along the rod may fluctuate during operation.

Other infrared sources include the high-pressure mercury arc within a silica envelope [92], a carbon arc, the incandescent lamp [55, pp. 14f], the tunable laser diode [104, pp. 24f] the CO_2 laser, useful only in the 1100 to 900 cm^{-1} region [94] [105, p. 4], a synchrotron, and the free-electron laser.

The maintenance of constant temperature of the source is critical. For most infrared spectrometers currently in use for routine analysis of samples, the source temperature can be fairly well controlled. For contemporary research quality sources, the temperature can be controlled to $\pm 0.1°C$ (with a concurrent increase in cost!). At 1500 K a change in temperature of 0.01°C between the collection of the background and the sample would correspond to a change in absorbance of 0.00004 absorbance units. Air currents, electrical power, and water temperature (if the source is water cooled) fluctuations are three contributors to any observed intensity variation. Such fluctuations affect the overall quality of a spectrum obtained, sometimes expressed as the *signal-to-noise* ratio. Energy flux from the source is the principal component in the "signal" of the signal-to-noise ratio [106, p. 22]

[94], realizing that the real signal is the removal of energy by the sample for transmission measurements.

Today sources in much of the literature are categorized either as air- or water-cooled. Of value would be to request that the vendor state the operating temperature and temperature stability of the source. Note also that since an air-cooled source does not need water, it is preferred overall if other stability issues are similar and if the intensities are comparable.

Optical Components

Mirrors, Windows, and Lenses

Mirrors are used in spectrometers to redirect the beam so as to make spectrometers more compact, to focus the beam appropriately, and to avoid problems such as those associated with the care necessary with transparent lens materials. Five types of mirrors may be used in a spectrometer: plane, parabolic, ellipsoidal, spherical, and toroidal. Parabolic mirrors focus parallel light to a point, and ellipsoidal mirrors can take light from a focused point and refocus it at a new point. In many cases the spherical mirror is used to mimic the more expensive parabolic mirrors. The use of toroidal mirrors eliminates astigmatism in the source images at the focal points of the two beams; it functions as a two-dimensional ellipsoid [107]. Chromatic aberration is absent in mirrors, since all wavelengths reflect alike.

The mirrors are most likely made of glass and covered with a mirror coating of aluminum [108]. Glass is chosen because of its dimensional stability with respect to temperature variations. Such mirrors are called *front-surface mirrors* or *first-surface mirrors*. Second-surface mirrors (coated on the back of the glass, as with bathroom mirrors) are unacceptable for infrared instruments because the glass absorbs most of the radiation [109]. Mirrors that are "off-axis" are used to make spectrometers more compact and to avoid spherical aberration.

Lenses made of crystalline salts such as KBr are used infrequently as spectrometer components, except as detector windows, due to the fact that they degrade upon exposure to moisture. Coatings are available for protection of the surfaces. Windows for sample cells will be discussed below in Section 2.4.

Wavelength Separation Components

Except for spectrometers with tunable laser sources, every spectrometer must include an optical component or device that is able to distinguish between the different wavelengths (frequencies) of radiation produced by the source so as to monitor the amount of energy impinged on the sample for each wavelength. The dispersive spectrometer uses a monochromator; some simply use filters, and the FT-IR spectrometer usually uses a Michelson interferometer.

Dispersive Monochromator

A monochromator for the infrared consists of an optical system of mirrors and a dispersing device, either a prism or a diffraction grating, that spreads or modifies the spectrum across the exit slit plane (see Fig. 2.8). Movement of the prism, grating, mirror, or filter will then "scan" the spectrum across the exit slit [92, p. 41]. The filter, if used, will also isolate a part of the spectrum either by absorbance or by interference. Optical configurations for the dispersive-type spectrometers include the Wadsworth system (older instruments) and the Littrow system [100, p. 192].

Some spectrometers still in use employ prisms made of NaCl or KBr. Although they are capable of separating discrete wavelengths and exhibit narrow bandwidths, NaCl and KBr are hygroscopic, making them vulnerable to degeneration; they are therefore the least rugged of all the choices. For this reason few spectrometers nowadays use such prisms.

Gratings are much more rugged than prisms. A diffraction grating consists of a repetitive array of apertures (slits) or obstacles (rulings) that has the effect of producing periodic alterations in the phase and thus the amplitude of an emergent wave via interference [110, p. 424]. Usually cast from a master grating, replica gratings are mounted on a glass optical flat and aluminized. The ruling diamond is angled so that the profile of each ruling is tilted in sawtooth fashion to increase energy in a given order (*blaze*).

In contrast to the prism, grating "dispersion" in terms of wavelengths is nearly constant over a substantial range of grating movements. Gratings are available with various combinations of blaze and ruling pitch, making it possible to obtain extensive spectral coverage [92, p. 5] [2] [111] [112]. However, gratings require an order sorter [94], which is often a filter made of silicon, germanium, indium arsenide, or indium antimonide and is used to eliminate unwanted higher orders of spectra that might ordinarily partially overlap with the first-order spectrum.

Optical null systems (which could be classified as part of the detector system) incorporate a moving wedge in the reference beam. The wedge moves until the intensity of light coming through it and impinged on the detector matches that of the beam passing through the sample. (The detector alternately is impinged with the reference and sample beams.) The distance the wedge moves is related to the percent transmittance of the sample at that wavelength. An electronic null system detects the detector voltage difference between the sample beam and the reference beam, and relates this difference to the percent transmittance.

In addition all infrared instruments have both an aperture stop and a field stop as a part of the optical system to control the intensity and field of view, respectively. The intensity of the energy passing the entire optical systems is called *etendue* or *throughput*.

The Michelson Interferometer

The Michelson interferometer is shown schematically in Fig. 2.9 [93, p. 85]. The interferometer consists of two perpendicular mirrors, one of which is stationary and the other moves at a controlled velocity in the directions shown. Between these mirrors is a semireflector, or beam splitter, usually positioned at 45°, at which the incoming beam is divided and later recombined after a path difference has been introduced between the two beams. The velocity of the moving mirror does not have to be controlled precisely, since the data are related to the number of laser fringes that the mirror passes. The types of mechanisms commonly found in commercial instruments for mirror movement are air bearings, voice coils, and flex-pivot, to name a few.

The design of the beamsplitter usually involves a KBr disk that has a germanium or more sophisticated coating which divides the source radiation into two parts and a compensating plate to maintain similar optical paths. Additional information can be obtained by examining an engineering report from Hughes Danbury Optical Systems and references therein [113].

Some interferometer systems allow the collection of the interferogram in both directions. Other modifications in design include having the angle of the beamsplitter in the interferometer at 60° (rather than 45°), which is said to provide greater throughput. In addition to the usual aperture and field stops, Perkin-Elmer instruments have a "B-stop," which is used to vary the attenuation while keeping the image size at the detector constant.

Detectors

Radiation impinging on the detector must eventually produce over the whole spectral range of the spectrometer an electrical signal that is proportional to the incident radiation intensity. In infrared spectroscopy, two types of detectors are employed: thermal detectors and quantum detectors [114].

Thermal Detectors

Thermal detectors include thermocouples, thermopiles, bolometers, pyroelectric, and thermomechanical devices [92, pp. 5–9] [115]. Thermal detectors employ materials that possess strong temperature-dependent properties. The radiation absorbed causes the detector temperature to rise, which causes some property of the detector material to change, such as its resistance, that is detectable in an electric circuit [115] [116].

For FT-IR work, the thermal detector made of deuterated triglycine sulfate (DTGS) is the most common. The DTGS detector is a pyroelectric bolometer ("pyroelectric detector" for short) [53, p. 21]. A pyroelectric (from the Greek *πiρο*, meaning "fire") material has a permanent alignment of dipole moments and thus a permanent electrical polarization. Upon heating, there is a change in electrical polarization due to absorbed radiation and also to thermal expansion. These polarization changes induce a change in the electric charge on the electrodes deposited on the surface of the detector which acts essentially as an electrical capacitor [117]. In fact DTGS is a special case of pyroelectricity, or rather, ferroelectricity, since it alters its orientation direction or quenches it at a transition temperature [17, p. 8].

Quantum Detectors

Well-known quantum detectors are based on materials with photoemissive, photoconductive, or photovoltaic properties [118]. Photoconductive detectors are semiconductors that undergo changes in the electronic distribution and hence a change in the conductivity on absorption of radiation [116]. Incident photons excite electrons from the valence band of a semiconductor into the conduction band (also called an *intrinsic detector*), producing two charge carriers—a free hole and a free electron. These will eventually recombine, but while they are separated, the material remains slightly more conductive. Under the influence of an electric field, the conduction electrons will more in one direction and the holes in the reverse direction, creating a current so that a load resistor will generate a voltage drop. Thus the photon flux is detected either by measuring the current or the voltage drop [118].

An example of a quantum detector is the MCT detector, which is a ternary semiconductor compound containing mercury, cadmium, and tellurium [116]. For photons with energy greater than the semiconductor band-gap energy, electrons will be excited into the conduction band, thus increasing the conductivity of the MCT. The ratio of mercury to cadmium to telluride may be changed to decrease the band gap and thus increase the range of detectable photons (bandwidth), but this decreases the sensitivity [53, p. 21] [118] [119]. A brochure entitled *Infrared Detectors 1995* published by EG&G (Judson-Optoelectronics) [119] is an excellent beginning source of information on this topic.

Detectors are characterized in part by D-Star (D^*), active size, cutoff wavelength, responsivity, and time constant. D^* is a relative sensitivity parameter used to compare the performance of different detector types. This quantity serves as a yardstick to assess the performance of detectors. However, since not all detectors depend on the detector area in the way specified by this definition, the concept of specific detectivity must be used with care [116] [119, pp. 20, 50] [100, p. 111].

Which detector is right? To choose a suitable infrared detector, factors to consider include the spectral range of the response, the response time, and other issues such as robustness, convenience, reliability, operating temperature (does it operate at room temperature or does it require a coolant?), and cost [120].

As with all photon detectors, the optimum system performance is achieved with the smallest size detector capable of collecting the available incident radiation. Photoconductive devices offer two advantages over thermal detectors. First, their response time is much shorter, typically of the order of 1 μs rather than 1 ms. Second, since they do not respond to such a wide band of the spectrum, the limiting background noise will be smaller. A disadvantage in addition to cost is that of their cooling requirements. See also Ewing's article [121] for an excellent summary of infrared detectors.

Electronics

Typically the signal from the detector is amplified by a preamplifier, and then either detected by a lock-in amplifier for dispersive instruments or changed to a digital signal by an A/D converter for FT-IR instruments. Some FT-IR instruments use lock-in amplifiers (SSFTIR). A filled 24-bit A/D converter, with one bit representing the sign, results in an uncertainty in the absorbance of about ± 0.0001 absorbance units for 4000 data points (2 cm^{-1} resolution). When signal averaging, the appropriate level of noise needs to be converted. For proper filling of the A/D converter, this level is typically at least 2 bits [122] [25, pp. 77, 144f]. For additional information on A/D converters, consult the *Design-In Reference Manual*, published by Analog Devices [123] or *Digital Signal Processing Applications with the TMS 320 Family: Theory, Algorithms and Implementations*, published by Texas Instruments [124].

Mounts

Considerable effort should be made to minimize vibrational noise. Vibration isolation may be of value for certain applications.

Noise Sources

The electronics of the instrument and the emission from the spectrometer components, environmental sources including physical vibration and temperature, design deficiencies, and the infrared source itself can all contribute to noise in the spectrum. However, the predominant noise in infrared spectroscopy is called *detector-limited noise* [137]. For detectors at room temperature, this is *Johnson noise* [125, p. 263] or *thermal noise* [126, p. 464] and results from random movement of electrons in the detector. For detectors cooled to very low temperatures, the dominant noise is from the random arrival of photons from the background to the detector, not by intrinsic detector noise [119], and is thus called *background-limited* noise [125, p. 264] [25, pp. 142–144] [55, p. 40] [121] [3, p. 365]. A detector is referred to as *BLIP* (background limited) when D^* is limited by the noise associated with photons from the background radiation.

Beside Johnson noise, two other types of noise [126, pp. 464–466] associated with electronic circuitry are shot noise, due to random emission of electrons from semiconductor junctions [127], and flicker noise (often called $1/f$ noise) [121] [127].

FFT Algorithms for FT-IR Spectrometers

An interferogram from the detector of an interferometer consists of a composite of cosinusoidal voltage signals with the maxmum occurring when the path lengths in the interferometer are the same (the *zero path difference*, or ZPD). Since the incoming radiation is polychromatic, the detector signal (the *interferogram*) is a mixture of maxima and minima for each infrared frequency, which occurs at different mirror positions for different infrared frequencies and phases, according to the expression

$$I(t) = 0.5H(v)\, I_0(v)\{1 + \cos(2\pi v V t + \theta(v))\} \tag{2.3}$$

In the expression above $I_0(v)$ is the intensity of the incident energy, v is the infrared frequency, V is the velocity of the mirror in cm/s, t is the time after mirrors A and B are equidistant, $\theta(v)$ is a phase shift introduced by the electronics and the beam splitter, and $H(v)$ introduces the departure from the ideal case due to the interferometer optics (with a value less than one).

The detector signal varies sinusoidally with a detector frequency $f_v = 2Vv$ [93, pp. 84–85] for each infrared frequency. Thus the interferogram, $I(t)$, is composed of the resultant signal from equation (2.3) for each frequency in the spectrum. Since the interferogram contains information on the intensity and phase of each infrared frequency in the spectrum, a mathematical operation known as the Fourier transformation is used to obtain the actual infrared spectrum.

Fast Fourier transform (FFT) algorithms are employed to convert the raw data, or interferogram (intensity or voltage vs. mirror position), into spectra (intensity vs. frequency). Programs consisting of about 150 lines of code are readily available and may be easily programed into a personal computer for the purposes of either studying the effects on the transform of apodization [128, p. 15] [129, p. 64], resolution, phase correction, zero fill, and sampling rate or simply obtaining the spectrum. In order for the transform to be a meaningful representation of the spectrum, information theory requires sampling points from the interferogram at mirror intervals of $1/(2\nu_{max})$ with ν_{max} being the bandwidth of the spectrum (usually $\nu_{max} = 4000\,cm^{-1}$) to acquire the desired spectral range. The resolution depends on the distance the mirror travels. The number of data points required are determined by these two variables [30].

Good conceptual introductions to Fourier transforms may be found in Griffiths, Foskett, and Curbelo [130], Hurley [131] and a series of articles by Perkins [132] [133] [134].

In most cases the interferogram collected is asymmetric and thus requires a phase correction ($\theta(\nu)$ above). This is a process of removing sine components from an interferogram or removing their effects from a spectrum in order that the intensities computed in the Fourier transform are correct [128, p. 27]. Two common phase correction methods are the Mertz and the Forman methods, with the Mertz being more common because it is easier to program. For the Mertz method, a Fourier transform is first performed on a small portion of the data to calculate the phase correction [128, p. 94], and then the file is apodized (the side lobes of peaks in the spectrum are suppressed) to minimize abrupt signal changes due to a finite mirror displacement. Finally the Fourier transform is performed on the whole data set and then phase-corrected to obtain the spectrum.

Advantages and Disadvantages of the Different Instrument Systems

A number of advantages of an FT spectrometer over a dispersive instrument have been put forth. "Jacquinot's advantage" arises from the fact that the circular apertures used in FTIR spectrometers have a larger area than the linear slits used in a prism or grating spectrometer, thus allowing for greater throughput of radiation, all other parameters being equal [25, pp. 352, 357] [135, pp. 19–23] [27].

"Fellgett's advantage," due to the fact that all frequencies emanating from the IR source impinge simultaneously on the FT detector, not sequentially as with a dispersive instrument, is a time advantage [135, pp. 23–25] [136] [27]. The signal-to-noise ratio (S/N) [137]) for an FT instrument is greater than that of a dispersive instrument by a factor of $(M)^{1/2}$, M being the number of resolution elements. The time advantage is greater than the sensitivity advantage, being directly proportional to M. All other parameters being the same, a $2\,cm^{-1}$ resolution spectrum that takes 30 minutes on a grating spectrometer should take about 1 second on an FT instrument [128, p. 275].

"Connes's advantage" results in more accurate wavenumber values and is realized due to the relationship between the infrared frequency and the accuracy of the moving mirror position as determined by the precision of the laser wavelength. Hirschfeld stresses that a Michelson interferometer exhibits a sharply reduced level of optical aberrations [138].

One advantage of a dispersive instrument over the FT is the capability of monitoring the intensity of the signal with time while observing a narrowband of wavelengths. A dispersive instrument ratios the background signal and the sample signal "continually" (typically every 1/15 s) compared to the FT instrument which takes a background scan (which may take up to 5 minutes) and then the sample scan. Zero percent transmittance is not easily calibrated on the FT instument.

2.4 EXPERIMENTAL CONSIDERATIONS

Instrument Performance Evaluation—How Do You Know if the Instrument is Functioning Properly?

A number of evaluative tests can be performed on a spectrometer to determine if it is functioning properly. A good summary of evaluative tests, mainly for FT instruments, is found in *Practice for Describing and Measuring Performance of Fourier Transform Infrared (FT-IR) Spectrometers: Level Zero and Level One* [139]. Some of these tests include the following:

1. S/N ratio test
2. Baseline linearity
3. 100% line test
4. Wavenumber precision and accuracy
5. Photometric precision and accuracy
6. Purge efficiency (the water band should contribute less than 0.001 absorbance units after a 5-minute nitrogen purge)
7. Interferogram reproducibility
8. Digitization noise (need at least 2 bits of noise for coaddition); this determines the number of practical scans one may meaningfully add (see [25])
9. Gain and gain ranging accuracy
10. Linearity of the detector; for MCT no spurious band may be present below the detector cutoff—Is the detector being flooded?

11. Instrument polarization (only if you are performing an orientation study)
12. Beam alignment (centerburst maximization), determined with by centerburst intensity or with a liquid crystal card
13. Temperature constancy for beam splitter and source (especially if water-cooled)
14. Long-term stability
15. Stray light (for FT-IR in many cases is negligible)
16. Do coadditions follow the $\sqrt{N}$ rule for S/N?

Wavelength Standards

Beginning in the 1960s, wavelength standards have been established [140] and revised [141] [142] for reference spectra. Secondary standards for wavelengths have been established and are useful for calibrating the spectrometer. Accepted values for 14 bands (most to $\pm 0.3\,\mathrm{cm}^{-1}$) in the spectrum of a film of polystyrene about 0.07 mm thick are listed in the *CRC Handbook of Spectroscopy* [143, p. 50]. Nyquist [144] has provided spectral data for the polystyrene standard at temperatures ranging from 30 to 210°C. An indene-camphor-cyclohexanone mixture has been used as a standard for many years [143, pp. 51–52] [105, pp. 607–646]. Ammonia is employed as a standard for gas phase spectra and is usually observed in a 10-cm cell at a pressure of 50 mm [146, pp. 605, 662f] [145, pp. 616–619]. *Tables of Wavenumbers for the Calibration of Infrared Spectrometers* [145] are available, which also include infrared emission bands for inert gases such as argon, for calibrating the range from 1 to 4 μm. For accurate work, a correction needs to be made for the wavelength in vacuum [146, pp. 556–557] [147, p. 109].

The National Institute of Standards and Technology (NIST) has a Standard Reference Database (number 26) available on computer disk that provides rapid access to experimental data on the vibrational and electronic energy levels of small, polyatomic transient molecules having from 3 to 16 atoms.

Intensity Standards

John Bertie and coworkers have been involved in the establishment of intensity standards (photometric accuracy standards), and their work should be consulted [148]. Polished calcium fluoride crystals have also been used as photometric standards [149] [150] [95, p. 15].

Types of Samples, Sample Preparation Methods

Transmission infrared analysis has more widespread applicability than any other instrumental analytical technique for the identification of unknown compounds, since the technique is amenable to samples that are solids, liquids (including solutions), or gases. However, to produce useful spectra, in many cases the analyst must engage in a careful and sometimes lengthy sample preparation. The effort expended in this task is usually the most important factor in obtaining meaningful results.

Choice of the sample compartment, or cell, is often also critical to the success of the analysis, and we have chosen to mention types of cells available for gaseous, liquid and solid samples.

Gaseous Samples

Gases are most easily manipulated by constructing a vacuum manifold to which both the infrared cell and the sample container are connected. The gaseous sample is allowed to enter an evacuated infrared cell to the desired pressure and sealed via a valve.

The shortest commercially available gas cells are about 50 mm in length, though shorter ones can be made if desired. A 10-cm cell is most practical for routine sampling of pure gases. For mixtures of gases, 25- to 115-meter cells may be of valuable for trace analysis (ppm or even ppb range) and for low vapor pressure materials. It is sometimes easier to run a volatile liquid sample as a gas by using an infrared gas cell. Note also that gas IR cells are available which may be heated to temperatures as high as 800°C.

Gaseous samples with high water vapor content should be run in cells with windows made of a water-insoluble crystalline material. For low water vapor content gas samples, KBr or NaCl windows can safely be used if there is no chance of water condensation, which would cause the windows to fog and thus scatter the IR beam. For gas cells, one has to be cautious that the material being introduced into the cells is easily removed because samples like amines may remain sorbed onto the walls and introduce impurity bands in the next samples. Other sample preparation methods include depositing the gas on a cold window or in an inert gas matrix ("matrix isolation"), as used in GC/LC IR.

For pollution studies of atmospheric gases or gases escaping from smokestacks, rivers, and toxic waste sites, the detector can be placed as far as 200 or more meters from the infrared source. Alternatively, a retroreflector may be used to send the signal from the source back through the environmental sample and back to the detector [151] [152].

Liquid Samples

Pure ("neat") liquids are first examined to determine whether their viscosity is low, medium, or high. If the viscosity is high, one approach is to smear the liquid onto a salt plate, most commonly KBr, assuming that the sample components do not attack the plate.

TABLE 2.4 Available IR Windows

Material	%T	Low λ Cutoff	Thickness	Upper λ Cutoff
KBr	50%	32 m	4 mm	Unavailable
KBr	10%	40 m	2 mm	0.25 m
NaCl	50%	17 m	10 mm	Unavailable
NaCl	50%	22 m	1 mm	Unavailable
NaCl	10%	26 m	2 mm	0.21 m
Csl	10%	80 m	2 mm	0.25 m
CsBr	10%	55 m	2 mm	0.3 m
CaF_2	10%	12 m	2 mm	0.13 m

Source: [153]

A second approach is to apply the liquid to a 3M Disposable IR Card™, a 5 by 10 cm card with a circular aperture containing a thin, microporous substrate that is the sample application area. An alternative is to apply the liquid to a Janos Screen Cell™, a polymer-coated fiberglass mesh substrate. Note that for an FT-IR instrument, an IR spectrum is usually taken of the same substrate as planned for the sample (for background) before the liquid is smeared onto the substrate. Or the substrate spectrum can be first run and then subtracted from the sample spectrum.

If the liquid sample has a medium or low viscosity, it is more convenient to place a drop of the liquid between two salt plates with a polyethylene terephthalate or teflon spacer, forming a thin film sandwiched between two windows. Table 2.4 lists a variety of materials available as windows [153].

If the viscosity is so low that the sample runs out from between the sandwiched plates too quickly, or if it is so volatile that all of it disappears before the spectrum can be run, one may have to use a liquid cell. For quantitative work the recommended thickness of sealed cells ranges from 0.005 mm for samples with carbonyl components to 0.050 mm for aliphatic materials. Microcells are available that allow for the measurement of samples of 10 μg or less. One disadvantage of using sealed cells is that they may have to be reopened and cleaned before they can be reused. And, because of their higher cost, they cannot simply be thrown away after use.

To perform quantitative analysis, the sample must either be analyzed in a cell with known pathlength or with the same cell to keep a constant path length. Table 2.5 lists the typical path lengths needed for a range of solute concentrations [154, p. 37].

The shortest fixed length liquid cell commercially available is 0.015 mm, and the longest about 100.0 mm (Spectra-Tech). Variable path-length liquid cells come with rotating windows and nonrotating windows with a lower pathlength specification of 0.005 mm. Demountable cells are available that

TABLE 2.5 Typical Pathlengths for Solution Analyses by IR

Analyte Concentration	Typical Path Length
>10%	0.05 mm
10–1%	0.1 mm
1–0.1%	0.2 mm
<0.1%	>0.5 mm

are easily cleaned, but they suffer from an inherent disadvantage which is irreproducible thickness (path length). Heated liquid cells [155, p. 13] [156] and flow-through cells are also available [53, p. 24] [157].

Since the spectrum of the unknown should ideally be comparable to a spectrum that can be found in an infrared library (reference spectra) in terms of band height compared to the baseline, the "thickness" of the liquid is really quite critical. The use of a variable path-length liquid cell is by far the best method to achieve a "library" quality spectrum most easily identifable by comparison to a reference spectrum. The disadvantage of such a cell is its high cost. Demountable cells are available that are easily cleaned, but they suffer from an inherent disadvantage which is irreproducible thickness (path length).

Solutions may be examined using the same methodology as liquids. An added complication is that the solvent itself contributes spectral peaks, which are "mixed in" with the peaks of the unknown solute, whose identity is being sought. Thus in most cases it is desired to remove the solvent bands from the spectrum, which can be accomplished either by having the solvent in the background or by computer subtraction of the solvent bands.

In many cases where the solute in a solution is to be identified, the solvent is evaporated, and the solute is examined as a solid, as described below. A convenient method for dealing with solutions is the 3M Disposable IR Card™ mentioned above. The microporous substrate allows volatile solvents to evaporate rapidly under ambient conditions. Drying the sample using a heat lamp or pulling a vacuum on the sample are two means of removing a solvent. Metal loops made of 20-gauge copper wire dipped into emulsions or latex mixtures can be employed in some cases to achieve thin films, where the loop is held horizontally until the solvent evaporates. If the thin film does not break, this method often works quite well.

Solid Samples

The most challenging sampling is that with solid samples. Depending on their physical properties, samples may have to undergo melting, smearing, freezing, filing, or sandpapering, dissolving, emulsifying, film casting, grinding, mulling with mineral oils, and/or compression with an alkali halide or a diamond cell.

a. Cast Films

If the solid is a polymer or compound with a low melting point (under 250°C), a common method is to place a small amount of it onto a salt plate, which is in turn laid on a microscope slide (to prevent thermal shock of the plate) on a heating unit. A warm spatula is used to smear the sample across the salt plate to achieve the appropriate thickness. Ideally the thickness of the material to be analyzed is in the 10-μm range. The use of a fine metal mesh as a substrate for a melted and resolidified polymer is also suggested.

If a suitable solvent can be found for the solid, a solution is made and the sample treated as with solutions described above using a 3M Disposable IR Card™, which is especially good for acids. Or the solution can be poured onto a salt plate and the solvent allowed to evaporate, producing what is known as a cast film. This latter process is often done stepwise with a buildup of very thin layers until the ideal thickness is achieved. The need to recast may be eliminated by casting a film in a wedge rather than a film of uniform thickness. A small wood stick is placed under one edge of the salt plate to accomplish this task, and thus the solution is thicker on the lower end than the upper.

b. Mulls

For materials that do not form films, such as powders, other techniques are employed. The most common are mulls, pellets, or wafers. These techniques require that the particles be one-tenth of the shortest wavelength or smaller to avoid distortions in the spectrum due to effects such as scattering with a concomitant sloping baseline and/or the Christensen effect [3, pp. 283f] [53, pp. 6f]. Since scattering is most pronounced when the cross section of particles is 0.4λ [158, p. 415], the particles should be less than 1 μm across for infrared analysis. If grinding cannot achieve small enough particles, then ideally the matrix material should have a refractive index in the infrared region similar to that of the powder to minimize these effects.

For the oil/mull method, about 10 mg of sample are ground with a mortar and pestle until a fine powder is obtained. A small amount of mulling oil such as Nujol (a highly refined mineral oil) is added, and the mixture is ground until a pastelike consistency is obtained. The mull paste is spread between two salt plates and mounted in the infrared instrument to obtain the spectrum. Mechanical grinders such as the Wig-L-Bug apparatus using a stainless steel vial and balls are faster but more difficult to clean. If the sample is difficult to grind, it can be ground with sodium chloride which acts as an abrasive, or if it is rubbery, it can be frozen with liquid nitrogen

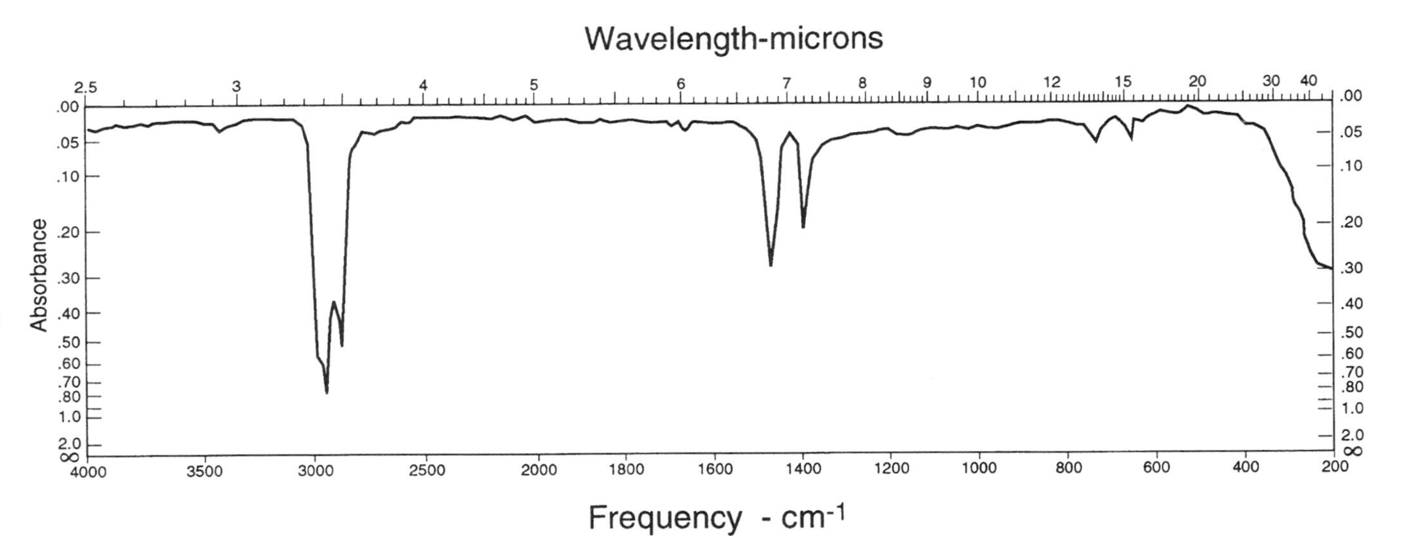

Fig. 2.10 Infrared spectrum of Nujol.

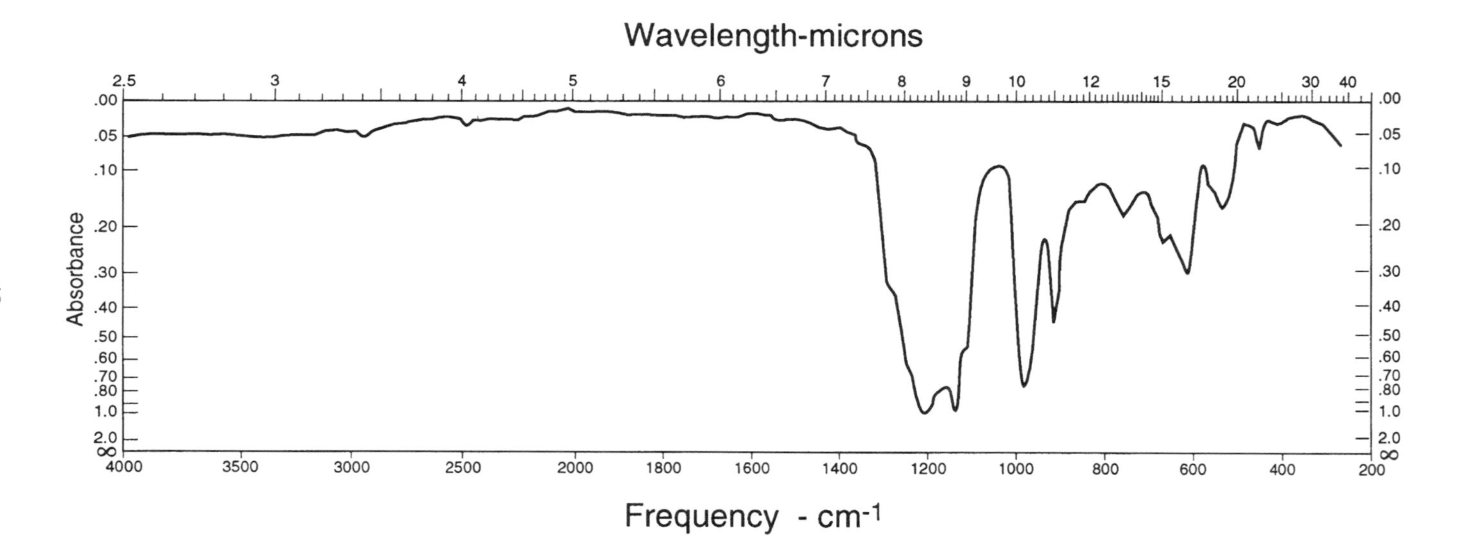

Fig. 2.11 Infrared spectrum of fluorolube.

to make it brittle before grinding. A small amount of solvent can be added to aid in grinding, but it should be vaporized before collecting the spectrum. The position of the solvent bands should be known. The spectroscopist needs to verify that they are missing.

Nujol exhibits significant infrared bands above 1300 cm^{-1} and below 750 cm^{-1} and thus can be used for samples having bands in the 1300- to 850-cm^{-1} region for which interferences generally are minimal. Fluorolube, a completely fluorinated hydrocarbon with about the same viscosity and refractive index as Nujol, may be used for samples absorbing in the region from 4000 to 1300 cm^{-1}, since Fluorolube shows no strong bands above 1300 cm^{-1}. Reference spectra of these two materials are shown in Figs. 2.10 and 2.11 [53, p. 15]. The process of preparing two mulls for a particular sample, one with Nujol and one with Fluorolube, to obtain a spectrum completely free of interference bands is called the *split mull technique*.

c. *Pellets and Wafers*

In many laboratories, the KBr pellet technique has essentially replaced the mull method. The sample is first ground in an agate mortar and pestle and then mixed with an alkali halide such as KBr, AgCl, KCl, or Csl and reground. The mixture is compressed under at least 10 tons of pressure over a period of 2 to 10 minutes into an infrared-transparent disk (they may not be optically transparent). For a 13-mm die, about 300 mg of KBr and 1.5 to 3 mg of sample are required.

The KBr Ultra-Micro Die from Perkin-Elmer allows for the pressing of either 1.5-mm or 0.5-mm diameter pellets, requiring only one to two micrograms of sample. Strong absorbers can be run at the submicrogram level. In this case a press capable of delivering 500 pounds of total force and a vacuum pump are required. A beam condenser is needed to obtain the best spectra with micropellets. The pellet produced may require a special holder for mounting the sample in the beam, or the pellet may be pressed into the holder itself.

As an alternative to KBr pellets with a die and press, two hardened steel blocks, about 1 by 1 by $\frac{1}{2}$ in., can be used to create KBr wafers. The blocks are sanded and washed with white soap and rinsed well. With the KBr and sample mixture between the steel blocks, 5000 pounds of force exerted for just a few seconds produce small oval wavers ranging from 20 × 5 mm to as small as 5 × 2 mm. The pellet produced is mounted over the holder with strips of tape. This method is preferred in some laboratories because it is much faster.

The advantages of the KBr pellet method over the mull method for sample preparation include the following:

- Spectra obtained are essentially free of interfering bands.
- Pellets can be stored in a dessicator.

- Spectra often show superior resolution compared to other preparatory techniques.

There are two disadvantages:

- For ionic samples, the bromide may exchange with the anion of the sample; for example, the KBr can react with amine hydrochlorides and produce amine hydrobromides.
- Water is almost always present, so the O–H stretching vibration near $3450\,cm^{-1}$ and the O–H bending vibration near $1640\,cm^{-1}$ are found in spectra of many samples prepared in this way. If the measurement of hydroxyl is critical, this method cannot be used.

The KBr abrasion method is useful for obtaining a spectrum of a thin coating such as the sticky side of a strip of tape or film. The sample is fastened to a smooth, hard surface. A small amount of KBr powder is sprinkled over the sample material, lightly rubbed into the surface with a steel plate, and the KBr scraped off with the edge of the steel plate. The KBr with minute particles from the sample surface can now be pressed into a wafer.

Insoluble, hard materials can be reduced to a useful sample size by using a flat mill bastard file. New files should be rinsed with heptane and acetone to remove any protective oils before using to file off tiny particles. Files should be used only once because it is nearly impossible to remove residual microparticles from the file surfaces. Sandpapering can also be done, but it is less satisfactory.

For very hard samples one can polish both sides of the material until the desired thickness is achieved, but some effort and care is required to ensure that the composition of the material is not altered. For samples that are extremely hard, techniques other than transmission infrared spectroscopy, such as diffuse reflectance, photoacoustic spectroscopy, or attenuated total reflectance (if one can achieve sufficient contact with the internal reflectance element) may be employed. For hard, carbon-filled polymers, pyrolysis is recommended. For extremely hard, small particles, an IR microscope may be required. As a matter of fact in many cases the infrared microscope is found to be superior for obtaining a useful spectrum of small defects, since it avoids matrix effects and is faster. This technique is the subject of Chapter 7 in this book.

A diamond compression cell (with pressures as high as 1000 psi) is available for transmission studies of single fibers and other microsamples [155] [156]. The orientation of functional groups in polymers as crystals may be of interest and is determined by employing a polarizer in the infrared beam.

Accessory and Methodology Choices

Since this chapter is concerned with transmission infrared spectroscopy, we will limit the accessories to those applicable for this type of analysis. Note that the advantages in sensitivity of Fourier transform infrared spectroscopy allow the spectroscopist to choose among a wider variety of sampling accessories than for dispersive infrared spectroscopy.

Harrick [155] gives a nice summary of the considerations that need to be addressed before the purchase of an accessory, namely the size of the sample compartment, the holder location and means of mounting, the direction of beam travel, the location of the focus, the beam shape, beam magnification or condenser, and the presence or absence of rail mounts.

The choice of a sample cell is critical to the success of many analyses. Harrick [155] lists ten considerations for choosing sample cells, one of which is the path-to-volume ratio for liquid and gaseous samples, since the amount of sample available may limit the type of accessory that can be used. The temperature range of the accessory may also be of interest. The phase of the sample is, naturally, one of the most important factors in the type of sample cell chosen, as we have mentioned above.

For solids, devices to hold salt discs vary mainly in the size of the aperture in the sample holder and the type of device to hold the sample, usually a spring clamp or some type of screw device. Since the beam is roughly circular in FT-IR, the aperture of the holder should be circular rather than rectangular in shape, as found for sample holders in dispersive IR. The location of the sample compartment is also important, since the location in many spectrometers depends on where the beam is focused, usually in the center of the sample holder or to either side. A beam condenser may be necessary for very small samples (specks or defects), unless an FT-IR microscope is available.

Two basic types of polarizers are available. One is based on wire grids and the other on the reflection off dielectric surfaces at critical angles. Discussion of these types of polarizers may be found in Harrick [155]. Other items often desirable in the laboratory include a mortar and pestle, a desiccator, a pellet press (which can be a heated press), a vacuum pump, a goniometer, and a variety of sample cells and holders. Usually a polishing kit is recommended if salt discs are used. The analysis of pyrolyzates (liquids and vapors) can be accomplished with modified test tubes or special chambers from Spectra-Tech [159. pp.11–13].

Setting up the Instrument for Sample Spectrum Collection

Before a spectrum is collected, the spectroscopist needs to make a number of choices, with the goal of achieving the most meaningful spectrum with the desired signal-to-noise ratio (S/N) in the shortest time, or providing the best match to library spectra.

Choosing the resolution is a important decision, since it typically ranges from 32 cm^{-1} (about 500 data points) to about 2 cm^{-1} (about 8000 data points). A resolution of 0.02 cm^{-1} or better is attainable but not from most currently available spectrometers. Obviously better resolution requires much more file space in the computer, but it may be desired for the spectral comparisons required, especially if the library includes spectra at 2 cm^{-1} resolution. The spectroscopist will have to choose (1) the spectral range covered, (2) the number of scans needed, (3) the signal gain, and (4) resolution [128, p. 98]. The advantage of accumulating multiple spectra is that the signal-to-noise ratio increases as the square root of the number of scans. This results from the fact the noise increases as the square root of the number of scans and the signal increases as the number of scans. This process of improving the S/N ratio by adding together corresponding data points from different scans is called *co-addition* (for "coherent" addition).

Certain trading rules must be considered such as the realization that more scans require more time. Greater resolution also requires more time to achieve the same signal-to-noise ratio. Once the parameters have been chosen, they should be the same for both the background spectrum and that of the sample.

Depending on the instrument design, the spectroscopist may (or may not) also be able to control (1) the size of the phase array file for the phase correction, (2) the level of zero fill (increasing the smoothness of the transformed data by calculating more spectral points), and (3) the apodization function. The 100%T and 0%T line on a dispersive instrument will have to be set using an open beam and a blocked beam, respectively.

In FT-IR, before the sample or background spectrum is run, usually the sample chamber is purged with nitrogen for about five minutes (depending on the efficiency of the purge system) to reduce the carbon dioxide and water peaks. Then the background spectrum is collected. (On one commercial system this process decreases the water absorption to about 37% of its original value.) The sample is then introduced into the chamber, and again the sample chamber is purged for five minutes; then, the spectrum of the sample is obtained and labeled. A convenient feature on many spectrometers is the option of collecting or co-adding more scans on an already collected spectrum.

Identifying Acceptable Spectra without Artifacts

Once the sample spectrum is obtained, it should be examined to determined if it is acceptable or if it exhibits any undesirable characteristics, such as distorted bands or extraneous features ("artifacts"). If the peaks are too weak, more sample should be introduced into the chamber; if the peaks are too strong, dilution of the sample is required. For some good spectral examples of both cases, see [102, pp. 144f].

What constitutes an "acceptable" spectrum? For general analytical work and for producing reference spectra, the major peaks should show a maximum between 5 and 25% transmittance, ideally around 10% transmittance. The background or baseline should be adjusted to between 90 and 95% transmittance [53, p. 14].

A badly ground sample prepared in a salt matrix will exhibit two artifacts, namely, a rising baseline in %T, from a low %T at higher wavenumber to a higher %T at lower wavenumber, and asymmetric band shapes, the latter being due to the Christiansen effect [53, pp. 6–7] [103, p. 520] [98, p. 93] [99, p. 234] [160] [161] [102, p. 147].

The Christiansen effect is due to particle size and differences in the refractive index of the matrix and the sample. Figure 2.12 (see [53, pp. 6–7] [162, p. 53]) illustrates how the refractive index of the sample decreases rapidly as we approach the absorption maximum of the sample from the

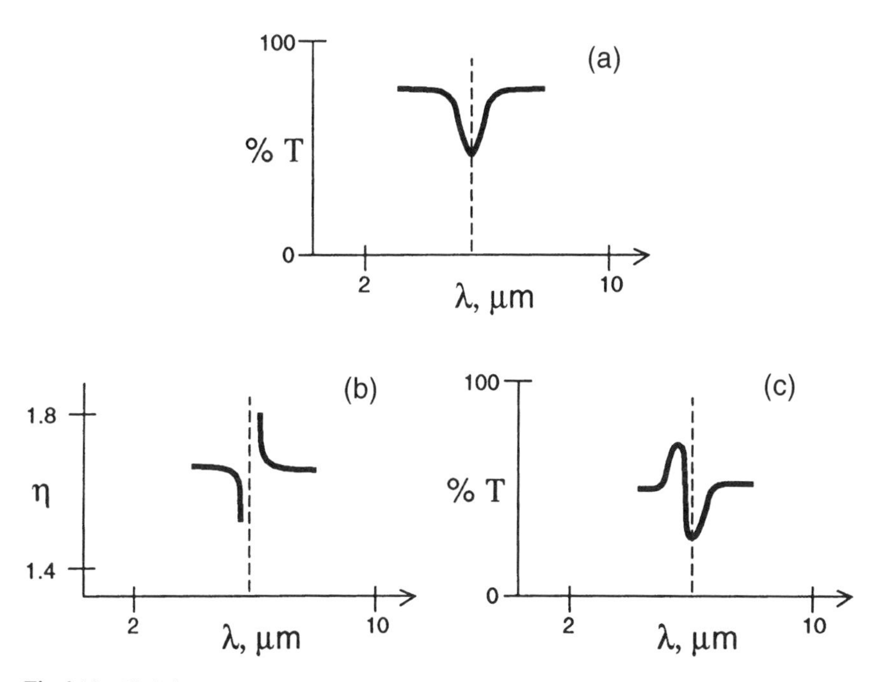

Fig. 2.12 Christiansen effect. The "true" band (*a*) shows a sharp peak in a plot of $\%T$ versus wavelength (μm). When plotting the refractive index versus wavelength for small particle samples in a matrix, a discontinuity may appear in the region of the absorption (*b*). The resulting distortion of the band (*c*) is shown in the $\%T$ versus wavelength plot.

shorter wavelength direction and approaches the refractive index of the matrix. Negligible differences in the refractive indexes of the sample and the matrix result in much less scattering from the particles in the matrix; thus more incident radiation reaches the detector. An increase in transmittance, rather than a decrease in transmittance, will occur just before the absorption maximum. In contrast, as we pass the absorption maximum and continue toward longer wavelengths, the refractive index of the sample also passes through a maximum. The significant difference between the refractive indexes of the sample and matrix results in much scattering and less radiation reaches the detector. The percent transmittance just after the absorption maximum is even less than expected. A sample exhibiting the Christiansen effect can be pressed into a disc of potassium iodide or cesium iodide rather than the typical potassium bromide, or reground and pressed into a new KBr disc [3, pp. 283–284]. The *Reststrahlen* effect results from the variation in refractive index of the sample as it goes through an absorption maximum.

A comprehensive article by Jones et al. [163] discusses photometric/instrumental errors more fully, including optical effects within the sample cell (reflection losses, beam convergence, polarization and nonparallelism of cell windows), and optical effects within the sample layer (reflection, interference, and anomalous dispersion). Hirschfeld [164] [165] and Wilson [166] offer in-depth discussions of instrumental errors. MacKenzie [167, pp. 31f] discusses the reaction of the sample with the matrix and the effect of nonuniform distribution of materials, including holes in the sample. Launer [168] discusses spurious bands from foreign gases, the effects of fringing or channeling, particle size, crystallized versus amorphous samples, and crystal orientation. Robinson [143] discusses stray radiation and instrument errors. The temperature of the sample may affect the spectrum [169], and low-boiling liquids in a badly filled or badly sealed liquid cell may show the effect of bubbles in the sample. Williams also covers artifacts and presents a number of examples [170]. Controlling errors in infrared spectroscopy is the subject of a series of articles in *Spectrochimica Acta* [171] [172].

Data Manipulations

A wide variety of data manipulations can be performed with the currently available software. Once collected, the spectrum can be examined in greater detail using massaging techniques such as the roll, zoom, cursor, and autoscale options.

A sloping baseline (see Fig. 2.13*a*) may distort important bands and make interpretation more difficult. To improve the spectrum, a baseline correction is performed (see Fig. 2.13). First, we must be sure the spectrum is in

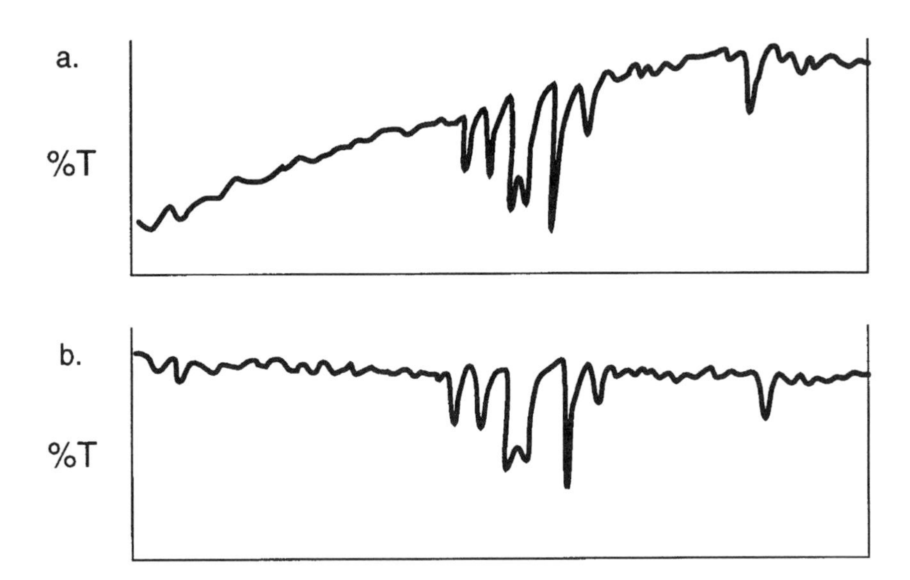

Fig. 2.13 Computer-controlled baseline corrections. An unacceptable sloping baseline in (*a*) may be corrected by computer so that the baseline is horizontal (*b*). Note that the corrections were made on the absorbance spectrum, then converted back to %T.

absorbance format rather than in percent transmittance format. One difficulty is the adjustment of the absorbance values after this correction, since some software packages don't set the baseline to zero absorbance after the correction (Fig. 2.13*c*) [102, pp. 146f].

Peak labeling is much more convenient than in previous decades, when the spectroscopist needed to identify specific peaks or bands manually. Most software will print out peak positions and intensities directly on the spectrum adjacent to a given peak. A means of setting the threshold for which peaks should be labeled is a necessity. Most software prints out the conditions at which the spectrum was run.

One very useful computer software capability is spectral subtraction. Multispectral display is a must for subtraction. If a solution spectrum is obtained, spectral subtraction allows the spectroscopist to computer-subtract the solvent bands quickly and efficiently, and thus obtain the spectrum of the solute. All contemporary software allows conversion of the spectrum quite readily from absorbance versus wavenumber to percent transmittance versus wavenumber, and vice versa.

The plot format should be easy to modify. Spectral compression above 2000 cm^{-1}, usually by a factor of two, is often desirable, since most of the critical information is typically found between 200 and 2000 cm^{-1}. Other capabilities should include a means of eliminating peaks in the interferogram so as to remove fringing in the spectrum. Access to the raw data, phase array, and apodization functions gives the spectroscopist additional freedom to massage the data. The data should be in a form that can be transferred to another computer.

Deconvolutions of the peaks, as well as the ability to assign the appropriate baseline and integration limits for peak areas, are other desired features. Software may be specifically written to provide peak ratios and more sophisticated calculations in some commercial systems.

It can be of value to have two implementations of spectroscopic software, one on the instrument and a second on another PC. This allows data manipulation without delay in data acquisitions, and accumulation of data into a common format from instruments having varying formats.

2.5. INTERPRETATION

Qualitative Analysis

As we have discussed in some detail above, each vibration and rotational transition in a molecule will have a specific energy associated with it. Transitions among each of these vibrational modes will usually result in absorption of infrared radiation in a specific region of the infrared spectrum. This fact is the basis for structural determinations by infrared spectroscopy. The "fingerprint region," which lies between 910 and 1430 cm^{-1} (7 to 11 μm) is generally accepted as unique for each molecule [173, p. 29].

Correlation Charts and Spectral Libraries

The analyst interested in qualitative identification of samples generally relies on accumulated empirical data, relating infrared absorption bands with structural features of molecules. A number of correlation charts have been published, which can be of great value, including Colthup, Daly, and Wiberley's *Introduction to Infrared and Raman Spectroscopy* [43] and Bellamy's *The Infrared Spectra and Complex Molecules* in two volumes [174] [175]. Colthup [176] published one of the earliest correlation charts in 1950. Other instructive books and articles on spectral interpretation include those written by Silverstein, Bassler, and Morrill [177] [178] and by Simons [179].

The best-known spectral library, consisting of more than 50,000 spectra, is the Sadtler Infrared Spectral Library (available in both computer and hard copy). The Perkin-Elmer Library, the Sprouse Collection, the Aldrich

TABLE 2.6 Regions of Absorptions for Various Functional Groups

Region	Frequency	Look for	Indication
1	3300	Broad, strong peak	O–H or N–H stretch
2	>3000	Any absorption	C—H where C is sp^2 or sp (often aromatic)
	<3000		C–H where c is sp^3
3	2250	Sharp peak	Triple bond to C
4	1650–1780	Strong, sharp peak	Carbonyl
5	1050–1230	Medium-strong peak	C–N or C–O stretch
6	690–840	Strong peak(s)	C—H out-of-plane bends

Library, and the Environmental Protection Agency library are other important spectral collections. An excellent, recently published bibliography of reference works for infrared and Raman spectral interpretation is located in an article by Coates [180].

The use of noncomputerized spectral libraries to find spectral matches is a time-consuming effort, however, and the analyst must have some idea of what important functional groups are present in order to narrow down the field of possibilities. To assist the beginner, McMinn [181] has listed six important regions in the infrared spectrum where some generalized types of vibrational transitions occur, given in Table 2.6. As an example, we discussed the various modes of stretching and deformation for infrared absorption by the $-CH_2-$ moiety in Section 2.2. The approximate spectral positions for those modes [53, p. 3] are listed in Table 2.7. As experience in spectral interpretation is gained, pattern recognition becomes the primary means of spectral identification.

TABLE 2.7 Infrared Spectral Positions for Vibrational Modes for the $-CH_2-$ moiety

	Spectral position	
Transition Type	cm^{-1}	μ
Asymmetric stretch	2926	3.42
Scissoring	1468	6.81
Twisting	1305	7.66
Symmetric stretch	2853	3.51
Wagging	1305	7.66
Rocking	720	13.89

Computer Searching Libraries and Spectral Searching Algorithms

Computer-assisted interpretation is now common with interactive software programs. *IRexpert* [182] and *EXPEC* [183] are two programs designed to assist in this effort. The actual libraries associated with these programs are limited to about 10,000 spectra. In contrast to computer-assisted interpretation, computer searching does not interpret the spectrum.

"Reverse" searching is most useful for spectra believed to be from samples likely to contain impurities or for spectra of mixtures. The computer determines which of the reference spectra have bands that are all contained in the sample spectrum. Thus the reverse search method ideally will identify all the components present, assuming that those components are in the reference library.

"Forward" searching is quicker and is useful if the sample is assumed to be "pure." Forward searching requests the computer to match all the bands in the unknown spectrum with those of a reference spectrum in the computer's spectral library. "Extra" bands in the unknown spectrum will cause the correct match to be rejected.

Computerized searching works best when the computer's task is to select a spectrum from a closed list of compounds, each one of which has a unique frequency of substantial intensity. The spectrum to be searched should be baseline-corrected and normalized before searching commercial databases, and the sample should be free of impurities and be prepared in exactly the same way that the material for the library spectrum was prepared. The state of the material (crystalline or amorphous, in solution, or neat, liquid or gas) should be the same as that of the reference material in the library. Instrument wavelength calibration must be the same as well as the instrument's handling of any scattering from the sample. Deresolving the spectrum can lessen the dependence on peak width variations resulting from sample preparation. Some computerized searching systems react with an error message if the wavelength range of the unknown spectrum is not the same as the wavelength range of the library spectra.

As one deviates from the requirements above, the probability of adequate sample identification is reduced. In this circumstance it is best to regard the search program as a tool for eliminating very unlikely matches and providing a list of 20 or so spectra that will be sorted through by hand and examined for a "match." Experience will allow the spectroscopist to sense if a match is likely to be an impurity or a true match. The experienced spectroscopist should never ever use a single band seriously to suggest the presence of a unique compound.

It can be helpful in assisting the computer search program to find useful results in less than ideal circumstances if one limits the search range to $100\,cm^{-1}$ down to the lower limit of the library spectra; this avoids most of the water and carbon dioxide interferences and mitigates the worst of the effects from scattering effects. If obvious impurities are present, all may not be lost; many search programs allow multiple mask regions to blot out the peaks from the impurities. Indeed the performance of some of the commercial searching programs in a protocol of identifying one component of a mixture, subtracting the reference spectrum of that compound, and searching the remainder to identify another component is phenomenal.

Hit values from the search need to be good, of course, but spectroscopic uniqueness is suggested by the degree of difference among the hit values in the search report. The more different they are, the more likely the hit is correct. The subject of spectroscopic orthogonality is treated well in Kalivas and Lang's text [184].

Even if the hit qualities are not good, there may be some information available. For example, if all the hits are aliphatic esters, one might suspect, after visual inspection of the data and the library spectra for counter-indications, that an aliphatic ester character is present. Peaks that do not fit this characterization are good candidates for other mixture components or impurities.

Sometimes mixtures can be parsed if some method of causing a separation, however slight, in the components can be effected. The various components can be spotted by seeing which peaks go up or down together in intensity.

Lastly, as always, confirmation of an identification from spectroscopy should be sought in other spectroscopies and in simple wet chemical tests, perhaps conducted under a microscope.

Quantitative Analysis

Beer-Lambert Law

The basis for any quantitative work in spectroscopy is the Beer-Lambert law* relating the absorbance A to the path length b and the concentration c with the proportionality constant a (called the "absorptivity," the "extinction coefficient," or other less commonly used terms) as

$$A = abc. \tag{2.4}$$

The absorbance is related to the transmittance by

$$A = -\log T = \log\left[\frac{I_0}{I}\right] \tag{2.5}$$

*This law assumes that the absorbed energy in an excited vibrational state is emitted before another photon relevant to this transition arrives.

in which I_0 is the intensity of the light beam before it enters the sample, I is the intensity after it passes through the sample, the T is the transmittance [185] [186]. We should bear in mind that the Beer-Lambert law is strictly valid for monochromatic light.

The path length b of a sealed or demountable cell can be determined only roughly by measuring the thickness of the spacer with a micrometer. A more accurate method is to insert the empty cell into the spectrometer and observe the interference pattern created by reflection of part of the incident beam from the internal faces of the cell [187, pp. 101f]. Calculation of the pathlength is accomplished by measuring the fringe spacings (see [187]). Cell lengths greater than 1 mm are difficult to measure by this method, because the fringe separation is small.

Other methods include calculating b from a measured absorbance when the cell is filled with a substance of known absorptivity. Benzene was formerly a common choice, since it has a large number of bands with widely differing absorbancies [188, p. 74], but it is used less frequently now because of its toxicity. As an example, the 850 cm^{-1} band of benzene has been found empirically to have an absorbance of 0.22 for each 0.1 mm of thickness [188, p. 74]. Another method is to match the absorbance of a sample of known absorptivity in the cell to be measured with the absorbance of the same material in a variable path-length cell. If the path length of the variable cell is calibrated, then the pathlength of the cell to be measured must be identical [187, p. 102].

Use of Intensity Measurements

For a given path length and concentration, the numerical value of the absorbance observed depends on the ability of the molecule to capture the electromagnetic radiation (i.e., it depends on the absorptivity or extinction coefficient a of the molecule). Molar intensity data have been obtained by Flett for several characteristic chemical groups and are reported both as peak extinction coefficients and as integrated band areas [188]. Absolute absorption intensities for several group vibrations may be found in [51, pp. 57f] and [99]; see Table 2.8.

To assist in converting among the different units, let us use the carbonyl group of an alkyl ketone as an example from Table 2.8. This group has an integrated absorption of about 8×10^{-7} cm^2/molecule second. The reader can easily convert to different units with the conversion 1 L/mole $cm^2 = 1.15 \times 10^{-10}$ cm^2/molecule second [188, p. 1541], leading to 7000 L/mole cm^2 or 7000 "darks" [3, p. 19]. One other common way to express the integrated band areas is 7.0×10^6 cm/mole. Finally the area can be expressed as Γ which is related to the absorbance through the frequency, usually in wavenumbers, via the approximation $A = \Gamma v$, where v is the observed band center [189].

TABLE 2.8 Absolute Intensities for Several Group Vibrations

Group	Position of bands, cm^{-1}	Integrated absorption, $\times 10^7$
C=O	~1720	
Esters		13.0
Alkyl ketones		8.0
C–O	~1200	
Esters		15.0
Alkyl ketones		2.5
N–H	~3400	
Dialkylamines		0.05
Alkylanilines		1.7
C≡N	~2250	
Alkylcyanides		0.25
CH_3	~2900	
Hydrocarbons		4.4
$-COCH_2-$		0.74
$-COOCH_2-$		2.54
CH_3	~1460	
Hydrocarbons		0.54
$-COCH_2-$		1.3
$-COOCH_2-$		2.0
CH_2	~2900	
Hydrocarbons		3.8
$-COCH_2-$		0.5
CH_2	~1460	
Hydrocarbons		0.23
$-COCH_2-$		1.17
$-COOCH_2-$		0.65

Source: [98, p. 262].

Note: In units of cm^2/molecules.

The theory of intensity in the infrared spectra of polyatomic molecules is delved into in some detail in books by Gribov and Orville-Thomas [63, pp. 335f] and by Painter, Coleman, and Koenig [73, pp. 167f].

Infrared intensity data can be used to determine intra- and intermolecular interactions [190]. Accurate intensity data allow the determination of static and dynamic charge distributions in molecules. Theoretical models have been developed to delineate these distributions.

Procedures for Infrared Quantitative Analysis

In quantitative work the use of calibration curves for one particular wavelength of visible light is a familiar procedure in most laboratories. Translating the method to the infrared region is not quite so simple. The difficulties and limitations of quantitative infrared spectroscopy have been discussed for years [191] [164] [192].

The simplest method is to select a region of the infrared spectrum that has a dominant peak characteristic of the solute (and not the solvent) and is free of other absorbing bands nearby. In quantitative work the detector should not be flooded; the study must be undertaken in a linear region of the detector. Spectra must be collected for a range of concentrations within which unknown concentrations are likely to lie.

The concentration of the unknown sample may be related to the peak height at the selected specific wavenumber. However, for some bands the wavenumber of the maximum absorbance changes as a function of concentration or even the band width due to interactions between the solute and solvent. Thus one needs to empirically examine the peaks and carefully check the wavenumber location of the analytical band in question, rather than automatically assume that the maximum will not move [139, p. 10]. Using peak areas may avoid this problem to some extent (see below).

A second and more commonly used method involves ratioing the peak areas of analyte and reference rather than simply measuring the peak heights for different concentrations of solute, with one peak being used as a reference so as to monitor thickness variations.

One primary question in determining either the areas of peaks or the heights of the peaks is, How do you determine *exactly* where the baseline is? Methods for determining the baseline are discussed in [55, pp. 168f] [153, pp. 275f] [193, pp. 50f] [194] [195] [196]. The decision as to where to draw the baseline is most likely to be the primary error source. Establishing a calibration curve relating peak heights or areas versus concentration is best done with a least squares regression program, provided that there are no overlapping peaks.

Typical precision for determination of the concentration of an unknown solution or mixture is 3 to 10% by weight [197], but standard deviations of 0.03% have been reported [198]. Note that the maximum precision in absorbance is said to occur at a transmittance value of 38.8% T [199, p. 980], but in some cases the optimum transmittance may be as low as 2% T or higher than 90% T [196]. A discussion of procedures to minimize the errors involved in quantitative IR work is found in Robinson [168, pp. 101f] and includes the following:

1. Choose absorption bands free from interfering bands.
2. Use cell windows that match the refractive index of the solvent.
3. Use solutions at a concentration of $<2\%$ if possible.
4. Be extraordinarily careful with cell positioning.
5. Work at sufficiently high resolution so that you are approximating monochromaticity, which is about 1/10 bandwidth.

For dispersive instruments,

1. Adjust the spectrometer for wide slits and low noise.
2. Set zero accurately.
3. Adjust all variables so the intensities lie between 20 and 60% *T*.
4. Scan slowly.

For FT instruments,

1. Run the background and the sample as close in time as possible.
2. Operate with narrow band passes to maintain more linear detector response.
3. Choose the optimum apodization function, which may be Happ-Genzel for close bands and "boxcar apodization" for isolated bands [196].

Chemometrics

Chemometric* methods are often used in quantitative IR analyses [202] [203]. Classical least squares analysis (CLS) using a *K*-matrix is useful for moderately complex mixtures, since peaks of interest may be overlapping [201]. Multiple linear regression (MLR) and inverse least squares (ILS) using a *p*-matrix is useful for very complex mixtures [201]. The most utilitarian aspect of these chemometric methods is that the spectroscopist is not only able to improve precision but also is able to extract spectra of unknowns from complex mixtures.

Principal component analysis (PCA) allows for partial modeling of nonlinearities. Partial least-squares methods (PLS) [202] and principal component regression (PCR) are full-spectrum multivariant calibration methods that are able to reduce the calibration spectral intensity data at many frequencies to a relatively small number of intensities in a transform full-spectrum coordinate system [202].

Calculated Spectra

The spectroscopist most interested in molecular structure determinations rather than qualitative or quantitative analysis of a sample may be involved in normal coordinate analysis, as described in the theory section. In some cases it may be possible to propose a structure, calculate the spectrum expected from this structure, and then compare it with that of the unknown. Calculation of several structures may provide enough insight to deduce the correct structure. This procedure is expensive and time-consuming at

*The term "chemometrics" literally refers to any type of mathematical manipulation of chemical data, but common usage limits the use of the term to predicting properties of materials using chemical data and linear algebra [200] [201].

present, but great strides have been made and are being made; small compounds can be calculated on a desktop IBM compatible or MacIntosh computer.

2.6. EXPERIMENTAL RESULTS AND CASE STUDIES

Water in the Gaseous State

The infrared spectrum of water vapor has probably been studied more extensively than that of any other molecule; thus it lends itself as an interesting example through which one can gain some comprehension of the characteristics of vibration-rotational spectra observed for gases and how that information can be used to determine rotational constants, moments of inertia, bond angles, and other molecular and thermodynamic information.

A spectrum of air containing water vapor and carbon dioxide is illustrated in Fig. 2.14, covering the mid-infrared range from 400 to 4000 cm^{-1}. The water vapor present exhibits a dense pattern of vibration-rotational peaks in the 1400 to 1800 and 3550 to 3900 cm^{-1} regions, and part of the high energy pure rotational spectrum in the region from 650 to 400 cm^{-1}. The carbon dioxide present exhibits a perpendicular band centered at 667.3 and a parallel band centered at 2349.3 cm^{-1}.

Relating that complex spectrum to structural parameters for the water molecule may involve significant effort, and it requires matching the actual spectrum with the computer-generated spectrum, based on educated guesses regarding some transitions and their vibrational frequencies as well as guesses as to the rotational constants, the centrifugal distortion, and Coriolis coupling constants.

Water is classified as a very asymmetric rotor having three different moments of inertia. The three different vibrational modes are symmetric stretching, bending, and asymmetric stretching, designated as v_1, v_2, and v_3. Both the ground and all of the excited vibrational states of each of these vibrational modes have a ground rotational state and a number of excited rotational states. Figure 2.15 illustrates an energy-level diagram generated for the v_2 bending vibration mode of water in the gaseous state (see [3, p. 455] [204, p. 144]).

If the molecule were perfectly rigid, the rotational constants A, B, and C for both ground and excited vibrational states would be the same, and the pattern and spacings of rotational levels in both the ground and excited vibrational states should be the same. However, water not only is not a rigid rotor; it exhibits extremely strong centrifugal distortion [205] [206]. As a matter of fact, the highest pure rotational energies ever deduced from available spectra are claimed to be for water molecules, up to 11,000 cm^{-1} [205].

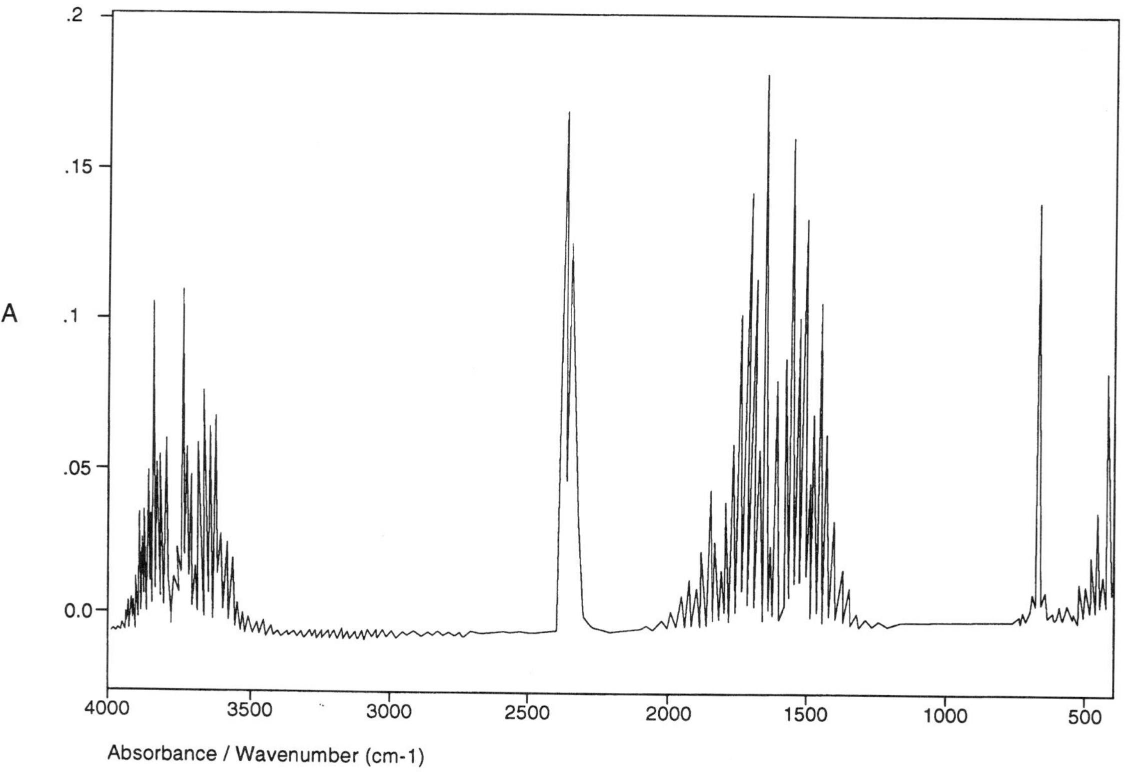

Fig. 2.14 Vibrational-rotational spectrum of water vapor. Plotted as absorbance versus wavenumber, the spectrum of water vapor in an IR gas cell with ambient air also shows some peaks from carbon dioxide present.

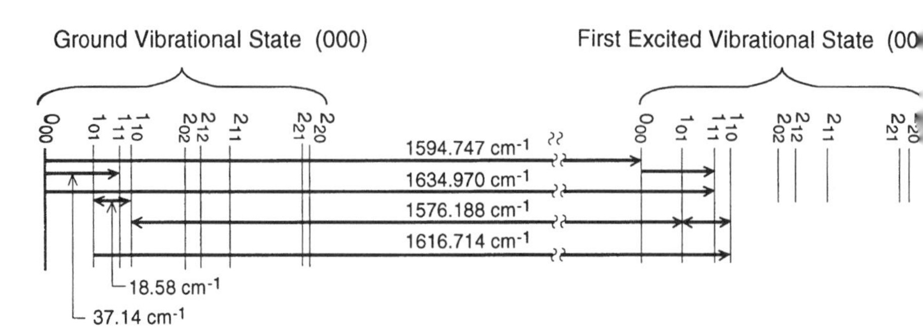

Fig. 2.15 Energy level diagram of the ground and first excited vibrational states of the bending mode for water vapor molecules. Some of the rotational-vibrational transitions and pure rotational transitions are indicated.

Calculation of A, B, and C are based on a series of equations incorporating the observed vibrational-rotational transition frequencies, almost always solved with the aid of a computer program [145, p. 192]. Note that the rotational constants for the other vibrational modes for water are usually slightly different. Thus the patterns are likewise distorted in different ways compared to that of the bending mode [207] due to these rotational constant differences and differences in the changing dipole moment direction. From the rotational constants, moments of inertia, I_i, can be calculated via equations such as

$$I_A = \frac{h}{8\pi^2 A}. \tag{2.6}$$

Using moments of inertia from a variety of molecules with different isotopic substitutions, we can calculate the distance from the center of mass to the center of the nucleus r, via equations based on (2.7), and from there determine bond distances:

$$I_A = \sum mr^2. \tag{2.7}$$

Force constants can be obtained from normal coordinate analysis [208] of the vibrational frequencies.

It is rather astonishing that such a few calculated constants can represent thousands of observed transitions, not only in the infrared, but also in the microwave and UV-visible range. The specific assignments of all the water peaks can be found in an extensive compilation by Fland et al. [209].

Supporting information may be obtained on the rotational states of water vapor from the far-infrared and microwave spectra, which are not illustrated in this chapter. The pure rotational spectrum of water vapor in the range 0 to 530 cm^{-1} can be found in [145, p. 631].

Of importance to the general spectroscopist is that water vapor and carbon dioxide bands are typically seen in many infrared spectra, even in some chambers purged with "dried" air. In addition water vapor plays an important role in atmospheric transmission. Precise determination of its absorption and variation of the absorption with temperature and concentration is required for the solution of many problems related to the physics and chemistry of the atmosphere [210].

Quantitative Analysis of Xylene Mixtures

PLS and PCR are two modern methods for executing quantitative infrared analysis of mixtures. The results from these methods are, in general, superior to previous results using peak heights or peak areas, since these methods may be applied to spectra exhibiting serious band overlaps. Since the mathematics of these methods are at best tedious, taking advantage of the software packages that are commercially available allows one to apply these newer techniques rapidly and with proficiency. The procedure for undertaking the analysis by using the commercially available software program *PLSplus for Grams/386* is well documented [211].

The application of these techniques to a three-component mixture of o-xylene, m-xylene, and p-xylene illustrates the determination of an unknown mixture to a precision of better than 5% by weight, and in some cases it suggests that an order of magnitude improvement is feasible. We chose this example because it can easily be reproduced in the laboratory. Ten calibration samples (a training set) were prepared including samples of the three xylenes and seven mixtures of the three components. Their infrared spectra were obtained and entered into the computer together with their compositions (see Table 2.9).

A mixture of *o*-xylene and *p*-xylene was prepared as a test for the method, so as to compare the calculated composition with the known composition. A spectrum of a sample with about a 25 μm path length was entered into the program. To perform the analysis, one must select the spectral region of preference and decide the number of factors needed to represent the data based on results from the ten calibration samples. We chose three factors and the spectral range from 1700 to 2000 cm^{-1} from spectra with a resolution of 2 cm^{-1} (157 total data points). We also chose options including sample averaging for more efficient processing and mean-centering.

The computer output using *PLSplus for Grams/386* for the quantitative analysis of the test mixture is listed in Table 2.10. The predicted composition of the test mixture appears to be acceptable compared to that of the

TABLE 2.9 Xylene Mixtures as Standards for Quantitative IR Analysis

Sample Label	o-Xylene[a]	m-Xylene[a]	p-xylene[a]
Pure o-xylene	1	0	0
Pure m-xylene	0	1	0
Pure p-xylene	0	0	1
Mixture 1	0.114	0.434	0.452
Mixture 2	0.444	0.112	0.444
Mixture 3	0.454	0.435	0.11
Mixture 4	0.661	0.339	0
Mixture 5	0.333	0.667	0
Mixture 6	0	0.669	0.331
Mixture 7	0	0.337	0.663

[a]Mass fractions.

TABLE 2.10 Computer Output of Quantitative Analysis of Mixture

	Prepared	Predicted	SPC Residual	F Ratio	Probability
o-Xylene	0.339	0.333	0.00947	1.2407	0.7108
m-Xylene	0	0.003	0.00946	1.2372	0.7103
p-Xylene	0.661	0.664	0.00946	1.2372	0.7103

prepared mixture. In the table the SPC residual is what is left of the spectral file when the spectrum is reconstructed and compared to original data. The F-ratio is the value of the PRESS (the sum of the squared differences between the predicted and known concentrations) for a specified number of factors divided by the minimum PRESS value obtained when the factors are varied up to the maximum chosen. The calculated probability indicates the significance of each factor up to the minimum of the PRESS.

Molecular Orientation Studies

Practical applications of polarized radiation in infrared spectroscopy include determining chain conformation and structure, estimating the degree of orientation and detecting conformation-sensitive bands and "crystallinity bands" [212, pp. 70f].

Molecular orientation studies may shed light on polymer deformation mechanisms when the material is stretched with a manual or motorized stretching device that allows symmetrical uniaxial deformation. The extension ratio is the ratio of the length of deformed material to that of undeformed material. Experimentally determined dichroic ratios are defined as

$$R_i = \frac{A_{\parallel}}{A_{\perp}}, \tag{2.8}$$

where $A_{\parallel}$ and $A_{\perp}$ are the absorbances of the same band measured with radiation polarized parallel and perpendicular to the stretching direction, respectively. The dichroic function F_i is defined as

$$F_i = \frac{R_i - 1}{R_i + 2}. \tag{2.9}$$

As an example, consider *cis*-1,4-polyisoprene (see Fig. 2.16). Three absorption bands must be found that are the result of motions whose transition moments form a set of three linearly independent vectors. Measurements of three such bands are sufficient to determine the angles between the transition moments of those motions and the local chain axis, in cartesian coordinates X, Y, Z as α_1, α_2, and α_3 [213].

The dichroic functions (F_i) for those specific infrared bands have been determined as a function of various extension ratios [213]. A graph of that function allows for estimates of F_i values. For an extension ratio of 5, for example, the F_i values are estimated as 0.04625, -0.0145, and -0.01475 for the 1663 cm^{-1} ν(C=C) band (v_1), the 1376 cm^{-1} $\delta_s(CH_3)$ band (v_2), and the 837 cm^{-1} γ(CH) band (v_3) of polyisoprene, respectively.

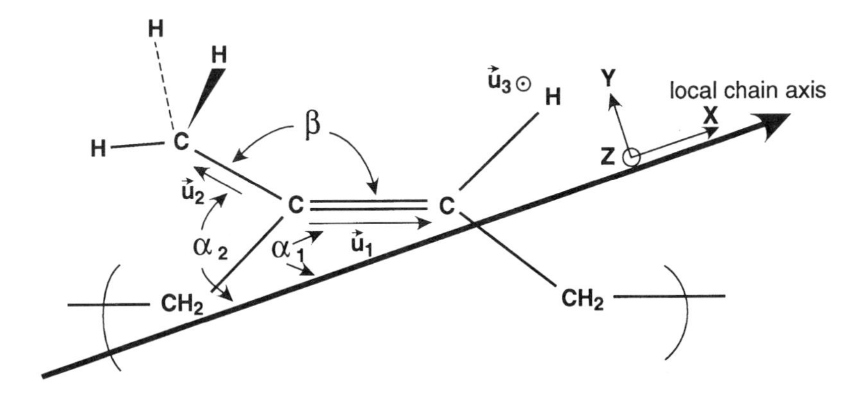

Fig. 2.16 *Cis*-1,4 polyisoprene. The local chain axis direction has been arbitrarily drawn.

Since the second moment of the orientation function $\langle P_2(\cos\theta)\rangle$ (defined for convenience as P_2) is the same for all three measured absorption bands, seven general equations can be formulated for the seven unknowns X, Y, Z, α_1, α_2, α_3, and P_2:

$$\delta_1 X = \cos\alpha_1, \qquad \delta_2(X\cos\beta + Y\sin\beta) = \cos\alpha_2,$$

$$\delta_3 Z = \cos\alpha_3, \qquad \frac{2}{3\cos^2\alpha_1 - 1}F_1 = P_2,$$

$$\frac{2}{3\cos^2\alpha_2 - 1}F_2 = P_2, \qquad \frac{2}{3\cos^2\alpha_3 - 1}F_3 = P_2$$

$$X^2 + Y^2 + Z^2 = 1.$$

Assuming that $\beta = 120^\circ$, then for the eight possibilities for δ_i of ± 1, with the subscript $i = 1$, 2, and 3, only one combination is physically meaningful after examining the calculations from a spreadsheet, namely with $\delta_1 = -1$, and $\delta_2 = \delta_3 = +1$. For this situation a value of the orientation function was determined to be 0.0637, and the values of $\alpha_1 = 25.0^\circ$, $\alpha_2 = 64.9^\circ$, and $\alpha_3 = 65.1^\circ$ were determined as well. Amram et al. [213] have gone beyond this and determined different orientational behaviors of the amorphous and crystalline parts in natural rubber. From the orientation function the average angle between the chain axis and the stretching direction was found to be 68°.

2.7 COURSES, USERS' GROUPS, AND MEETINGS

Regional and International FT-IR conferences have been held for many years. A variety of seminars on the subject are offered, particularly through instrument companies such as Perkin-Elmer, Digilab, Bomen, Spectra-Tech, MIDAC, Bruker Mattson, and Nicolet. Some of these companies have formed Users' Groups that meet periodically in order for spectroscopists to keep abreast of instrument updates, accessories available, and new developments in methodologies and techniques. We recommend that such companies be contacted for information on what is offered. Miami University of Ohio (513-529-2874) offers a short course on Microspectroscopy, and Arizona State University (602-965-4476) offers a course on Applied Molecular Spectroscopy.

Currently there are a number of World Wide Web sites that deal with infrared spectroscopy. We envision that such sites will be expanded and improved and that electronic communications will be significantly expanded among spectroscopists in the future.

ACKNOWLEDGMENTS

The authors would like to thank Richard Jackson of Bruker Instruments for his contributions to the quantitative analysis example, Liliane Bobozka at the Ecôlé Supérieure de Physique et de Chimie Industrielles (Paris) for permission to use information in her article for our chemometrics example, and Robert A. Toth, Jet Propulsion Laboratory, Pasadena, CA, for assistance with the water vapor example. Our thanks also go to Robert Pransis of 3M for contacts that led to inclusion of the chemometrics example.

We thank J. Stevens and D. Misemer for a private communication that we referenced. We thank Karen Palanzo of Hughes Danbury Optical Systems, Inc., Jerry Auth of MIDAC, Mike Eastwood of the Jet Propulsion Laboratory, and Pete Brendal for information about beam splitters. We would like to thank 3M Graphics Department for assistance with the figures, the 3M Technical Library staff, especially Mary Hanson and Debra Yndstad, for extensive reference assistance and Colleen Spicer, Rebecca Dittmar, Gerald Lillquist and James Westberg of 3M and George Cleary of Abbott Laboratories for contributing improvements to the text. And finally, we thank the scores of spectroscopists who shared valuable information with us. We welcome comments to Marilyn.d.duerst@uwrf.edu.

REFERENCES

1. William C. Dampier, *A Shorter History of Science*, New York: Meridian Books, 1957.
2. *Diffraction Grating Handbook*, Bausch & Lomb, Inc., Rochester, New York, 1970.
3. George W. Chantry, *Long-wave Optics: The Science and Technology of Infrared and Near-millimetre Waves*, Vol. 1: *Principles*, New York: Academic Press, 1984.
4. Nicolas C. Thomas, "The Early History of Spectroscopy," *J. Chem. Educ.*, **68**, 631 (1991).
5. Lovell, "Sir William's Invisible Light," *Optical Spectra*, Apr 1970.
6. W. Abney and E. R. Festing, *Phil. Trans.* **172**, 887 (1882).
7. W. H. Julius, *Verhandl. Koniki. Acad. Wetenschappen Amsterdam*, **1**, 1 (1892).
8. A. A. Michelson, "Visibility of Interference Fringes in the Focus of a Telescope," *Phil. Mag.*, 31:**5**, 256 (1891).
9. A. A. Michelson, *Phil. Mag.*, 34:**5**, 280 (1892).
10. W. D. Perkins, "Fourier Transform-Infrared Spectroscopy; Part I: Instrumentation," *J. Chem. Educ.*, *63*, **A5** (1986).
11. J. W. Cooley, and J. W. Tukey, "An Algorithm for the Machine Calculation of Complex Fourier Series," *Math. Comput.*, **19**, 297 (1965).
12. M. L. Forman, "Fast Fourier-Transform Technique and Its Application to Fourier Spectroscopy," *J. Opt. Soc. Amer.*, **56**, 978 (1966).
13. Michael MacRae ed., "Perspectives on Ten Years in Spectroscopy: A Look Back and a Look Ahead," *Spectroscopy*, **10(8)**, 52 (1995).
14. Richard W. Duerst, William L. Stebbings, Gerald Lillquist, James W. Westberg, William

E. Breneman, Colleen K. Spicer, Rebecca M. Dittmar, Marilyn D. Duerst, and John A. Reffner, "Depth Profiling and Defect Analysis of Films and Laminates: An Industrial Approach," in *Practical Guide to Infrared Microspectroscopy*, H. Humecki, Ed., New York: Marcel-Dekker, 1995, p. 137f.

15. Linda Crabtree, Ed., "What's New in Spectroscopy?'," *Spectroscopy*, **7(4)**, 16 (1992).
16. *International Light* (brochure), 17 Graf Road, Newbury MA 01950.
17. R. Feynmann, *Lectures on Physics*, Vol. II, 6th printing, Reading, MA: Addison-Wesley, 1977.
18. Max Born and Emil Wolf, *Principles of Optics*, 3rd ed., London: Pergamon, 1964.
19. David S. Falk, Dierter R. Brill, and David G. Stork, *Seeing the Light*, New York: Harper and Row, 1986.
20. William L. Wolfe, *Handbook of Military Infrared Spectroscopy*, Washington, D.C.: Office of Naval Research, 1965.
21. Max Bender, "The Use of Light Scattering for Determining Particle Size and Molecular Weight and Shape," *J. Chem. Educ., 26*, **15** (1952).
22. Francis A. Jenkins and Harvey E. White, *Fundamentals of Optics*, 4th Ed., New York: McGraw Hill, 1976.
23. John G. Foss, "Absorption, Dispersion, Circular Dichroism and Rotary Dispersion," *J. Chem. Educ.*, 40, **592** (1963).
24. T. G. Goplen, D. G. Cameron, and R. N. Jones, "Absolute Absorption Intensity and Dispersion Measurements on Some Organic Liquids in the Infrared," *Appl. Spectr.*, **34**, 657 (1980).
25. Alan G. Marshall and Francis R. Verdun, *Fourier Transforms in NMR, Optical and Mass Spectrometry: A User's Handbook*, New York: Elsevier, 1990.
26. S. G. Lipson and H. Lipson, *Optical Physics*, 2nd Ed., Cambridge, England: Cambridge University Press, 1981.
27. Alan G. Marshall, Ed., *Fourier, Hadamard, and Hilbert Transforms in Chemistry*, New York: Plenum, 1982.
28. Deane B. Judd and Gunter Wyszecki, *Color in Business, Science and Industry*, 3rd Ed., New York: Wiley, 1975.
29. J. R. Partington, *An Advanced Treatise on Physical Chemistry, Vol. 4: Physico-Chemical Optics*, London: Longmans, Green, and Co., 1953.
30. Richard T. Weidner and Robert L. Sells, *Elementary Classical Physics*, volume 2, Boston: Allyn and Bacon, 1973.
31. Bernd Jahne, *Digital Image Processing: Concepts, Algorithms and Scientific Applications*, 3rd ed., Berlin: Springer-Verlag, 1995.
31. P. W. Kruse, L. D. McLaughlin, and R. B. McQuistan, *Elements of Infrared Technology: Generation, Transmission and Detection*, New York: Wiley, 1962.
33. Hans Herbert, "Exercises with an Optical Diffractometer," *Phys. Educ.*, **17**, 124 (1982).
34. N. A. Dodd, "Computer Simulation of Diffraction Patterns," *Phys. Educ., 18*, 294 (1983).
35. A. C. Thompson, G. Kozimer, and David G. Stockwell, "A Light Scattering Experiment for Physical Chemistry," *J. Chem. Educ., 47*, 828 (1970).
36. Franklin S. Harris, Jr., "The Physics of Light Scattering," *Opt. Spectra*, July/August 1970, p. 52f.
37. Wen-Chung Tsai and Ronald Pogorzelski, "Eigenfunction Solution of the Scattering of the Beam Radiation Fields by Spherical Objects," *J. Opt. Soc. Amer.*, **65**, 1457 (1975).

38. Eugene D. Olsen, *Modern Optical Methods of Analysis*, New York: McGraw-Hill, 1975.
39. R. M. Drake and J. E. Gordon, "Mie Scattering," *Am. J. Phys.*, **53**, 955 (1985).
40. John C. Whitmer, "Normal Vibrational Modes—A Simple Demonstration," *J. Chem. Educ.* **48**, 134 (1971).
41. H. W. Kroto, *Molecular Rotation Spectra*, New York: Wiley, 1975.
42. S. Califano, *Vibrational States*, New York: Wiley, 1976.
43. N. B. Colthup, L. H. Daly, and S. E. Wiberley, *Introduction to Infrared and Raman Spectroscopy*, 3rd ed., New York: Academic Press, 1990.
44. L. H. Little, ed., *Infrared Spectra of Absorbed Species*, New York: Academic Press, 1966.
45. Marlin D Harmony, *Introduction to Molecular Energies and Spectra*, New York: Holt, Rinehart and Winston, 1972.
46. J. L. Hollenberg and David Dows, "Measurement of Absolute Infrared Absorption Intensities in Crystals," *J. Chem. Phys.*, **34**, 1061 (1961).
47. T. Fujuyama, J. Herrin, and Bryce Crawford, Jr., "Vibrational Intensities XXV" *Appl. Spectroscopy*, **24**, 9 (1970).
48. L. Light, J. Huebner, and R. Vergenz, "How does Light Absorption Intensity Depend on Molecular Size?," *J. Chem. Educ.*, **71**, 105 (1994).
49. A. L. K. Aljibury, "Numerical Method for the Direct Determination of Interated Absorption Intensities from Transmission Curves," *Appl. Spectroscopy*, **23**, 72 (1969).
50. A. E. Martin, "The Accuracy of Infrared Intensity Measurements," *Trans. Faraday Soc.*, **47**, 1182 (1951).
51. Michael C. Hair, *Infrared Spectroscopy in Surface Chemistry*, New York: Marcel Dekker, 1967.
52. D. Steele and D. H. Whiffen, "Infrared Absorption Intensities of Hexafluorobenzene," *J. Chem. Phys.*, **29**, 1194 (1958).
53. *Infrared Spectroscopy for Use in the Coatings Industry*, Philadelphia, PA: Federation of Societies for Paint Technology, 1969.
54. Dieter O. Hummel, *Atlas of Polymer and Plastics Analysis*, 2nd Ed., vol. 2, New York: VCH, 1988.
55. Clifton E. Meloan, *Elementary Infrared Spectroscopy*, New York: Macmillan, 1963.
56. J. Ziomek, "Group Theory," in *Progress in Infrared Spectroscopy*, vol. 1, Szymanski, Herman A., Ed., New York: Plenum, 1962.
57. Robert A. Faltynek, "Group Theory in Advanced Inorganic Chemistry," *J. Chem. Educ.*, **72**, 20 (1995).
58. K. Yamasaki, "Simple Way of Labeling Rotational Levels with Respect to Full Symmetry Point Groups," *J. Chem. Educ.*, **68**, 574 (1991).
59. E. B. Wilson, Jr., J. C. Decius, and Paul C. Cross, *Molecular Vibrations*, New York: McGraw-Hill, 1955.
60. Harry C. Allen, Jr., and Paul C. Cross, *Molecular Vib-Rotors*, New York: Wiley, 1963.
61. R. F. Barrows, *Introduction to Molecular Spectroscopy*, New York: McGraw Hill, 1962.
62. J. C. D. Brand and J. C. Speakman, *Molecular Structure: The Physical Approach*, London: Edward Arnold, 1960.
63. L. A. Gribov and W. J. Orville-Thomas, *Theory and Methods of Calculation of Molecular Spectra*, New York: Wiley, 1988.
64. V. V. Rossihin and V. P. Morozov, "Potential Constants and Electrooptical Parameters of Molecules," Energoatomizdat, Moscow (1983) (in Russian).

65. Arthur Barlow and Max Diem, "Normal Coordinate Calculations as a Classroom Computer Project," *J. Chem. Educ.*, **68**, 35 (1991).
66. M. Tasumi and M. Nakata, "Test Data for Normal Coordinate Calculations," *Pure & Appl. Chem.*, **57**, 121 (1985).
67. Takehiko Simanouti, "The Normal Vibrations of Polyatomic Molecules as Treated by Urey-Bradley Field," *J. Chem. Phys.*, **17**, 245 (1949).
68. E. Fermi, "Infrared and Raman Spectra," *Z. Physik*, **71**, 260 (1931).
69. Robert S. Mullikan, "Solved and Unsolved Problems in the Spectra of Diatomic Molecules," *J. Phys. Chem.*, **41**, 5 (1937).
70. G. Herzberg, *Molecular Spectra and Molecular Structure, II. Infrared and Raman Spectra of Polyatomic Molecules*, New York: Van Nostrand, 1945.
71. E. F. H. Brittain, W. O. George, and C. H. Wells, *J. Introduction to Molecular Spectroscopy: Theory and Experiment*, New York: Academic Press, 1970.
72. A. D. Cross, *An Introduction to Practical Infra-Red Spectroscopy*, London: Butterworths, 1960.
73. Paul C. Painter, Michael M. Coleman, and Jack L. Koenig, *The Theory of Vibrational Spectroscopy and its Application to Polymeric Materials*, New York: Wiley, 1982.
74. George R. Harrison, Richard C. Lord, and John R. Loofburrow, *Practical Spectroscopy*, Englewood Cliffs, NJ: Prentice-Hall, 1948.
75. R. Chang, *Basic Principles of Spectroscopy*, New York: McGraw-Hill, 1971.
76. D. Papousek and M. R. Aliev, *Vibrational-Rotational Spectra—Theory and Applications of High Resolution Infrared, Microwave and Raman Spectroscopy of Polyatomic Molecules*, Amsterdam: Elsevier, 1992.
77. Alan Carrington and Andrew D. MacLaughlin, *Introduction to Magnetic Resonance*, New York: Harper and Row, 1967.
78. V. Thomsen, "Why do Spectral Lines Have a Linewidth?," *J. Chem. Ed.*, **72**, 616.
79. L. Petrakis, "Spectral Line Shapes," *J. Chem. Educ.*, **44**, 432 (1967).
80. K. S. Seshadri and R. N. Jones, "The Shapes and Intensities of Infrared Absorption Bands—a Review," *Spectrochimica Acta*, **10**, 1012 (1963).
81. David Park, "The Classical Theory of Pressure Broadening," *Amer. J. Phys.*, **39**, 682 (1971).
82. C. H. Townes and A. L. Schawlow, *Microwave Spectroscopy*, New York: McGraw-Hill, 1955.
83. M. M. Maricq and Szente, "Anomalous Pressure Dependence of Infrared Spectra Due to the Combination of Collision Broadening and Instrument Resolution," *J. J., Appl. Spectr.*, **46**, 1045 (1992).
84. D. A. Ramsay, "Intensities and Shapes of Infrared Absorption Bands of Substances in the Liquid Phase," *J. Amer. Chem. Soc.*, **74**, 72 (1952).
85. A. M. Thorndike, "Indirect Measurement of Spectral Line Breadth in the Infra-Red," *J. Chem. Phys.*, **16**, 211 (1948).
86. W. G. Richards and P. R. Scott, *Structure and Spectra of Molecules*, New York: Wiley, 1985.
87. W. O. George, I. W. Griffiths, B. Minty, and R. Lewis, "Enhancement of the Rotation-Vibration Bands of HCl," *J. Chem. Educ.* **71**, 621 (1994).
88. D. Welti, *Infrared Vapour Spectra*, London: Heyden & Son, 1970.

89. R. Zbinden, *Infrared Spectroscopy of High Polymers*, New York: Academic Press, 1964.
90. B. P. Straughan and S. Walker, Eds., *Spectroscopy*, Vol. 2, New York: Wiley, 1976.
91. R. F. Barrow, D. A. Long, and D. J. Millon, *Molecular Spectroscopy*, volume 1, London: Burlington House, 1973.
92. D. P. C. Thackeray, "The Infrared Spectrometer," in R. G. J. Miller and B. C. Stace, eds., *Laboratory Methods in Infrared Spectroscopy*, 2nd ed., New York: Heyden, 1972.
93. P. A. Griffiths, "Fourier Transform Infrared Spectroscopy," in R. G. J. Miller and B. C. Stace, eds., *Laboratory Methods in Infrared Spectroscopy*, 2nd ed., New York: Heyden, 1972.
94. Emil W. Ciurczak, "Building the Perfect Spectrometer," *Spectroscopy*, **9(2)**, 12 (1994).
95. Nelson L. Alpert, William E. Keiser, and Herman A. Szymanski, *IR: Theory and Practice of Infrared Spectroscopy*, New York: Plenum, 1970.
96. Chris J. Georgopoulos, "Interference Sources in Infrared Systems Operating in Closed Spaces," *Opt. Eng.*, **24**, 62 (1985).
97. J. F. Waymouth and R. E. Levin, *Designers Handbook: Light Source Applications*, GTE Products Corp, Danvers, MA, 1980, p. 72.
98. Armand Hadni, *Essentials of Modern Physics Applied to the Study of the Infrared*, London: Pergamon, 1967.
99. Werner Bruegel, *An Introduction to Infrared Spectroscopy*, London: Methuen, 1962.
100. Antonin Vasco, *Infra-red Radiation*, (Engl transl P. S. Allen, ed.), London, 1968.
101. R. A. Smith, F. E. Jones, and R. P. Chasmer, *The Detection and Measurement of Infra-Red Radiation*, Oxford: Clarendon Press, 1968.
102. B. W. Cook and K. Jones, *A Programmed Introduction to Infrared Spectroscopy*, London: Heydon and Son, 1972.
103. James E. Stewart, *Infrared Spectroscopy: Experimental Methods and Techniques*, New York: Marcel-Dekker, 1970.
104. Ivan Simon, *Infrared Radiation*, Princeton, NJ: Van Nostrand, 1966.
105. John R. Ferraro, *Low Frequency Vibrations of Inorganic and Coordination Compounds*, New York: Plenum, 1971.
106. P. A. Wilks, "In Quantitative Analysis, 'Signal-to-Noise' is the Name of the Game," *Spectroscopy*, **6(9)**, 22 (1991).
107. Perkin-Elmer manual for the model 137B Infrared Spectrometer, 1967.
108. Robert H. Leshne, "A Method of Producing Aluminum Front Surface Mirrors," *Optical Spectra*, Jan/Feb, 1968, p. 15.
109. Donald C. O'Shea, *Elements of Modern Optical Design*, New York: Wiley, 1985.
110. Eugene Hecht, *Optics*, Reading MA: Addison-Wesley, 1989.
111. *Diffraction Gratings: Ruled and Holographic Handbook*, J and Y Diffraction Gratings, Inc., Metuchen, NJ.
112. *Diffraction Gratings*, Milton Roy Co., Rochester, NY.
113. Engineering Report No. 8421, April 1966, Hughes Danbury Optical Systems.
114. "Infrared Detectors," Phillips Components Ltd. Technical Publications, No. 199206, London, 1989.
115. T. V. Higgins, "Thermal Detectors feel the heat of light," *Laserfocus World*, Nov. 1994, p. 65.
116. E. H. Putley, "Solid State Devices for Infra-red Detection," *J. Sci. Instrum.*, **43**, 857 (1966).

117. E. L. Dereniak and F. G. Brown, "Pyroelectric Detector Evaluation," *Infrared Physics*, **15**, 39 (1975).
118. T. V. Higgins, "Quantum Detectors Sense Light Directly," *Laserfocus World*, vol **30**, 93 (1994).
119. *Infrared Detectors 1995*, EG&G Optoelectronics, Inc./Judson, 1995.
120. E. H. Putley, "Modern Infrared Detectors," *Physics in Technology*, **4**, 202 (1973).
121. Galen W. Ewing, "LIX. Infrared Detectors," *J. Chem. Educ.*, **48**, A521 (1971).
122. James W. Cooper, "Errors in Computer Data Handling," *Anal. Chem.*, **50**, 804A (1978).
123. *Design-in Reference Manual: Analog Devices*, 1994, (800) 262–5643.
124. *Digital Signal Processing Applications with the TMS 320 Family: Theory Algorithms, and Implementations*, Texas Instruments, 1986.
125. John Chamberlain, *The Principles of Interferometric Spectroscopy*, New York: Wiley, 1979.
126. A. James Diefenderfer, *Principles of Electronic Instrumentation*, Philadelphia: Saunders, 1972.
127. T. Coor, "XXXIX. Signal to Noise Optimization in Chemistry—Part One," *J. Chem. Educ.*, **45**, A533 (1968).
128. P. R. Griffiths and J. A. DeHaseth, *Fourier Transform Infrared Spectroscopy*, New York: Wiley, 1986.
129. E. Oran Brigham, *The Fast Fourier Transform*, Englewood Cliffs, NJ: Prentice-Hall, 1974.
130. P. R. Griffiths, C. T. Foskett, and R. Curbelo, "Rapid-Scan Infrared Fourier Transform Spectroscopy," *Appl. Spectr. Rev.*, **6**, 31 (1972).
131. William J. Hurley, "Interferometric Spectroscopy in the Infrared," *J. Chem. Educ.*, **43**, 236 (1966).
132. W. D. Perkins, "Fourier Transform-Infrared Spectroscopy; Part II: Advantages of FT-IR," *J. Chem. Educ.*, **64**, A269 (1987).
133. W. D. Perkins, "Fourier Transform-Infrared Spectroscopy; Part III: Applications," *J. Chem. Educ.*, **64**, A296 (1987).
134. W. D. Perkins, "Fourier Transform Infrared Spectroscopy Part I: Instrumentation," *J. Chem. Educ.*, **63**, A5 (1986).
135. Robert John Bell, *Introductory Fourier Transform Spectroscopy*, New York: Academic Press, 1972.
136. Galen W. Ewing, "LXV. Multiplex Spectrophotometry," *J. Chem. Educ.*, **49**, A377 (1972).
137. Alan G. Marshall and Melvin B. Comisarow, "Fourier Transform Methods in Spectroscopy," *J. Chem. Educ.*, **52**, 638 (1975).
138. T. Hirschfeld, "An FT-IR Advantage Counter to the Hacquinot or Throughput Advantage," *Appl. Spectr.*, **39**, 1086 (1985).
139. *ASTM E 1421–94*, Standard Practices for Describing and Measuring Performance of Fourier Transform (FT-IR) Spectrometers: Level Zero and Level One Tests. Approved Dec. 15, 1994.
140. Coblentz Society Board of Managers, "Specifications for evaluation of infrared reference spectra," *Anal. Chem.*, **38**, 27A (1966).
141. Coblentz Society Evaluation Committee, "Specifications for Infrared Reference Spectra of Materials in the Vapor Phase Above Ambient Temperature," *Appl. Spectr.*, **33**, 543 (1979).
142. Coblentz Society, "Specifications for Evaluation of Research Quality Analytical Infrared Spectra," *Anal. Chem.*, **47**, 945A (1975).

143. J. W. Robinson, Ed., *CRC Handbook of Spectroscopy*, Vol. II, Cleveland: CRC Press, 1974.
144. R. A. Nyquist, "Infrared Study of Polystyrene at Variable Temperature", *Appl. Spectr.*, **38**, 264 (1984).
145. A. R. H. Cole, R. N. Jones, and R. C. Lord, compilers, *Tables of Wavenumbers for the Calibration of Infrared Spectrometers, parts III and IV, 600 to 1 cm^{-1}*: London: Butterworths, 1961.
146. A. R. H. Cole, R. N. Jones, and R. C. Lord, compilers, *Tables of Wavenumbers for the Calibration of Infrared Spectrometers, part I*; London: Butterworths, 1961.
147. K. N. Rao, C. J. Humphreys, and D. H. Rank, *Wavelength Standards in the Infrared*, New York: Academic Press, 1966.
148. John Bertie, C. Dale Keefe, R. Norman Jones, H. Mantsch, and D. Moffatt, "Infrared Intensities of Liquids. Part VII: Progress Towards Calibration Standards", *Appl. Spectr.*, **45**, 1233 (1991).
149. James E. Stewart, "A Rotating-Sector Attenuator of Adjustable Transmittance for Precise Spectrophotometry", *Appl. Optics*, **1**, 75 (1982).
150. James E. Stewart, "Photometric Accuracy of Infrared Spectrophotometry", *Appl. Optics*, **2**, 1303 (1963).
151. George M. Russwurm, and Jeffrey W. Childers, "FT-IR Spectrometers Perform Environmental Monitoring", *Laserfocus World*, Apr 95, p. 79.
152. M. R. Witkowski, C. T. Chaffin, T. L. Marshall, M. L. Spartz, J. H. Fateley, R. M. Hammaker, W. G. Fateley, R. E. Carter, and D. D. Lane, "Testing and Development of a Mobile Fourier Transform Infrared (FT-IR) Spectrometer System for the Analysis of Atmospheric Pollutants", *Proc. SPIE-Int. Soc. Opt. Eng.*, 1992, 1575 (8th Int'l Conf. Fourier Transform Spectroscopy, 1991), p. 340–341.
153. Herman A. Szymanski, *Progress in Infrared Spectroscopy*, vol. 2, New York: Plenum Press, 1964.
154. "Sampling Techniques for Infrared Analysis", Graseby-Specac, 301 Commerce Dr., Fairfield CT, p. 37.
155. Harrick IR-UV-Vis Accessories Catalog, 1994.
156. Specac Infra-red Sampling Accessories Catalog.
157. Buck Scientific IR/FTIR Accessories and Sampling Catalog.
158. Deane Judd, and Gunter, Wyszecki, *Color in Business, Science and Industry*, 3rd ed., New York: Wiley, 1975.
159. Spectra-Tech Catalog.
160. R. L. Henry, "The Transmission of Powder Films in the Infra-red", *J. Opt. Soc. Amer.*, **38**, 775 (1948).
161. G. Duyckaerts, "The Infrared Analysis of Solid Substances: A Review", *J. Soc. Anal. Chem.*, **84**, 201 (1959).
162. G. K. T. Conn, and D. G. Avery, *Infrared Methods: Principles and Applications*, New York: Academic Press, 1960.
163. R. N. Jones, D. Escolar, J. P. Hawranek, D. Neelakantan, and R. P. Young, "Some Problems in IR Spectrophotometry", *J. Mol. Structure*, **19**, 21 (1973).
164. Tomas, Hirschfeld, "Quantitative FT-IR: A Detailed Look at the Problems Involved" in *Fourier Transform Infrared Spectroscopy*, vol. 2, New York: Academic Press, 1970.
165. T. Hirschfeld, "Ideal FT-IR Spectrometers and the Efficiency of Real Instruments", *Appl. Spect.* **40(8)**, 1239 (1986).
166. P. A. Wilson, "FTIR in Quantitative Analysis and Automation", in *Computer Methods in UV, Visible and IR Spectroscopy*, W. O. George and H. A. Willis, eds., London: Royal Society of Chemistry, 1990.

167. MacKenzie, ed. *Advances in Applied Fourier Transform Infrared Spectroscopy*, New York: Wiley, 1988.

168. P. J. Launer, "Tracking Down Spurious Bands in Infrared Analysis" in Miller, R. G. J. and Stace, B. C., eds., *Laboratory Methods in Infrared Spectroscopy*, 2nd ed., New York: Heyden, 1972, p. xvii to xxi.

169. Perkin-Elmer Instrument News, 24, #2, 1974.

170. Robert C. Williams, "Errors in Spectral Interpretation Caused by Sample-preparation Artifacts and FT-IR Hardware Problems", in *Practical Sampling Techniques for IR Analysis*, Coleman, Patricia B., ed., Boca Raton, FL: CRC Press, 1993.

171. R. N. Jones, R. Venkataraghavan, and J. W. Hopkins, "The Control of Errors in Infrared Spectroscopy, I. The reduction of finite spectral slit distortion by the method of pseudo-deconvolution", *Spectrochim. Acta*, **23A**, 925 (1967).

172. J. P. Hawranek, P. Neelakantan, R. P. Young, and R. N. Jones, "The Control of Errors in I.R. Spectrophotometry-III. Transmission Measurements Using Thin Cells," *Spectrochim. Acta.*, **32A**, 75 (1976).

173. John R. Dyer, *Applications of Absorption Spectroscopy of Organic Compounds*, Englewood Cliffs, NJ: Prentice-Hall, 1965.

174. L. J. Bellamy, *The Infrared Spectra of Complex Molecules*, vol. 1, 3rd ed. New York: Chapman and Hall, 1975.

175. L. J. Bellamy, *The Infrared Spectra of Complex, Molecules*, vol. 2, 3rd ed., New York: Chapman and Hall, 1975.

176. N. B. Colthup, "Spectra-Structure Correlations in the Infra-Red Region," *J. Opt. Soc. Am.*, **40**, 397 (1950).

177. Robert M. Silverstein, G. Clayton Bassler, and Terence C. Morrill, *Spectrometric Identification of Organic Compounds*, 4th Ed., New York: Wiley, 1981.

178. Robert M. Silverstein and G. Glayton Bassler, "Spectrometric Identification of Organic Compounds," *J. Chem. Educ.*, **39**, 546 (1962).

179. J. H. Simons, *Fluorine Chemistry*, vol. II, New York: Academic Press, 1954.

180. John P. Coates, "Resources and References for Spectral Identification," *Spectroscopy*, **10(7)**, 14 (1995).

181. Dennis McMinn, "Introductory Use of Infrared Spectra," *J. Chem. Educ.*, **61**, 708 (1984).

182. D. Cabrol, J. P. Rabine, D. Ricard, M. Rouillard and T. P. Forrest, "Computer-Assisted Learning," *J. Chem. Educ.*, **70**, 120 (1993).

183. Hendrik Luinge, "EXPEC, A Knowledge-based System for Interpretation of Infrared Spectra," *Anal. Proc.*, **27**, 266 (1990).

184. John H. Kalivas and Patrick M. Lang, *Mathematical Analysis of Spectral Orthogonality*, Practical Spectroscopy Series, Vol. 17, New York: Marcel-Dekker, 1994.

185. Fred H. Lohman, "The Mathematical Combination of Lambert's Law and Beer's Law," *J. Chem. Educ.*, **29**, 155 (1955).

186. D. F. Swinehart, "The Beer-Lambert Law," *J. Chem. Educ.*, **39**, 33 (1962).

187. W. J. Price, "Sample Handling Techniques," in R. G. J. Miller and B. C. Stace, eds., *Laboratory Methods in Infrared Spectroscopy*, 2nd ed., New York: Heyden, 1972, p. 97f.

188. Robert T. Conley, *Infrared Spectroscopy*, Boston: Allyn and Bacon, 1966.

188. M. St. C. Flett, "Intensities of Some Group Characteristic Infrared Bands," *Spectrochim. Acta*, **18**, 1537 (1962).

189. R. C. Golike, I. N. Milles, W. B. Person, and Bryce Crawford, Jr., "Vibrational Intensities. VI. Ethylene and its Deuteroisotopes," *J. Chem. Phys.*, **25**, 1266 (1956).

190. G. Zerbi, E. Galbiati, and C. Castiglioni, "Intensity Spectroscopy and FTIR as New Tool for the Study of Polymer Blends," *SPIE, vol. 553, Proc. 1985 FCISC.*

191. B. A. Neff, D. J. Weissert, R. B. Lam, J. J. Leary and T. N. Gallagher, "Studies of Precision in FTIR Spectroscopy," *Amer. Lab.*, Nov. 1985, p. 72.

192. T. Fujiyama, John Herrin, and Bryce Crawford, Jr., "Vibrational Intensities. XXV. Some Systematic Errors in Infrared Absorption Spectrophotometry of Liquid Samples," *App. Spectr.*, **24**, 9 (1970).

193. V. J. I. Zichy, "Quantitative Infrared Analysis of Polymeric Materials," in R. G. J. Miller and B. C. Stace, eds., *Laboratory Methods in Infrared Spectroscopy*, 2nd ed., New York: Heyden, 1972, p. 48f.

194. D. M. Haaland and R. G. Easterling, "Improved Sensitivity of Infrared Spectroscopy by the Application of Least Squares Methods," *Appl. Spectr.*, **34**, 539 (1980).

195. D. M. Haaland and R. G. Easterling, "Application of New Least Squares Methods for the Quantitative Analysis of Multicomponent Systems," *Appl. Spectr.*, **36**, 665 (1982).

196. B. A. Neff, D. J. Weissert, R. B. Lam, J. J. Leary and T. N. Gallaher, "Studies of Precision in FTIR Spectroscopy," *Amer. Lab.*, Nov. 1985, p. 72f.

197. R. N. Jones, D. Escolar, J. P. Hawranek, P. Neelakantan, and R. P. Young, "Some Problems in Infrared Spectrophotometry," *J. Molec. Str.*, **19**, 21 (1973).

198. Perkin-Elmer, "Quant" chapter in *Introduction to Fourier Transform Infrared Spectroscopy Instrumentation-Selected Applications*, 1986.

199. I. M. Kolthoff, E. B. Sandell, E. J. Meehan, and Stanley Bruckenstein, *Quantitative Chemical Analysis*, New York: Macmillan, 1969.

200. Edmund R. Malinkowski and Darryl G. Howery, *Factor Analysis in Chemistry*, New York: Wiley, 1980.

201. James Duckworth, "Using PLSplus/IQ from the laboratory to On-Line," short course, Montreal, Canada, August 6, 1995, NIR-95 Meeting.

202. David M. Haaland and Edward V. Thomas, "Partial Least Squares Methods for Spectral Analysis. 1. Relation to Other Quantitative Calibration Methods and the Extraction of Qualitative Information," *Anal. Chem.*, **60**, 1193 (1988).

203. David M. Haaland and Edward V. Thomas, "Partial Least Squares Methods for Spectral Analysis. 2. Application to Simulated and Glass Spectral Data," *Anal. Chem.*, **60**, 1202 (1988).

204. G. W. Chantry, *Submillimetre Spectroscopy*, New York: Academic Press, 1971.

205. V. G. Tyuterev, V. I. Starkov, S. A. Tashkun, and S. N. Makheilenko, "Calculation of High Rotation Energies of the Water Molecule Using the Generating Function Model," *J. Molec. Spectroscopy*, **170**, 38 (1995).

206. Ron Woods and Giles Henderson, "FTIR Rotational Spectroscopy," *J. Chem. Educ.*, **64**, 921 (1987).

207. Byron T. Darling and David M. Dennison, "The Water Vapor Molecule," *Phys. Rev.*, **57**, 128 (1940).

208. Anthony R. Lacey, "A Student Introduction to Molecular Vibrations," *J. Chem. Educ.*, **64**, 756 (1987).

209. J. M. Fland, C. Camy-Peyret, and R. A. Toth, *Water Vapor Line Parameters from Microwave to Medium Infrared*, London: Pergamon Press, 1981.

210. A. Bauer, M. Godon, J. Carlier, and Q. Ma, "Water Vapor Absorption in the Atmospheric Window at 239 GHz," *J. Quant, Spectrosc. Radiat. Transfer*, **53**, 411 (1995).

211. *PLSplus for Grams/386* is available from Galactic Industries Corporation, 395 Main St., Salem NH 03079, (603) 898–7600.

212. Arthur Elliott, *Infra-red Spectra and Structure of Organic Long-Chain Polymers*, London: Arnold, 1969.

213. B. Amram, L. Bokobza, P. Queslei, and L. Monnerie, *Polymer*, **27**, 877 (1986).

GLOSSARY

The reader is urged to consult ASTM E131-91a for an extensive list of pertinent definitions

Amplitude (A_0) is the height of a crest for any electromagnetic wave.

Coherence refers to multiple waves that "match" up in terms of phase, wavelength, time and/or space. Coherence may be spatial or temporal.

Dielectric, from "dia" meaning "through" or "opposite of", a material through which electrical lines of force pass; a non-conductor.

Frequency (v) is the number of oscillations at a given point in space, in units of cycles per second or Hertz (Hz). Frequency is independent of the medium in which the wave is propagated.

Infrared Range:

Mid-infrared wavelength range includes electromagnetic radiation with wavelengths from about 2.5 to 50 μm (4000 to 200 cm^{-1}).

Near-infrared range borders the mid-infrared range on the high energy side from about 0.75 to about 2.5 μm (13000 to 4000 cm^{-1})

Far-infrared range borders the mid-infrared range on the low energy side from 50 to 1000 μm (200 to 10 cm^{-1}).

Intensity, S, is the flow of energy per unit area per second or in units of watts/sr and is related to the square of the amplitude of the wave.

Interference results when two or more waves meet and form new wave patterns.

Destructive interference occurs with waves of the same wavelength that are $\lambda/2$ out of phase (the crest of one wave matches up with the trough of the other), and no amplitude is observed for equal inputs.

Constructive interference results when the two waves are in phase or differ by λ (the crests match up); the resultant amplitude is doubled, for equal inputs.

Mathematical form used to represent a wave may be written in various ways, such as $A(t) = A_0 \cos[\omega t - kx - kA]$. Hundreds of theoretical texts deal with the mathematics of waves.

Phase refers to the position on the sinusoidal curve where the radiation is located with respect to some conveniently defined reference point.

Polarization refers to the plane in which the electric field or magnetic field of the wave is orientated, with respect to the propagation direction and a defined axis system in space.

Wavelength (λ) is the distance from crest to crest of either the electric or magnetic field associated with the radiation or, more generally, between corresponding points on consecutive waves, usually expressed in microns (μm) in the infrared region. The wavelength of a photon may change depending on the medium in which it is propagated.

3 Specular Reflection Spectroscopy

Robert J. Lipert, Brian D. Lamp, and Marc D. Porter

3.1 INTRODUCTION

This chapter on specular reflection spectroscopy focuses on applications in the mid-infrared region of the spectrum. Applications in other spectral regions or using other spectroscopic techniques, such as Raman spectroscopy, generally follow similar patterns as discussed in this chapter.

In the past few decades, infrared spectroscopy (IRS) has developed into an invaluable diagnostic tool for unraveling details about the bonding and molecular structure at surfaces [1, 2, 3, 4, 5, 6]. The importance of IRS derives from five major factors, The first and most critical factor is the role of interfacial phenomena in a vast number of emerging material and surface technologies. Examples include adhesion, catalysis, tribiology, microelectronics, and electrochemistry [3, 7, 8, 9]. The second factor stems from the

Modern Techniques in Applied Molecular Spectroscopy, Edited by Francis M. Mirabella.
Techniques in Analytical Chemistry Series.
ISBN 0-471-12359-5

information-rich content of an IRS spectrum of a surface. That is, the frequencies of the spectroscopic features can be used to identify the chemical composition of a surface and the magnitudes and polarization dependencies of the features used to determine average structural orientations. The third factor arises from advances in the performance of IRS instrumentation. The most important of these advancements are the high throughput and multiplex advantages of Fourier transform interferometry, the development of high sensitivity, low-noise IR detectors, and the improvement in the computational rates of personal computers and their adaptation to the operation of chemical instrumentation. The fourth factor results from the compatibility of IRS with a variety of sample environments. This situation opens the door to a wide range of *in situ* studies, including characterizations in high-pressure environments and in condensed phases. The fifth factor rests with the application of classical electromagnetic theory to IRS characterizations in a reflection mode. Such considerations provide a basis to delineate the experimental conditions requisite for optimal detection as well as to separate shifts and distortions of spectral bands induced by optical effects from those due to bonding and surface-induced structural changes.

As is evident from the general theme of this monograph, there are many different experimental modes for probing the composition and structure of interfaces with infrared spectroscopy. This chapter focuses on the application of infrared spectroscopy to the characterization of smooth surfaces. We define smooth surfaces as those that can be generally described as having a "mirrorlike" or specular finish, that is to say, surfaces with irregularities that are small in comparison to the wavelength of incident light. Armed with this definition, an overview of the reflection experiment within the context of classical electromagnetic theory is presented in the following section. We then utilize the formulations from these considerations to gain insight into the dependence of the reflection spectrum on the optical properties of the film, substrate, surrounding medium, and on the angle of incidence and polarization of the incoming radiation. This section is followed by a discussion of various experimental approaches for the collection of a reflection spectrum at a specular surface, an approach often referred to as IRS in a specular, external, or reflection-absorption mode (Fig. 3.1). This chapter concludes with brief descriptions of recent applications selected to demonstrate the overall utility of this technique.

3.2 THEORETICAL CONSIDERATIONS

Considerable insight has been gained in describing the phenomena associated with specular reflectance spectroscopy using classical electrodynamic theory [10, 11, 12, 13]. In this section we will first develop the wave

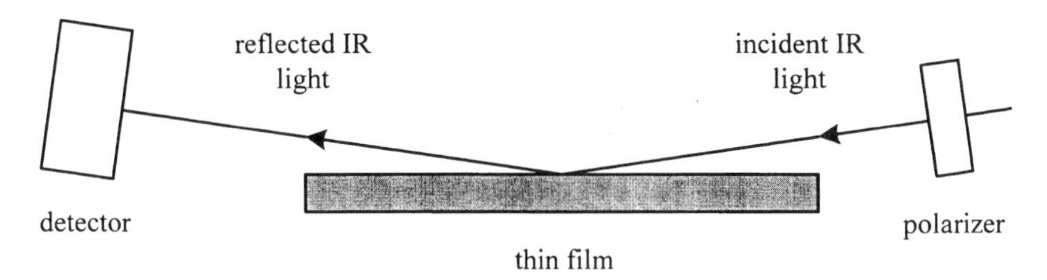

Fig. 3.1 Schematic diagram for a specular reflectance measurement.

equations for the propagation of light in both transparent and absorbing media. We will then consider the effects on a reflection measurement that occur when light encounters an interface of differing optical properties. Since the absorption of light by a modified surface is proportional to the mean-square electric fields at the interface, insights into the factors that effect these fields foster an understanding of the conditions optimal for the detection of coatings at different types of reflective substrates. We will also show how the optical properties of the film and substrate can give rise to shifts and distortions of the spectral band shapes of the modifying film and how considerations of such spectra can be exploited to obtain orientational information about the modifying film.

Propagation of Electromagnetic Radiation through a Nonconducting Homogeneous Medium

For electromagnetic radiation propagating through a transparent uniform isotropic medium, manipulation of Maxwell's equations results in the following expression for the electric field vector **E** [14]:

$$\nabla^2 \mathbf{E} = \frac{\varepsilon \mu_p}{c^2} \frac{\partial^2 \mathbf{E}}{\partial t^2}. \tag{3.1}$$

In this expression ∇^2 is the Laplacian operator, ε is the dielectric constant of the medium, μ_p is the relative magnetic permeability of the medium, c is the speed of light, and t is time. Equation (3.1) has the form of the classical wave equation

$$\nabla^2 f = \frac{1}{v^2} \frac{\partial^2 f}{\partial t^2} \tag{3.2}$$

where f is the function that describes the wave and v is its velocity. The velocity of the electromagnetic wave is therefore

$$v = \frac{c}{\sqrt{\varepsilon \mu_p}} \tag{3.3}$$

The plane traveling wave solution to (3.1) is given by

$$\hat{\mathbf{E}} = E^0 \exp[i(\mathbf{K} \cdot \mathbf{r} - \omega t)], \tag{3.4}$$

where E^0 is the maximum amplitude of the electromagnetic wave, $\mathbf{r}$ is the vector distance from the origin to the point of measurement, $\mathbf{K}$ is the propagation vector with units of inverse length, $\omega = 2\pi\nu = 2\pi c/\lambda$ and is the angular frequency with units of rad/s, and ν and λ are the frequency and vacuum wavelength, respectively, of the electromagnetic radiation. Substituting (3.4) into (3.1) shows that

$$K = (\varepsilon \mu_p)^{1/2}\left(\frac{\omega}{c}\right) = (\varepsilon \mu_p)^{1/2}\left(\frac{2\pi\nu}{c}\right) = (\varepsilon \mu_p)^{1/2}\left(\frac{2\pi}{\lambda}\right). \tag{3.5}$$

In a nonmagnetic medium, $\mu_p \cong 1.0$. Also the refractive index n of a medium is given by ratio of the velocity of light in the medium to the velocity in vacuum, namely $n = c/v$. Thus we see from (3.3) that

$$n = (\varepsilon)^{1/2}. \tag{3.6}$$

Combining and rearranging (3.4), (3.5) and (3.6) gives

$$\hat{\mathbf{E}} = E^0 \exp\left[i\left(\frac{2\pi n}{\lambda}\mathbf{s} \cdot \mathbf{r} - \omega t\right)\right], \tag{3.7}$$

where $\mathbf{s}$ is a unit vector in the direction of propagation.

In an absorbing medium, as depicted in Fig. 3.2, the plane traveling wave undergoes an exponential attenuation with increasing propagation distance. The electric field vector can then be represented as

$$\hat{\mathbf{E}} = E^0 \exp\left[i\left(\frac{2\pi n}{\lambda}\mathbf{s} \cdot \mathbf{r} - \omega t\right)\right] \exp\left(-\frac{2\pi k}{\lambda}\mathbf{s} \cdot \mathbf{r}\right), \tag{3.8}$$

where k is referred to as the absorption index or extinction coefficient [13]. This k can be related to the absorption coefficient α of Lambert's law, which describes the attenuation of the intensity of light as it propagates through an absorbing medium. Neglecting reflection at the surface of the absorbing

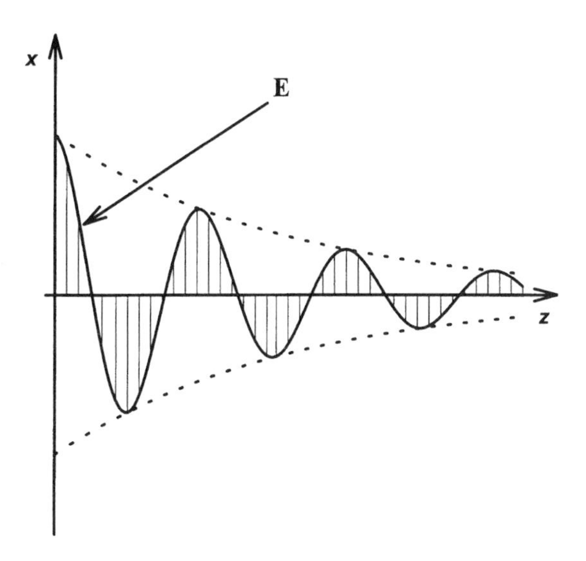

Fig. 3.2 Propagation of an electromagnetic wave into an absorbing medium (z-direction).

medium, the Lambert law is given by

$$I = I^0 \exp(-\alpha z), \tag{3.9}$$

where I^0 is the initial radiation intensity and z is the distance of propagation into the absorbing medium. The radiation intensity, I, is proportional to the mean-square electric field. Therefore I is proportional to $\exp\left(-\frac{4\pi k}{\lambda}\mathbf{s}\cdot\mathbf{r}\right)$ and thus

$$\alpha = \frac{4\pi k}{\lambda}. \tag{3.10}$$

It is therefore possible to obtain k for a medium from a transmission spectrum of a thin film of the material.

If we define a complex refractive index as

$$\hat{n} = n + ik, \tag{3.11}$$

Equation (3.8) can be recast into the form of (3.7) with $\hat{n}$ replacing n:

$$\hat{\mathbf{E}} = E^0 \exp\left[i\left(\frac{2\pi\hat{n}}{\lambda}\mathbf{s}\cdot\mathbf{r} - \omega t\right)\right]. \tag{3.12}$$

The complex refractive index of a medium is related to its complex dielectric constant by

$$\hat{n} = (\hat{\varepsilon})^{1/2}. \tag{3.13}$$

Therefore,

$$\hat{\varepsilon} = \hat{n}^2 = n^2 + i2nk - k^2 = \varepsilon' + i\varepsilon'', \tag{3.14}$$

where

$$\varepsilon' = n^2 - k^2 \qquad \varepsilon'' = 2nk. \tag{3.15}$$

In general, $\hat{n}$ or $\hat{\varepsilon}$ are functions of the radiation frequency. We can therefore write

$$\hat{n}(\bar{v}) = n(\bar{v}) + ik(\bar{v}) \quad \text{and} \quad \hat{\varepsilon}(\bar{v}) = \varepsilon'(\bar{v}) + i\varepsilon''(\bar{v}), \tag{3.16}$$

where $\bar{v} = 1/\lambda$ and has the units of wavenumber. These functions completely describe the optical properties of a medium. For example, the complex refractive index $\hat{n}$ describes both refraction, through n, and absorption, through k, of light by the medium. Importantly, the real and imaginary parts of these optical functions are not independent but are related through the Kramers-Kronig transformation. Using the refractive index as an example, the real part of the refractive index at $\bar{v} = \bar{v}_i$ can be calculated from $k(\bar{v})$ using the following expression [15]:

$$n(\bar{v}_i) = n_\infty + \frac{2}{\pi} P \int_0^\infty \frac{\bar{v}k(\bar{v})}{(\bar{v}^2 - \bar{v}_i^2)} d\bar{v}. \tag{3.17}$$

In this equation the constant n_∞ represents the contribution to the refractive index of regions far removed from any absorption band and P indicates that the Cauchy principal value of the integral must be taken because of the singularity at $\bar{v} = \bar{v}_i$. In practice, measurements are performed across an absorption band from $\bar{v}_a$ to $\bar{v}_b$. The transformation is then usually approximated as

$$n(\bar{v}_i) = \bar{n} + \frac{2}{\pi} P \int_{\bar{v}_a}^{\bar{v}_b} \frac{\bar{v}k(\bar{v})}{(\bar{v}^2 - \bar{v}_i^2)} d\bar{v}, \tag{3.18}$$

where $\bar{n}$ is either the baseline refractive index or the mean refractive index across the band [15]. As was noted earlier, the function $k(\bar{v})$ can be obtained from an absorption spectrum of a thin film.

Reflection of Electromagnetic Radiation at a Boundary between Homogeneous Media of Different Optical Properties

Reflectivity

The optical effects that arise when recording absorption spectra using specular reflection can have a large impact on the magnitudes and shapes of the spectra. To interpret the absorption spectra obtained using a specular reflectance geometry, we must understand the influence of the optical constants of the media on the electric fields at the reflecting interfaces. It will be shown that when an absorption band is present in a medium, not only is the light intensity affected by absorption but also by changes in the reflectivity of the interface. That is, the intensity of the reflected light is affected by changes in n as well as changes in k across the absorption band. Changes in reflectivity across an absorption band can result in a distortion of the band shape and a shifting of the measured absorption maximum, compared to a thin film transmission spectrum. In addition, because of the phase shift in the light that occurs upon reflection, the mean-square electric fields present near an interface are dependent on the polarization of the incident radiation. This situation gives rise to a surface effect, often referred to as the *surface selection rule*, that can be used to obtain information on the structural orientation of modifying films [10, 16].

The development of the equations for the electric fields present in a three-phase, optically isotropic medium [10] treats the system as a series of planar, parallel optical boudaries. Figure 3.3 presents a schematic of this model. In general, the propagating radiation can pass through n phases, with the optical functions of phase j given by

$$\hat{n}_j = n_j + ik_j. \tag{3.19}$$

Reflection occurs at the boundary between phases. It is usually the case that phase 1 is air (i.e., $n \approx 1.0$, $k_1 \approx 0$), phase 2 is the sample thin film, and phase 3 is a supporting medium of effectively infinite thickness. The z-direction is normal to the interfaces with the positive direction extending into phase 2 from phase 1. The degree to which radiation is reflected or transmitted at an interface depends on the polarization and angle of incidence of the incoming radiation as well as the optical functions of the two adjacent phases. Parallel-polarized (p-polarized) light is polarized in the plane of incidence and has x- and z-components, and perpendicular-polarized (s-polarized) light has only a y-component. The reflectivity of the interface between phases j and k, R_{jk}, is given by the Fresnel coefficients for reflection, r_{jk} [17, 18]:

$$R_{jk} = |r_{jk}|^2 = r_{jk}r_{jk}^*. \tag{3.20}$$

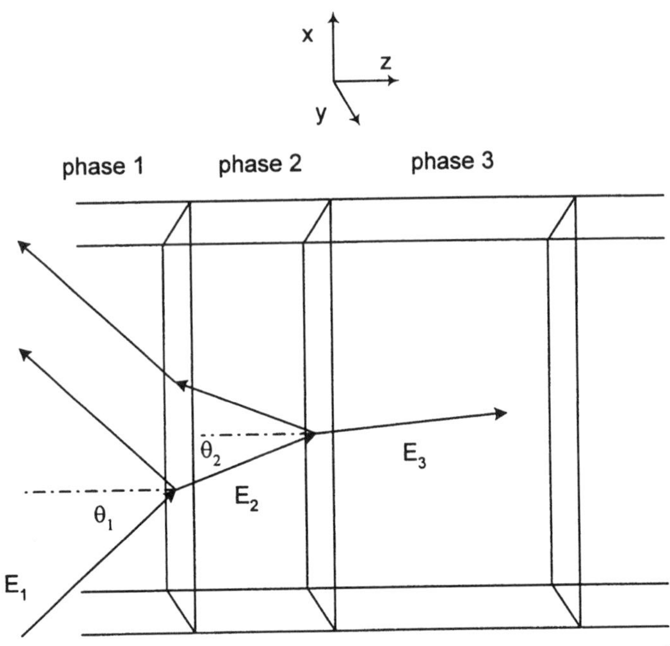

Fig. 3.3 Electric field vectors for plane-polarized light incident on a three-phase medium. The plane of polarization is the *x-z* plane.

The Fresnel coefficients are defined as the ratios of the complex amplitudes of the electric field vectors of the incident waves to those of the reflected waves. Similarly the intensity of transmitted light is given by the Fresnel coefficients for transmission, t_{jk}, which are the ratios of the complex amplitudes of the electric field vectors of the incident waves to those of the transmitted waves. Both sets of coefficients are obtained from Maxwell's equations by applying the continuity requirements for $\hat{\mathbf{E}}$ and $\hat{\mathbf{H}}$ across the phase boundaries. To simplify the expressions for r_{jk}, it is customary to define a refractive coefficient for phase j as

$$\xi_j = \hat{n}_j \cos \hat{\theta}_j, \tag{3.21}$$

where $\hat{n}_j$ and $\hat{\theta}_j$ are the complex refractive index and complex angle of incidence, respectively, in phase j [11]. The Fresnel coefficients for reflection and transmission of radiation polarized perpendicular and parallel to the plane of incidence are then given by

$$r_{\perp jk} = \frac{\xi_j - \xi_k}{\xi_j + \xi_k}, \tag{3.22}$$

$$t_{\perp jk} = \frac{2\xi_j}{\xi_j + \xi_k}, \tag{3.23}$$

$$r_{\parallel jk} = \frac{\hat{n}_k^2 \xi_j - \hat{n}_j^2 \xi_k}{\hat{n}_k^2 \xi_j + \hat{n}_j^2 \xi_k}, \tag{3.24}$$

$$t_{\parallel jk} = \frac{2\hat{n}_j \hat{n}_k \xi_j}{\hat{n}_k^2 \xi_j + \hat{n}_j^2 \xi_k}. \tag{3.25}$$

Thus the reflectivities at the interface between phases 1 and 2 for the two polarizations are

$$R_{\perp 12} = \left| \frac{\xi_1 - \xi_2}{\xi_1 + \xi_2} \right|^2, \tag{3.26}$$

$$R_{\parallel 12} = \left| \frac{\hat{n}_2^2 \xi_1 - \hat{n}_1^2 \xi_2}{\hat{n}_2^2 \xi_1 + \hat{n}_1^2 \xi_2} \right|^2. \tag{3.27}$$

It is instructive to evaluate the reflectivity for the ideal case where the incident light is normal to the phase boundary, phase 1 is air, and phase 2 is a very strongly absorbing medium whose refractive index can be approximated as purely imaginary. This situation is similar to the reflection of infrared radiation at a metal surface [19]. In this example, $\theta_1 = 0$, $\xi_1 = 1$, and $\xi_2 = ik$. The reflectivity then becomes

$$R = \frac{(1 - ik)(1 + ik)}{(1 + ik)(1 - ik)} = 1. \tag{3.28}$$

We therefore see that metals are highly efficient reflectors of infrared radiation because of their very strong absorption of the incident light. This situation arises from the high density of the free electron gas of metals, which strongly absorbs infrared radiation and results in a highly efficient reflection of the incident beam.

For the common situation of a three-phase system (e.g., the case where an organic film is supported on a smooth substrate and is in contact with the ambient environment), the perpendicular and parallel Fresnel coefficients r_{123} are

$$r_{\perp} = \frac{r_{\perp 12} + r_{\perp 23} e^{-2i\beta}}{1 + r_{\perp 12} r_{\perp 23} e^{-2i\beta}}, \tag{3.29}$$

$$r_{\parallel} = \frac{r_{\parallel 12} + r_{\parallel 23} e^{-2i\beta}}{1 + r_{\parallel 12} r_{\parallel 23} e^{-2i\beta}}, \tag{3.30}$$

where β represents the attenuation of the beam as it travels through phase 2 of thickness d and is given by

$$\beta = 2\pi \frac{d}{\lambda} \xi_2. \tag{3.31}$$

When calculating the electric field at an interface at which reflection is occurring, it is also necessary to take into account the phase change in the electric field that occurs upon reflection. This phase change, δ^r_{jk}, is given by the real and imaginary parts of the Fresnel coefficients as follows:

$$\delta^r_{jk} = \arg(r_{jk}) = \tan^{-1}\left[\frac{\mathrm{Im}(r_{jk})}{\mathrm{Re}(r_{jk})}\right], \tag{3.32}$$

where $\arg(r_{jk})$ is the angle that is the argument of the complex number r_{jk}. By convention, when the electric field vectors for incident and reflected light are antiparallel and equal in magnitude, $\delta^r_\perp = -180$ and $\delta^r_\parallel = 0$ [20]. The reflectivity of a three-phase system is then [11]

$$R_\perp = \frac{R_{\perp 12} + R_{\perp 23} e^{-4\mathrm{Im}\beta} + R^{1/2}_{\perp 12} R^{1/2}_{\perp 23} e^{-2\mathrm{Im}\beta} 2\cos(\delta^r_{\perp 23} - \delta^r_{\perp 12} + 2\mathrm{Re}\beta)}{1 + R_{\perp 12} R_{\perp 23} e^{-4\mathrm{Im}\beta} + R^{1/2}_{\perp 12} R^{1/2}_{\perp 23} e^{-2\mathrm{Im}\beta} 2\cos(\delta^r_{\perp 23} + \delta^r_{\perp 12} + 2\mathrm{Re}\beta)}. \tag{3.33}$$

The same expression holds for parallel polarization, which, with appropriate substitutions, is

$$R_\parallel = \frac{R_{\parallel 12} + R_{\parallel 23} e^{-4\mathrm{Im}\beta} + R^{1/2}_{\parallel 12} R^{1/2}_{\parallel 23} e^{-2\mathrm{Im}\beta} 2\cos(\delta^r_{\parallel 23} - \delta^r_{\parallel 12} + 2\mathrm{Re}\beta)}{1 + R_{\parallel 12} R_{\parallel 23} e^{-4\mathrm{Im}\beta} + R^{1/2}_{\parallel 12} R^{1/2}_{\parallel 23} e^{-2\mathrm{Im}\beta} 2\cos(\delta^r_{\parallel 23} + \delta^r_{\parallel 12} + 2\mathrm{Re}\beta)}. \tag{3.34}$$

We note that equations (3.33) and (3.34) can be viewed conceptually as the reflection analogs to the Beer-Lambert law for transmission spectroscopy.

Mean-Square Electric Fields

The change in the intensity of the reflected light because of absorption by a thin film is proportional to the mean-square electric field, $\langle E^2 \rangle$, in the thin film. For insight into the differences between infrared absorption spectra of free-standing thin films and reflection spectra of thin films supported on a reflective surface, it is worthwhile to examine the mean-square electric fields in each of the phases that constitute the reflecting stratified medium.

The electric fields of incident and specularly reflected light add vectorially to produce a standing-wave electric field in each phase except the last phase. Because a standing wave is formed above a reflecting surface, the mean-square electric fields vary with distance from the reflecting surface and are different for parallel and perpendicular polarizations. With the relationships derived above, the expressions for the mean-square fields for the various polarization components in phase 1, relative to the incident mean-square field ($\langle E_1^{ot2} \rangle$), are given by [11]

$$\frac{\langle E_{\perp 1}^2 \rangle}{\langle E_{\perp 1}^{ot2} \rangle} = (1 + R_\perp) + 2R_\perp^{1/2} \cos\left[\delta_\perp^r - 4\pi\left(\frac{z}{\lambda}\right)\xi_1\right], \tag{3.35}$$

$$\frac{\langle E_{\parallel 1x}^2 \rangle}{\langle E_{\parallel 1x}^{ot2} \rangle} = \cos^2\theta_1 \left\{(1 + R_\parallel) - 2R_\parallel^{1/2} \cos\left[\delta_\parallel^r - 4\pi\left(\frac{z}{\lambda}\right)\xi_1\right]\right\}, \tag{3.36}$$

$$\frac{\langle E_{\parallel 1z}^2 \rangle}{\langle E_{\parallel 1z}^{ot2} \rangle} = \sin^2\theta_1 \left\{(1 + R_\parallel) + 2R_\parallel^{1/2} \cos\left[\delta_\parallel^r - 4\pi\left(\frac{z}{\lambda}\right)\xi_1\right]\right\}. \tag{3.37}$$

Plots of the reflectivity, phase change, and mean-square electric fields at 2000 cm^{-1} are shown for various two-phase systems in Fig. 3.4 as a function of θ_1. The optical constants for the different media used in the calculations are given in Table 3.1. The selected surfaces—gold, glassy carbon, glass, and silicon—illustrate the effect of varying optical constants on the reflective properties of materials often encountered in a specular reflection characterization. We see that because gold has a high reflectivity for both polarizations at nearly all angles of incidence, the x- and y-components of the mean-square electric fields at the gold surface have negligibly small values at all angles of incidence. However, the z-component of the mean-square electric field at the gold surface has a significant intensity, with a maximum

TABLE 3.1 Optical Constants Used for Calculating the Optical Properties of the Various Media

Medium	n	k
Air	1.0	0.0
Gold[a]	2.7	28.5
Glassy carbon[a]	2.9	1.3
SiO_2 (glass)[b]	1.342	0.0
Si[b]	3.426	0.0

Note: Values of n and k are valid for 2000 cm^{-1} radiation.

[a]See [22].

[b]See [19].

that is nearly a factor of four stronger than the incident field when θ_1 is slightly less than 80°. It is the difference in the intensities of the electric field components that gives rise to the surface effect that is exploited to determine the spatial orientation of the modifying film (see below).

In contrast to gold, glassy carbon, which has a similar n to gold but is a much weaker absorber of infrared radiation (i.e., a smaller k), has a much lower reflectivity, and hence the z-component of the mean-square electric field at the surface is smaller. In this case the maximum in the z-field at glassy carbon occurs at a 59° angle of incidence. The lower reflectivity of glassy carbon also results in x- and y-components with much greater surface electric fields than at gold.

The last two materials, glass and silicon, are transparent at 2000 cm^{-1} ($k = 0$) but have different values for n. We see that silicon, with the larger n, is more reflective than glass; see equations (3.26) and (3.27). The maximum in the z-field occurs at a 58° angle of incidence for glass and 59.5° for silicon. In addition the relatively large values of the x- and y-components of the mean-square fields leads to a breakdown of the surface selection rule operative at metal surfaces. The variation of the mean-square electric fields as a function of distance from the surface will be examined in more detail next.

The plots of the mean-square electric field shown in Fig. 3.4 are diagnostic of the fields present at the interface between two media. As mentioned above, a standing-wave electric field is set up in the incident medium. Inspection of equations (3.35)–(3.37) shows that the period p of the standing wave is given by

$$p = \frac{1}{2\bar{\nu} n_1 \cos\theta_1}. \tag{3.38}$$

Figure 3.5 is a plot of the relative mean-square electric fields as a function of distance from the same four reflecting media—gold, glassy carbon, glass, and silicon—examined in Fig. 3.4. The periodic nature of the mean-square electric fields in the incident medium is apparent. Also note that changing the angle of incidence from 78.5° for gold to 59° for glassy carbon results in a standing wave of shorter period via equation (3.38). We also see that for a highly reflective surface like gold, if the thickness of the phase-2 film is much less than the period of the standing wave, then the film will not interact with s-polarized light. Note also how the phase change occurring on reflection from glass has resulted in the z-field having a minimum at its surface. In the case of glass, and to a lesser extent silicon, the mean-square electric fields extend into the second phase and are constant because these two media are transparent and there is no interference from a reflected wave.

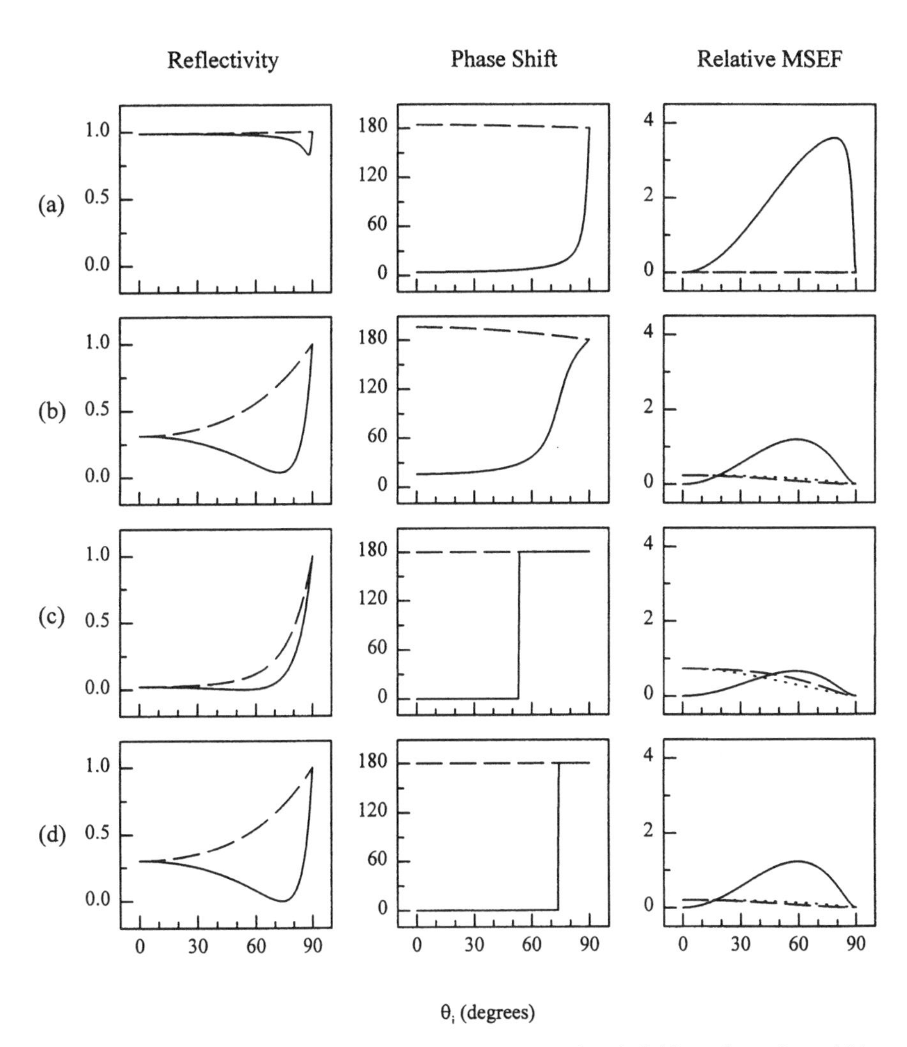

Fig. 3.4 Reflectivity, phase shift, and relative mean-square electric fields at the surface of (a) gold, (b) glassy carbon, (c) glass, and (d) silicon as a function of the angle of incidence. The incident medium is air. For reflectivity and phase shift, values are shown for parallel (—) and perpendicular (– – –) polarization of the incoming light. The relative mean-square electric fields are shown for the *z*-(—), *x*-(----), and *y*-(– – –) components.

Band Distortions

It has long been recognized that the experimentally determined band shapes of infrared reflection spectra of thin films can be shifted and distorted by the dispersion in the real part of the refractive index [10, 21]. However, a priori

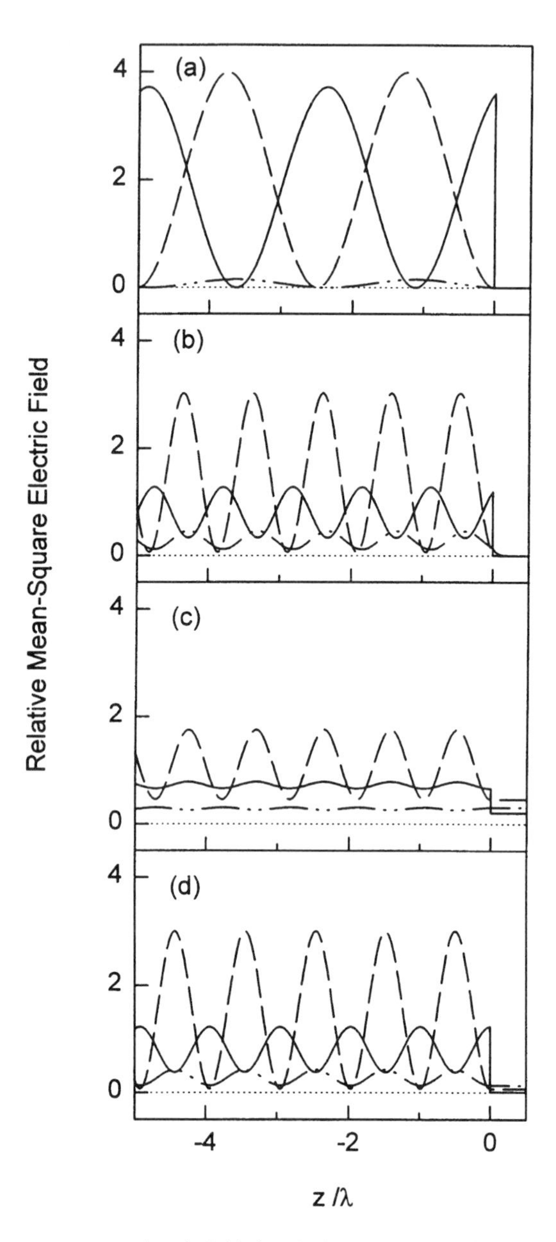

Fig. 3.5 Relative mean-square electric fields in a 2-phase system as a function of distance from the interface. Phase 1 is air and phase 2 is (a) gold, (b) glassy carbon, (c) glass, and (d) silicon. The mean-square electric fields are shown for the *z*-(—), *y*-(– – –), and *x*-(----) components. Distances have been normalized to the wavelength of the incident light, which corresponds to 2000 cm^{-1}.

predictions of band shapes are hindered by a complicated interplay of mean-square electric fields, angles of incidence, polarizations, and the reflectivities between different phases in a stratified medium. It is therefore valuable to examine a series of simulated spectra of absorbing films under differing conditions. As a starting point, Fig. 3.6 presents simulated data for $\hat{n}$ for a thin organic film. The absorption band has an absorption index $k(\bar{\nu})$ with a Gaussian line shape and a maximum value of 0.25 at $1730\,\text{cm}^{-1}$. These properties are similar to those for the carbonyl stretch of an ester functionality in a polymer like poly(methyl methacrylate). Using this $k(\bar{\nu})$, $n(\bar{\nu})$ can be calculated using the Kramers-Kronig relation given in equation (3.18). A plot of the resulting $n(\bar{\nu})$ is also shown in Fig. 3.6. It is seen that $n(\bar{\nu})$ has a derivative shape, decreasing on the high-energy side of the absorption band and increasing on the low-energy side.

The extent to which band shapes are dependent on angle of incidence is illustrated in Fig. 3.7. This figure shows a series of simulated spectra for a 1000-Å thick film, whose optical functions are as shown in Fig. 3.6, coated onto a glassy carbon substrate. To generate the spectra, the reflectivities for coated (R) and uncoated (R_0) substrates were calculated as a function of wavelength. The reflection analogue to absorbance is then given by $A = -\log(R/R_o)$. As is evident, the shapes of these spectra are reminiscent of that of $n(\bar{\nu})$, though the sign of the distortion changes in going from an angle of incidence of 20° to 80°.

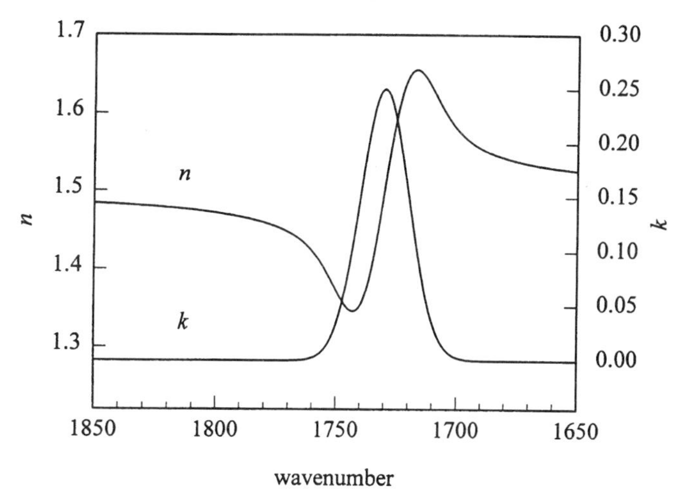

Fig. 3.6 Absorption index k and refractive index n of a simulated infrared absorption band centered at $1730\,\text{cm}^{-1}$.

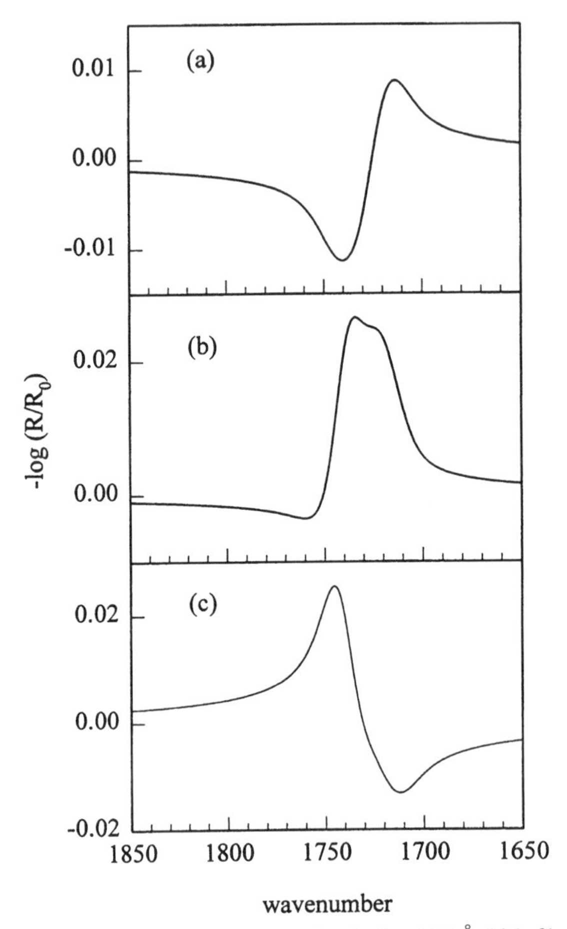

Fig. 3.7 Simulations of a reflection absorption band of a 1000 Å-thick film whose optical functions are displayed in Figure 6. The incident medium is air and the substrate is glassy carbon whose optical functions are assumed to be constant ($n = 2.9$, $k = 1.3$) across the spectral region. This figure illustrates the changes that occur in the band profile as the angle of incidence of *p*-polarized light is changed from (a) 20°, to (b) 59° to (c) 80°.

To diagnose the role of $n(\bar{\nu})$ in determining these lineshapes, the calculations were repeated with a hypothetical film with $k(\bar{\nu})$ set to zero over all wavelengths, while the dependence of $n(\bar{\nu})$ in Fig. 3.6 was maintained. The results of these calculations are displayed in Fig. 3.8. Comparing these spectra with those in Fig. 3.7 reveals that at both high and low angles of incidence, the band shape is dominated by $n(\bar{\nu})$, whereas the spectrum at a 59° angle of incidence contains contributions from both $n(\bar{\nu})$ and $k(\bar{\nu})$.

Figure 3.9 shows the effect of varying the substrate on the band shapes and peak positions. As in Fig. 3.4, the substrates include gold, glassy carbon,

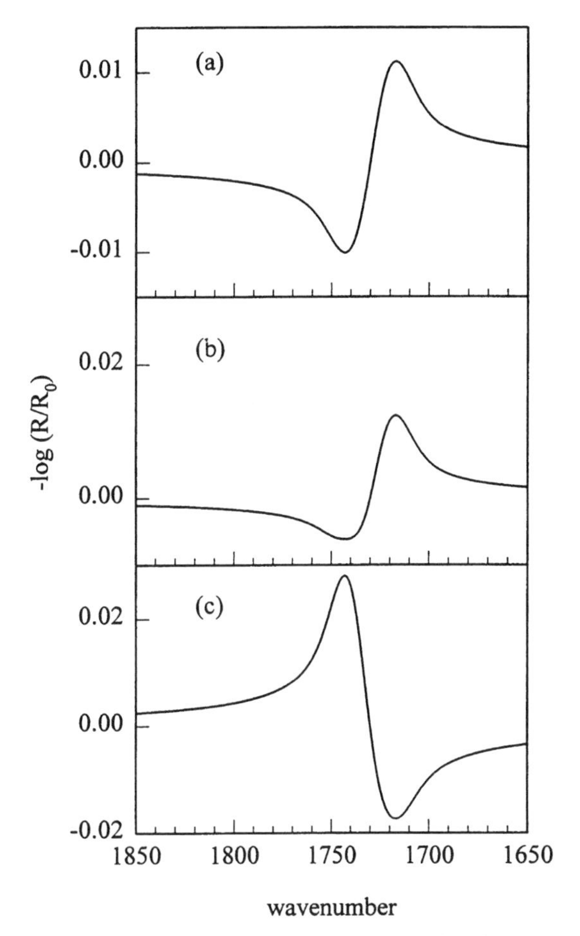

Fig. 3.8 Spectral simulations under the same conditions as in Figure 7 except that the absorptivity of the film has been set to zero while retaining the same dispersion in *n*.

silicon, and glass. In all cases the angle of incidence was set at a value to produce the largest mean-square electric field at the surface of the substrate. Because of the convolution of optical effects, the distortion of the band shapes differ depending on the substrate. In some instances a maximum is observed at higher energies than the maximum for $k(\bar{\nu})$, and in some at lower energies. The following shows that these distortions are accurately predicted based on comparisons of observed and calculated spectra.

The distortions discussed above are evident in the spectra of poly(methyl methacrylate) (PMMA) films on glassy carbon. Figure 3.10 [22] shows both the observed and calculated spectra for three different PMMA film thicknesses (i.e., 3270 ± 100, 362 ± 30, and 78 ± 15 Å) in the region of the

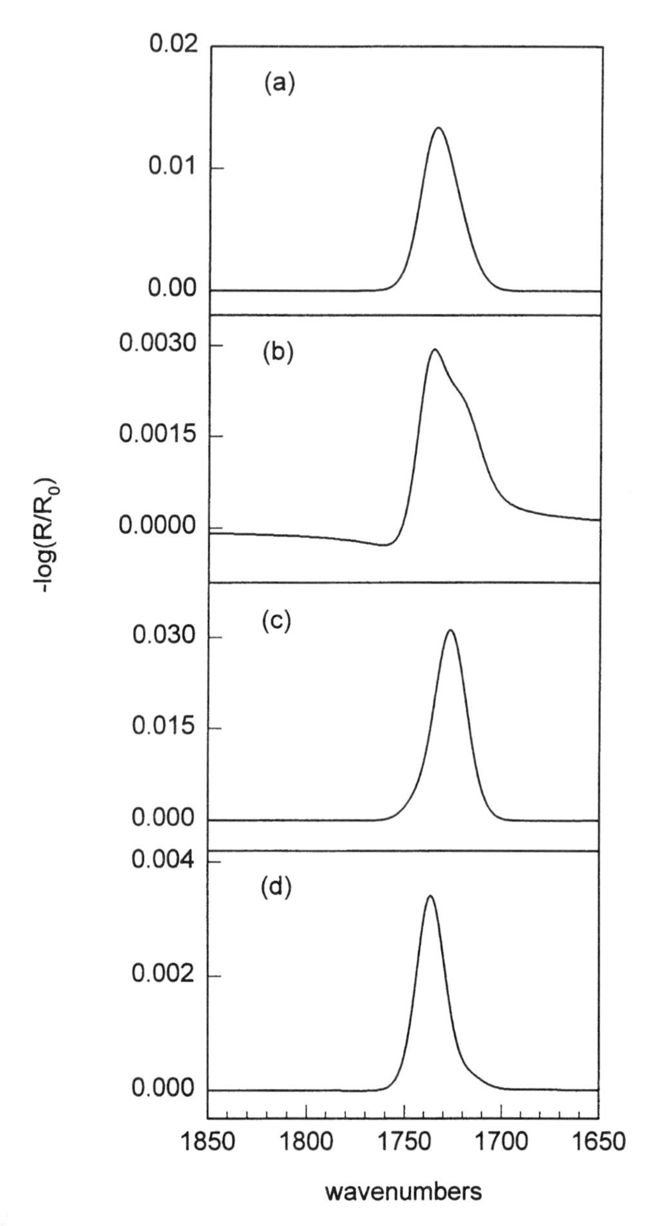

Fig. 3.9 Simulated reflection spectra illustrating the effect of different substrates on band profiles. The spectra are for 100 Angstrom-thick films with optical constants given in Figure 6. The substrates and angle of incidence are: (a) gold, 70°; (b) glassy carbon, 59°; (c) glass, 58°; and (c) silicon, 59.5°. These angles are those that result in the largest relative mean-square electric field in the *z*-direction at the substrate interface.

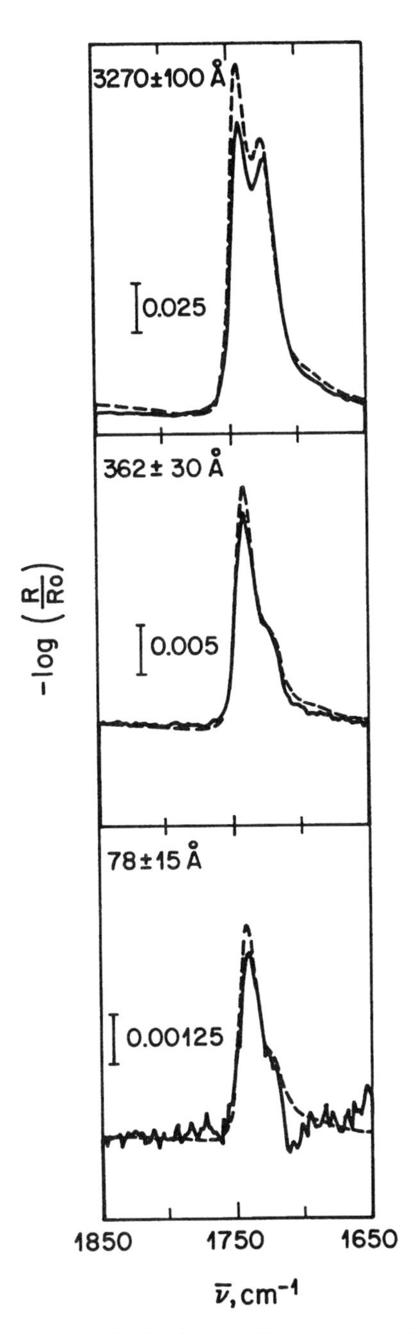

Fig. 3.10 Experimental (—) and calculated (----) reflection spectra of the carbonyl stretching mode of poly(methyl methacrylate) films of various thicknesses on glassy carbon. The *p*-polarized light is incident at 60° (reproduced with permission from reference 22).

carbonyl stretching vibration obtained with p-polarized light incident at 60°. The calculated spectra were produced by starting with a transmission spectrum to generate an initial approximation for $k(\bar{\nu})$. This $k(\bar{\nu})$ was used to generate $n(\bar{\nu})$ through the Kramers-Kronig transformation. The resulting $k(\bar{\nu})$ and $n(\bar{\nu})$ data set was then used to calculate the transmission spectrum. Differences between the calculated and observed spectra were used to adjust $k(\bar{\nu})$, and the calculation cycle repeated until no further improvement in the fit was obtained. The final $k(\bar{\nu})$ and $n(\bar{\nu})$ data set was used to calculate the specular reflection spectrum at each film thickness. As discussed for the data in Fig. 3.8, optical effects can lead to large alterations in the shapes of reflection spectra compared to those found in transmission spectra. Furthermore these distortions are strongly dependent on film thickness as well as the polarization and angle of incidence of the incoming light. Figure 3.11 shows spectra of the same films as in Fig. 3.10 taken with a 20° angle of incidence. These alterations in peak location and shape, which are caused by the optical properties of the film, must be accounted for before assigning differences between reflection and transmission spectra to surface induced changes in the structure or chemical bonding of the film.

Orientation Analysis

As was noted earlier, the anisotropy of the mean-square electric fields at highly reflective surfaces can be used to determine the spatial orientation of an organic film [23]. The analysis begins by recognizing that the magnitude of an infrared absorption is given by

$$A \propto |M \cdot E|^2, \tag{3.39}$$

where M is the dipole moment derivative of the vibrational mode with respect to the normal coordinate of the vibration. Since maximum detection at a highly reflective surface occurs when p-polarized light is used at near-grazing angles of incidence, the electric field vector is oriented perpendicular to the surface (i.e., in the z-direction). It follows that the magnitude of the absorption will be given by

$$A \propto |M \cdot z|^2 \propto \cos^2 \theta, \tag{3.40}$$

where θ is the angle of orientation of the vibrational mode with respect to the surface normal. Thus the most strongly excited vibrations will be those with dipole transitions that are oriented perpendicular to the surface. This orientation dependence of the absorption intensity leads to observable differences between the spectra of isotropically and anisotropically oriented films, a situation that be exploited to deduce the orientation of molecules in

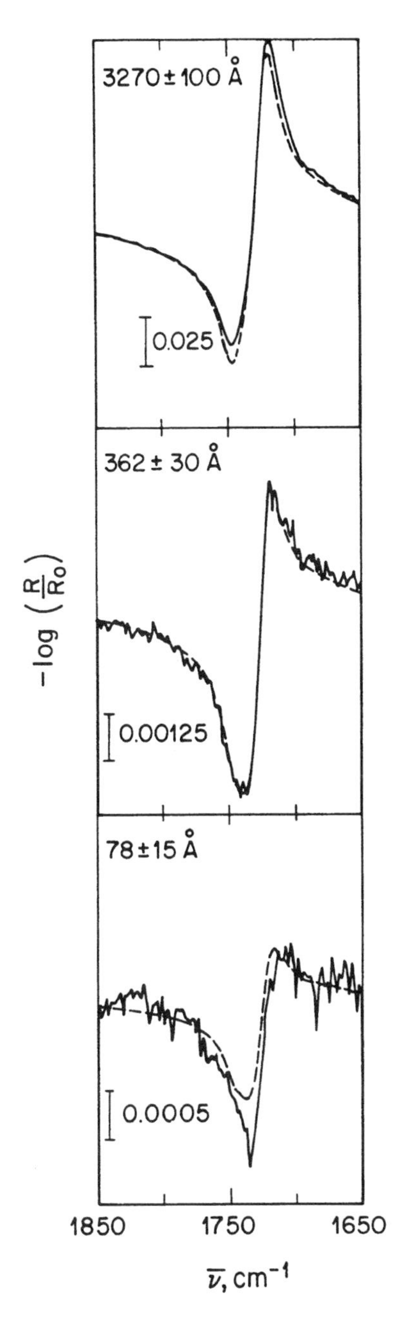

Fig. 3.11 Experimental (—) and calculated (----) reflection spectra of the carbonyl stretching mode of poly(methyl methacrylate) films of various thicknesses on glassy carbon. The *p*-polarized light is incident at 20° (reproduced with permission from reference 22).

an ordered thin film. The average spatial orientation of a given vibrational mode can be calculated via

$$\cos^2 \theta = \frac{A_{\mathrm{obs}}}{3A_{\mathrm{calc}}}, \tag{3.41}$$

where A_{obs} is the observed absorbance and A_{calc} is the calculated absorbance for an isotropic film of equivalent thickness.

The above analysis has proved invaluable in the characterization of a variety of thin-film systems [1, 2, 3, 4, 5, 6]. There are, however, two important limitations that merit consideration. The first stems from the breakdown of the strong anisotropy of the electric fields that are present at surfaces with high reflectivities when using surfaces with low reflectivities. This situation, which was discussed earlier, results from the much larger presence of the x- and y-components of the electric fields at surfaces with low reflectivities (see Fig. 3.4). Thus the observed reflection spectrum for films coated on surfaces like glassy carbon contains contributions from a "mixed" surface electric field.

The second complication results from $\hat{n}$ being a tensor for an anisotropic medium. Most treatments do not consider $\hat{n}$ as a tensor but rather a scalar. Recently, however, an approach that treats the calculation of the reflection spectrum of a sample via a matrix transform method to account for the tensor quality of $\hat{n}$ had been developed [24]. The essence of the approach involves the iterative calculation of a reflection spectrum by varying the spatial orientation of the ordered film until reaching an optimal fit with the experimental spectrum. This approach represents an important refinement of orientation determinations, as will be demonstrated for one example in the next section. It is nevertheless important to note that use of all orientational analyses are only valid if the vibrational modes of the adsorbate are not strongly perturbed as a consequence of immobilization.

3.3 EXPERIMENTAL CONSIDERATIONS

IRS has long been an important method for the investigation of structure and bonding in organic materials. The extension of this technique to the investigation of the structure of thin films as surface coatings has generally been limited by the ability of infrared instrumentation to detect small amounts of material on a surface. Recent instrumental developments, such as advances in Fourier transform infrared (FT-IR) technology, make such characterization much more tractable.

Given the small amounts of material present in most thin films and the resulting low magnitude of their spectra, one of the limiting factors in the application of IRS to specular reflectance has been the attainment of the requisite signal-to-noise (S/N) ratio sufficient to provide tractable spectra. For instance, in the case of spontaneously adsorbed monolayers, absorbances are generally on the order of 10^{-4} or less, requiring a S/N ratio often unattainable without specialized detectors. Recent improvements in detector design and electronics have led to high-quality, "off the shelf" detectors that provide more than adequate S/N for most experiments.

The advent of computer technology has also been valuable to the development of reflectance methods in FT-IR spectroscopy. Improvements in computer design have allowed the replacement of vendor-built minicomputer systems with commercially available systems as well as a notable decrease in the calculation time required for the Fourier transformation analysis.

The general experimental design for external reflectance infrared spectroscopy varies only slightly from experiment to experiment. In its most basic form, the experiment requires four components. These include a stable source of broadband infrared radiation, a method for mounting the sample in the infrared beam, a method for separating the incident light into its component polarizations, and a detector, all of which are coordinated through a computer-driven data acquisition and control system. By far the most common instrumental arrangement utilizes Fourier transform technology by combining a Globar type blackbody emission source, Michelson interferometer, and liquid-nitrogen-cooled HgCdTe or InSb detector. Commercial versions of these instruments are now capable of providing the stability and low noise levels necessary for reflectance experiments at submonolayer levels. This being true, judicious choice of components is still a necessity in the construction of such an instrument.

The selection of an appropriate IR source has become much less problematic in recent years. The development of Globar sources, which function as stable blackbody generators, has reached the point where stability is not a major concern. A remaining consideration currently involves the external cooling method for the source. Recently ceramic technology has allowed the more combersome liquid-cooling methods to be replaced by much simpler air-cooled sources that provide stable outputs over a fairly long lifetime.

Coincident with the progress in computer technology has been a general increase in the availability of high-performance detectors. For most applications, a liquid-nitrogen-cooled HgCdTe broad- or narrowband detector provides the requisite low noise levels.

The convoluted dependence of specular reflection spectra on polarization, angle of incidence, and substrate optical properties is a challenging problem for potential users of the technique. Insights into the physical optics of the measurement must be developed in order to optimize sampling configuration. Moreover these complications must be accounted for when using the spectra obtained in this mode for quantitative measurements or for structural interpretations based on peak positions and band shapes. For example, the nature of the standing wave electric field near the substrate surface results in a nonlinear dependence of absorption on film thickness.

There are several sampling approaches for specular reflectance measurements; all are designed to maximize detection capabilities while minimizing contributions from background interference. The vast majority of specular reflectance experiments involves the characterization of thin organic films deposited onto metallic (i.e., highly reflective) substrates. As was discussed in the last section, the experimental configuration for maximum detectability has an incoming p-polarized beam at grazing angles of incidence. Using this configuration with high-performance instrumentation, it is possible to characterize submonolayer quantities of an adsorbate. Orientational information can also be derived from such spectra via the infrared surface selection rule [10, 16].

External reflection spectra can be measured in several ways with commercially available, fixed or variable angle accessories that readily can be mounted into the sample compartment of an FT-IR spectrometer. In Fig. 3.12, the optical layout of the Versatile Reflection Attachment (VRA, Harrick Scientific Corp.) is shown. This accessory allows the angle of incidence to be varied continuously from 15° to 85° without realignment. This is accomplished by rotation of the sample-mirror combination, which has a fixed angle of approximately 100° between them, about an axis that lies along the intersection of the sample and mirror planes. This accessory can accommodate a range of sample sizes and is convenient for examining thin films supported on a smooth metal substrate, such as gold or silver, vapor deposited on a microscope slide. While the optimum angle of incidence for characterizing films on gold in the mid-infrared region of the spectrum is 88°, it is often advantageous to use a lower incidence angle because a circular IR beam incident at 88° is elongated to nearly 30 times over its original diameter. Thus, unless very large samples or small beam diameters are used, much of the light will not impinge on the sample. In contrast, an 80° angle of incidence spreads the beam by only a factor of six, simplifying the measurement. Larger samples, such as computer disks, are more easily handled if mounted horizontally. Accessories optimized for such purposes have been developed. However, changing the angle of incidence is not as convenient as with the VRA, and most accessories are set up to provide a fixed angle of incidence.

The spreading of the IR beam across the sample at high angles of incidence limits the spatial resolution of the technique. Importantly, acces-

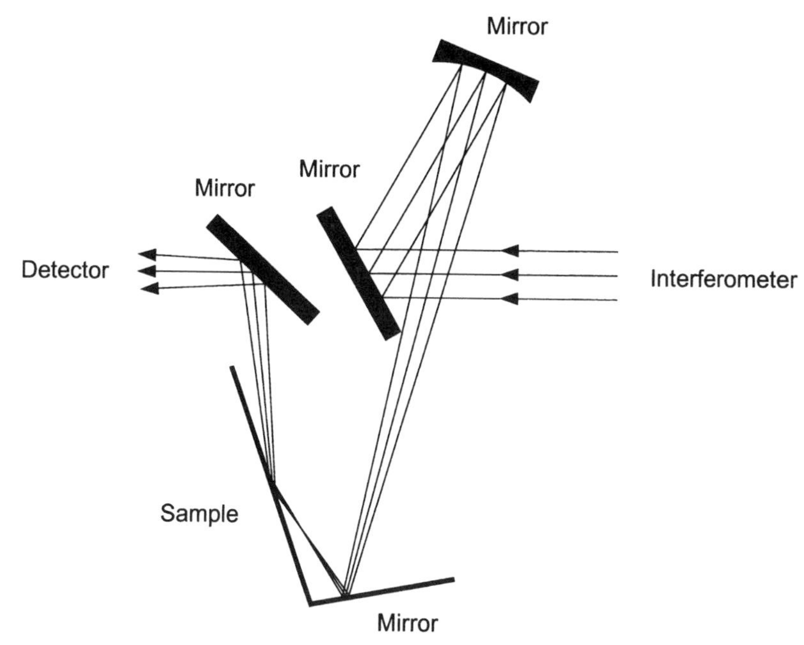

Fig. 3.12 Optical layout of an FT-IR spectrometer accessory for performing specular reflection measurements.

sories are available for use with an infrared microscope [25] that allow measurements to be made at grazing angles of incident with a spatial resolution of $\sim 50\,\mu$.

A more challenging problem involves the analysis of films deposited on nonmetallic substrates, such as silicon or carbon surfaces. Maximum detectability in these cases occurs at an incident angle of $\sim 70^\circ$ with respect to the surface normal. In comparison to characterizations at highly reflective surfaces, those at surfaces with low reflectivities suffer from two complications. The first is the accentuation of the distortion of spectral band shapes that results from the dispersion in the refractive index of the film in the vicinity of absorption features (see Section 3.2). The second is the "breakdown" in the surface effect that gives rise to the surface selection rule at metallic surfaces (see Section 3.2). Both complications arise from the lower reflectivity of substrates like silicon and carbon.

Perhaps most important in reflectance measurements at substrates coated with thin films is the choice of a material for a reference spectrum. The most common reference is an uncoated sample substrate that has been cleaned using acidic or organic solvents immediately prior to use. These "bare" references are sufficient for use with samples having large absorbances but can be troublesome for samples with low signal strengths (i.e., ultrathin films). Contaminants on reference substrates have serious consequences in such cases by introducing artifacts into the resulting spectrum. To counteract these effects, reference samples are often prepared by use of a protective film that minimizes adventitious contaminant adsorption and does not have spectral features that overlap with those of the sample film. Examples of such films include perdeuterated [26] or fluorinated [27] organothiolates adsorbed at gold. Both materials have low surface tensions and are thereby not highly susceptible to contaminant adorption. However, where perdeuterated thiolates are of use over a significant portion of the mid-IR spectral region, the fluorinated organothiolates have strong spectral features associated with C–F stretching modes that limit their usefulness largely to the C–H stretching region.

Another critical consideration in the design of infrared specular reflectance characterizations involves an understanding of the environmental requirements for a particular experiment. A significant source of interference in low signal measurements is the strong absorbance of water vapor and carbon dioxide in critical regions of the infrared spectrum. At the typically small absorbance levels for thin films, inadequate removal of these interferences can result in a considerable masking of spectral details. Masking by water vapor can be especially problematic because of the overlap of its envelopes of stretching and bending modes with the C–H and C=O stretching regions. The use of vacuum techniques has been quite successful in dealing with this interference, but is available on only a few commercial instruments. Moreover the direct translation of the results of *in vacuo* characterizations to actual *in situ* conditions may prove problematic.

A more common method for dealing with water vapor and carbon dioxide is through the use of efficient and consistent purge methods combined with spectral subtraction techniques to remove signals from ineffectively compensated interferences. Instrument designs currently contain the requisite "plumbing" to purge efficiently the spectrometer with either dry air or nitrogen. Purging the bench, for example, with a clean, dry purge gas (e.g., boil-off from liquid nitrogen) can diminish water interference to a level where the resulting spectroscopic features are negligible compared to those of the sample. Spectral subtraction with the necessary reference spectrum can then be used to compensate for any residual interference signal, resulting in a high-quality sample spectrum.

Another approach for discrimination against the "background" signals of the surrounding media employs polarization modulation (PM) techniques [28]. These techniques take advantage of the anisotropy of the mean-square electric field near a highly reflective surface for *p*- versus *s*-polarized incident light. Because only the *z*-component of the electric field is present near the surface, immobilized films absorb only *p*-polarized light. The surrounding, isotropic medium, however, absorbs both *p*- and *s*-polarized light.

In PM experiments the polarization of the infrared light is modulated at kilohertz frequencies between the two polarization states through the use of a photoelastic modulator in conjunction with a polarizer. A photoelastic modulator is an IR transparent crystal with an isotropic refractive index when unstressed. The application of a resonant periodic strain along one axis of the crystal with a piezoelectric actuator induces an anisotropy in the refractive index of the crystal, which results in a phase retardation and hence, a rotation of the polarization of the incident beam. The addition of a fixed polarizer to the optical path that is aligned to pass either *s*- or *p*-polarized light completes the setup, with the beam impinging on a surface modulated between the two polarization states. The signal from the reflection experiment is then demodulated using phase-sensitive detection to generate a difference spectrum $(I_p - I_s)$ between the *p*- and *s*-polarized components of the signal. The differential reflectance spectrum (S) of the sample is obtained as the ratio of the difference spectrum to the sum $(I_p + I_s)$:

$$S = \frac{(I_p - I_s)}{(I_p + I_s)}. \tag{3.42}$$

A block diagram of this instrumentation is shown in Fig. 3.13. Further details of the basic principles of this technique, which has also been successfully applied to compensate for the background of *in situ* electrochemical studies [5, 29, 30], as well as recent experimental advances are available [31, 32].

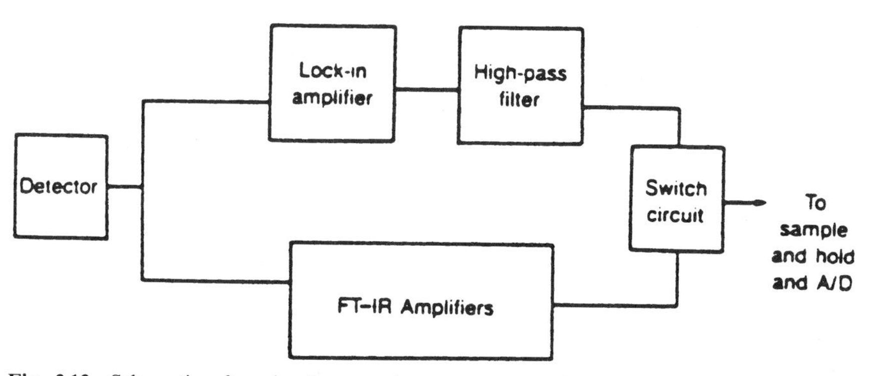

Fig. 3.13 Schematic of a signal processing arrangement for a polarization modulation measurement. The signal for the detector amplifier is split into two channels, with each channel demodulated separately (reprinted with permission from reference 5).

3.4 APPLICATIONS

The previous discussions lay a foundation for the application of IRS to a wide range of surface characterizations. Applications have ranged from the identification of the products from a plasma-induced modification of the surface of a polymeric film [33]; the determination of the average spatial orientation of ordered organic films at metal [23, 34, 35, 36, 37, 38, 39, 40, 41], carbon [24], and aqueous [41, 42] interfaces; the delineation of the effects of applied potential on the composition of a redox-transformable organic adsorbate at electrode surfaces [43, 44, 45]; to the reactions of gas phase species at surfaces under ultrahigh vacuum [46]. In view of the breadth of these applications, the following discussion is limited to three types of characterization problems. The first example deals with the structural characterization of organized organic thin films. This example illustrates the use of IRS as a probe of the spatial orientation of monolayer and multilayer films. The second example deals with the dependence of the composition and structure of an organic monolayer films with a pendent redox-transformable functionality. This example demonstrates the capability of IRS as an *in situ* structural probe of liquid-solid interfaces. The third example describes a study of the composition of thin films of aluminum nitride. This example shows the utility of IRS as a probe of the effects of preparation and handling on the composition of this material important to microelectronics research.

Organized Thin Films

In selecting representative applications of IRS characterizations of samples in a specular mode, we first discuss the results of a series of characterizations of organized monolayer and multilayer films of organic compounds. These examples demonstrate the strengths of this characterization approach by unraveling details concerning both the composition and average spatial arrangement of the thin film. The first example examines the composition and spatial orientation of the monolayer formed by the spontaneous adsorption of a long alkyl chain length alkanoic acid (i.e., arachidic acid, $CH_3(CH_2)_{18}COOH$) at ambient silver surfaces [47]. The interest in such systems stems from the high relative degree of order of the organic structures which are formed by this and other adsorbate-substrate combinations (e.g., organosulfur compounds at gold surfaces [48, 49]) in comparison to the morphological and compositional heterogeneity of functionalized polymer surfaces. This attribute, coupled with ease of preparation, has lead to an explosive growth in the use of these systems as models for examining the structural origins of a host of interfacial phenomena (e.g., electron transfer, wettability, lubrication, and biocompatibility) [49] and as platforms for the construction of chemical sensors and other forms of analytical transduction schemes [48].

Figure 3.14 gives details about both the mode of interaction between arachidic acid and the ambient silver surface and the packing density of the alkyl chains. The presence of the strong band at $1400\,cm^{-1}$ and the weak, barely detectable band at $1514\,cm^{-1}$, coupled with the absence of a band around $1730\,cm^{-1}$, reveals that the chemisorption of the carboxylic acid group and the thin layer of silver oxide present at ambient silver surfaces transforms the acid to a carboxylate salt. A dispersion of the arachidic acid in KBr has an absorption band near $1730\,cm^{-1}$, which is the $\nu(C{=}O)$ mode for a protonated carboxylic acid. As judged by comparison to spectra for silver carboxylate salts, the bands at ~ 1514 and $1400\,cm^{-1}$ are assigned to the $\nu_a(COO^-)$ and $\nu_s(COO^-)$ modes, respectively. Importantly, the much larger absorbance of the $\nu_s(COO^-)$ mode with respect to the $\nu_a(COO^-)$ mode argues that the carboxylate group is symmetrically bound as a bridging ligand to the silver surface, an assertion consistent with the peak positions for bulk metal carboxylate salts of bridging ligands.

The monolayer spectrum in Fig. 3.14 also contains information that serves as a basis for the development of a conformational description of the alkyl chains. Earlier studies of crystalline hydrocarbons [50, 51], which were undertaken in part to aid in the characterization of the chain structure of the bilayer structure of cell membranes, have found that the coupling of the $\omega(CH_2)$ and $\gamma(CH_2)$ modes is a signature of alkyl chains composed of all-trans conformational sequences. Thus the envelope of the $\omega(CH_2)$ and

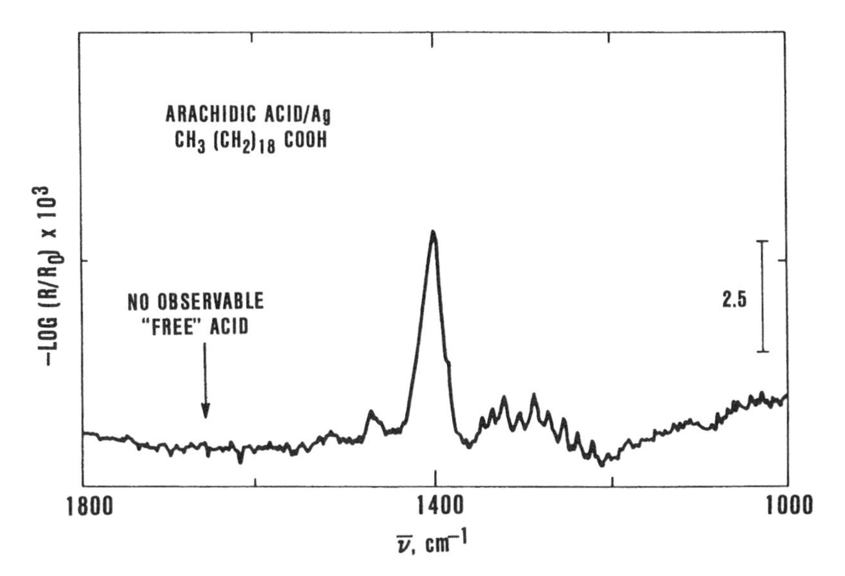

Fig. 3.14 Infrared specular reflectance spectrum between 1800 and $1000\,cm^{-1}$ for a spontaneously adsorbed monolayer from aracidic acid at silver (adapted with permission from reference 47).

$\gamma(CH_2)$ between 1350 and 1200 cm^{-1} is diagnostic of an extended chain conformation. Though unclear as to the extent of conformational defects, it follows that the resulting interfacial structure exists in a densely packed environment.

Further insight into the structure of the alkyl chains of the monolayer formed from arachidic acid can be gleaned from the spectrum in Fig 3.15. This figure presents both an experimental and calculated spectrum in the C–H stretching region. The bands at 2917 and 2851 cm^{-1} are assigned to the $\nu_a(CH_2)$ and $\nu_s(CH_2)$ modes, respectively. Bands attributed to the methyl modes are also evident, with the in-plane $\nu_a(CH_3)$ mode at 2965 cm^{-1} and two bands for the Fermi-resonance couplet of the $\nu_s(CH_3)$ mode at 2938 and 2879 cm^{-1}. Importantly, the positions of the methylene stretching modes are indicative of a densely packed array of alkyl chains, which is consistent with the presence of the envelope of the $\omega(CH_2)$ and $\gamma(CH_2)$ modes in Fig. 3.14.

Details about the spatial orientation of the alkyl chains can be quantitated by an examination of the absorbances of the methylene modes for the calculated and experimental spectra. In this case both the average tilt (θ) and twist (ϕ) of the alkyl chains (see Fig. 3.16) can be determined by

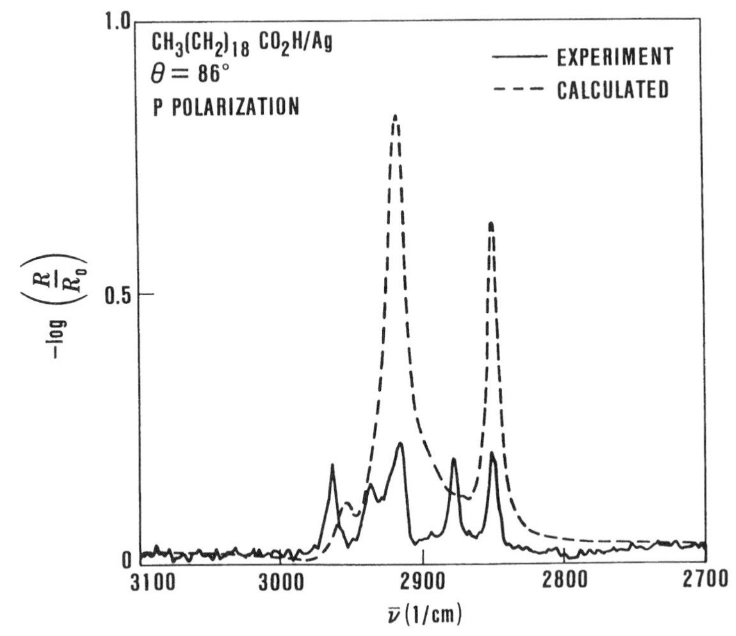

Fig. 3.15 Observed (—) and calculated (-----) infrared specular reflectance spectrum between 3100 and 2700 cm^{-1} for a spontaneously adsorbed monolayer from aracidic acid at silver (adapted with permission from reference 47).

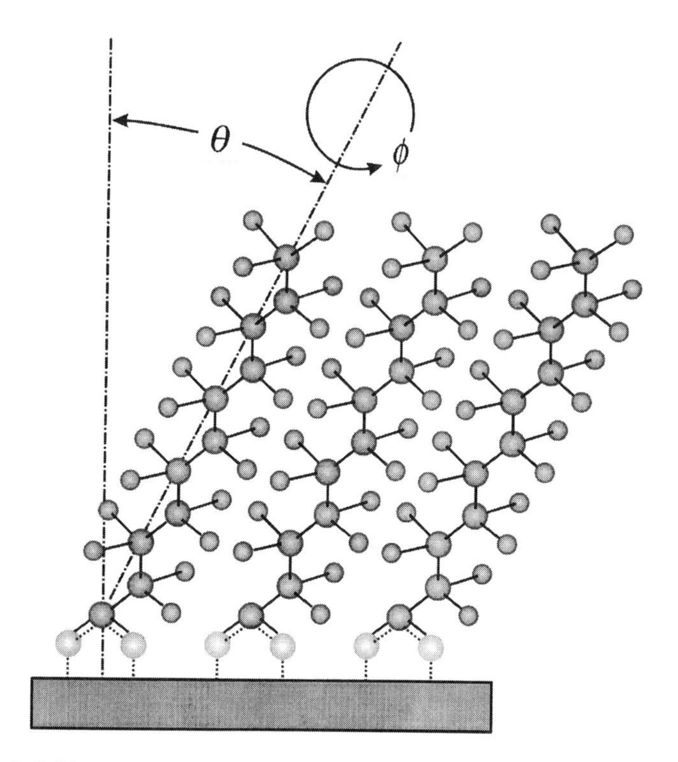

Fig. 3.16 Definition of the tilt angle θ and twist angle ϕ of an all-trans alkyl chain with respect to a planar surface.

recognizing that the transition dipoles for the $\nu_a(CH_2)$ and $\nu_s(CH_2)$ modes are orthogonal to each other as well as to the molecular axis of an extended alkyl chain. For such a system, equation (3.41) can be recast as

$$\sin^2\theta\cos^2\phi = \frac{A_{\text{obs}}^{\nu_s}}{3A_{\text{calc}}^{\nu_s}}. \tag{3.43}$$

$$\sin^2\theta\cos^2\phi = \frac{A_{\text{obs}}^{\nu_a}}{3A_{\text{calc}}^{\nu_a}}. \tag{3.44}$$

This treatment yields an average value for θ of 24° and of 48° for ϕ. These angles are consistent with a carboxylate head group bound to silver as a symmetric bridging ligand. That is, a θ of 24° follows from the expectation of a fully extended chain conformation connected to a symmetrically bound carboxylate group.

An analysis of the chain structure for organized thin films supported on surfaces with low reflectivity can be performed as well. An example of such a situation is presented in Fig. 3.17 [1]. This figure shows an observed and calculated spectrum in the C–H stretching region for a five-monolayer Langmuir-Blodgett film of cadmium arachidate supported on a glassy carbon substrate. Based on the low reflectivity of the support, the optimal angle of incidence for detection is at a much lower value than at a metal substrate (see Fig. 3.5). The derivative-shape of the spectra arises from the importance of the reflectivity at the air-film interface along with the order of the chain structure. For this multilayer system the comparison of the observed and calculated spectrum, where the calculated spectrum was obtained by the iterative process briefly described in Section 3.2, yields an average θ of 15° and an average ϕ of 45°.

Redox Transformations at Electrochemical Interfaces

In contrast to the above example, the application of IRS to probe processes at electrochemical interfaces has two major obstacles. The first is the strong absorption of the electrolytic solution. The second is the discrimination of very small absorbance changes (i.e., absorbances of 10^{-5}–10^{-4}) in the presence of the large (and sometimes overwhelming) background absorbance of the electrolytic solution. A solution to the first obstacle is the use of

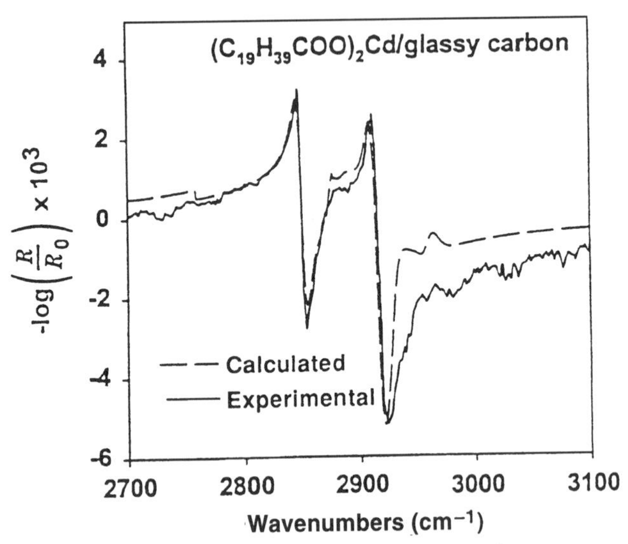

Fig. 3.17 Observed (—) and calculated (-----) infrared specular reflectance spectrum between 3100 and 2700 cm^{-1} for a five-monolayer Langmuir-Boldgett film of cadmium arachidate on a glassy carbon substrate. The *p*-polarized light is incident at 55° (adapted with permission from reference 1).

an electrochemical cell in which the contacting solution layer is held at thicknesses of a few microns for aqueous electrolytes and tens of microns for some nonaqueous media [29, 30]. Solutions to the second obstacle often employ electrochemical [29, 30] or optical (e.g., polarization; see Section 3.3) modulation schemes that are coupled with phase-sensitive detection techniques.

Figures 3.18 and 3.19 give an example of the design and optical layout of a thin layer cell. Such cells are designed for operation primarily in a specular reflection mode with the incoming beam impinging on an IR-transparent hemispherical window, passing through the thin layer of electrolytic solution, and reflecting off the electrode. The use of a hemispherical window ensures that light reflected from the air-window interface does not impinge on the detector. This design, coupled with the "beam-steering" peripheral mirrors in Fig. 3.19, facilitates optimization of the angle of incidence at the solution-electrode interface, which is a strong function of the optical properties of both the window material and electrolytic solution.

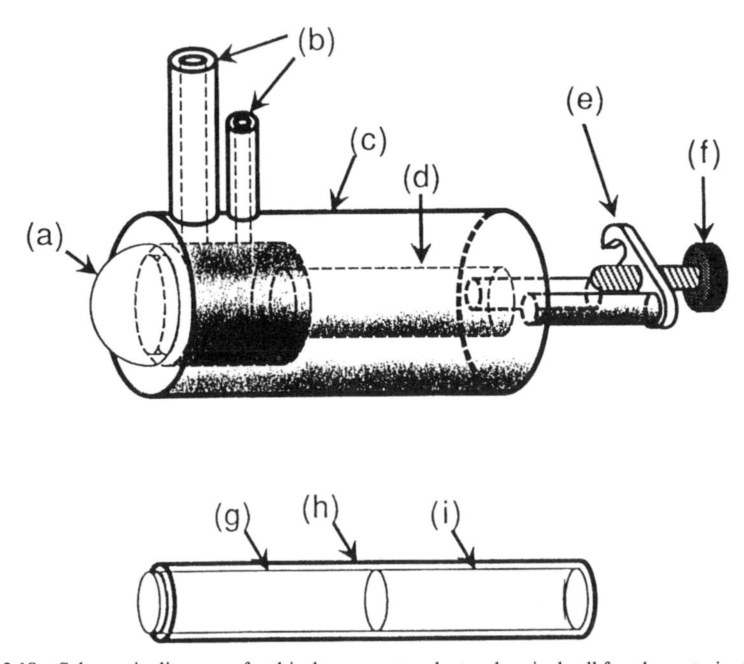

Fig. 3.18 Schematic diagram of a thin-layer spectroelectrochemical cell for characterization of organic coating at electrode surfaces by infrared specular reflectance spectroscopy. Top: (a) hemispherical window (retaining window bracket not shown); (b) solution inlet and outlet ports, (c) Kel-F cell body, (d) barrel, (e) plunger retaining screw, and (f) retaining screw. Bottom: (g) borosilicate glass core, (h) Kel-F jacket, and (i) brass core (reproduced with permission from reference 42).

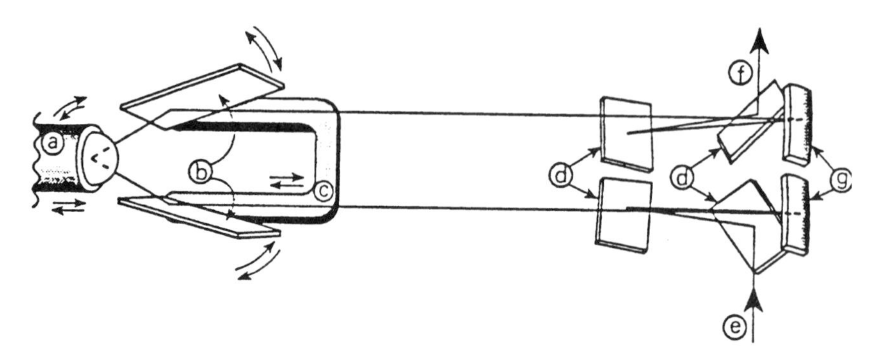

Fig. 3.19 Illustration of an optical layout for an *in situ* spectroelectrochemical characterization by specular reflectance spectroscopy. The cell assembly (a–c) and the focus projection accessory (d–f) are designed to provide flexibility to select the optimal irradiation geometry as dictated by the optical properties of the experimental components. Left: (a) front of cell (see Figure 18), (b) peripheral mirrors, and (c) arm of peripheral mirrors (which translates parallel to the optical axis). Right: (d) plane mirrors, (e) incoming beam from spectrometer, (f) outgoing beam to detector, and (g) concave mirrors to project focal plane (reproduced with permission from reference 42).

The unique structural features of spontaneously adsorbed monolayers from organosulfur compounds has also attracted the imagination of the electrochemical community [9, 49]. These systems have opened new avenues for explorations of heterogeneous electron-transfer mechanisms, electrical double-layer theories, and the creation of new types of electrochemical-based sensors [9, 49]. In each of the above cases, it is important to determine if the electrochemical transformation of a pendent redox moiety induces a structural change in the monolayer. For example, redox transformations may cause reversible/irreversible changes in the structure of the underlying polymethylene chain and/or in the double-layer structure at the interface formed by the solution-monolayer contact.

Figures 3.20 through 3.24 present the results of a study that examines the affects of a redox transformation on the structure of an organic monolayer film [43]. These results are for a monolayer prepared by the chemisorption of 11-mercaptoundecyl ferrocenecarboxylate $[(\eta^5 - C_5H_5)Fe(\eta^5 - C_5H_4)-COO(CH_2)_{11}SH$ (abbrev. $FcC_{11}SH$), where Fc denotes the ferrocenyl group] at a gold electrode. Figure 3.20 shows a set of cyclic voltammetric current-potential ($i - E$) curves for the as-formed monolayer. The wave observed as the applied voltage is scanned from +0.20 to +0.85 V reflects formally the oxidation of the Fe(II) of the ferrocenyl moiety to Fe(III), whereas the wave found upon the change in scan direction arises from the reduction Fe(III) to Fe(II). Both the shapes of the $i - E$ curves and the linear dependence of peak currents with scan rate are consistent with the

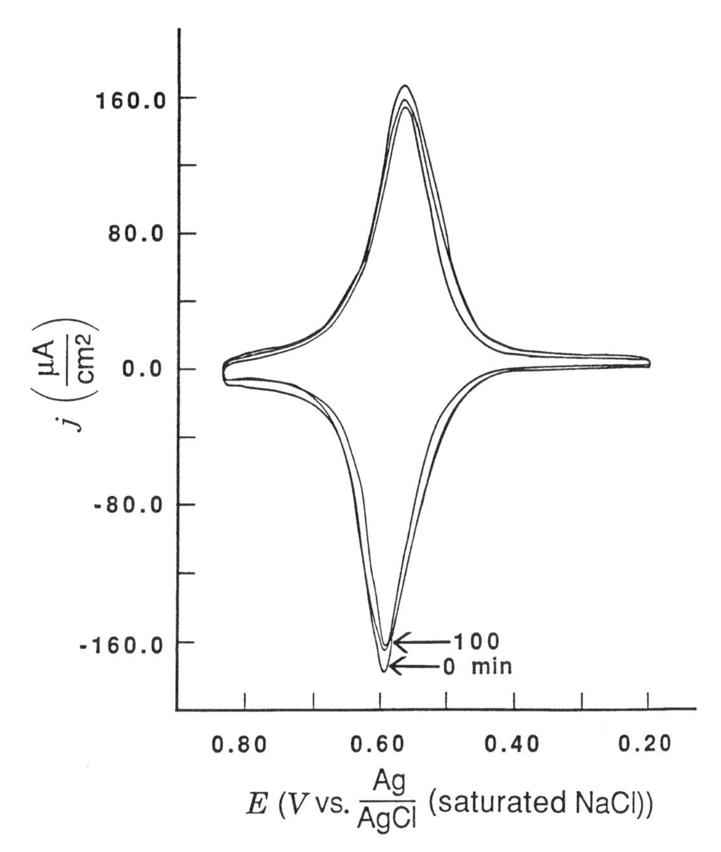

Fig. 3.20 Cyclic voltammetric curves for a monolayer formed from $FcC_{11}SH$ at gold. The scan rate is 400 mV/s, with each scan recorded after holding the applied voltage at +0.20 V for the noted time period (adapted with permission from reference 43).

electrolysis of an immobilized redox species [9]. The general repeatability of the $i - E$ curves demonstrates the electrochemical stability of the immobilized species.

The electrochemical data provide two additional insights into the structure of the $FcC_{11}SH$-based monolayer. First, the charge passed to electrolyze the ferrocenyl moiety can be used to determine the surface concentration of the monolayer. Based on such an analysis, the surface coverage corresponds to a packing density limited largely by the size of the ferrocenyl end group. Second, the charging current that flows before and after oxidation of the ferrocenyl group changes markedly. This change reflects one of two (or a combination of both) possibilities. One possibility is a change in the capacitance of the interfacial structure that arises from the

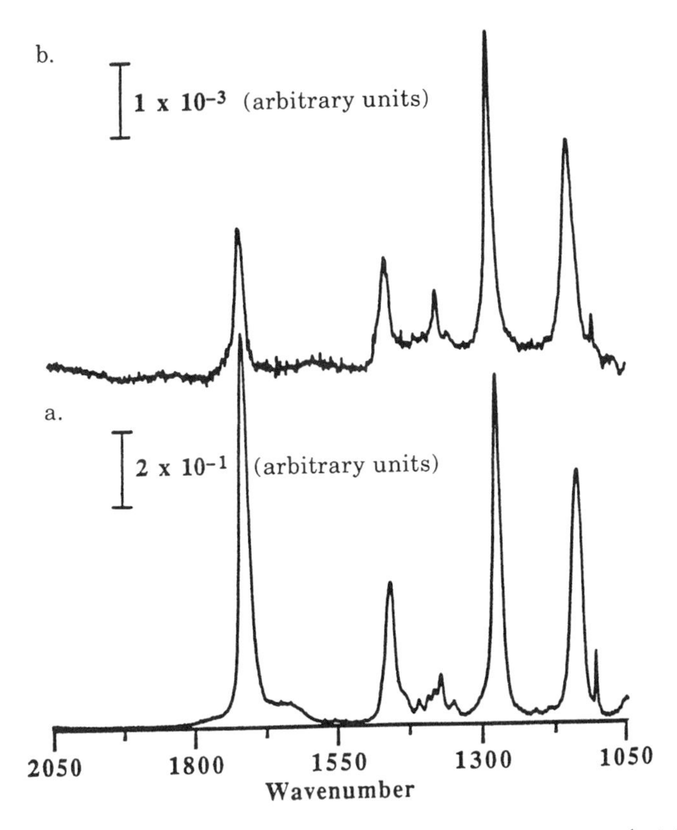

Fig. 3.21 Infrared specular reflectance spectrum between 2050 and 1050 cm^{-1} of (a) a KBr pellet of $FcC_{11}SH$ and (b) of a spontaneously adsorbed monolayer from $FcC_{11}SH$ at gold. The monolayer spectrum was collected at an incident angle of 82° using *p*-polarized light (adapted with permission from reference 43).

creation of a cationic ferrocenyl group; the other is a change in the structure of the monolayer because of the repulsion of closely spaced, charged end groups. Gaining insight into the viability and relative importance of both possibilities is critical to the interpretation of data from studies aimed at the use of this system as a model interfacial system.

Figures 3.21 and 3.22 presents the IRS spectrum for a monolayer from $FcC_{11}SH$ at gold. The former is the spectrum in the low-energy region, and the latter in the high-energy region. Both figures include spectra for the precursor dispersed in a KBr pellet. Bands diagnostic of the pendant group on the monolayer include the carbonyl stretching modes at 1714 cm^{-1}, the two C–O stretching modes at 1147 and 1283 cm^{-1}, the ferrocene ring C–H stretching modes at $\sim$3100 cm^{-1}, and the ferrocene ring modes between

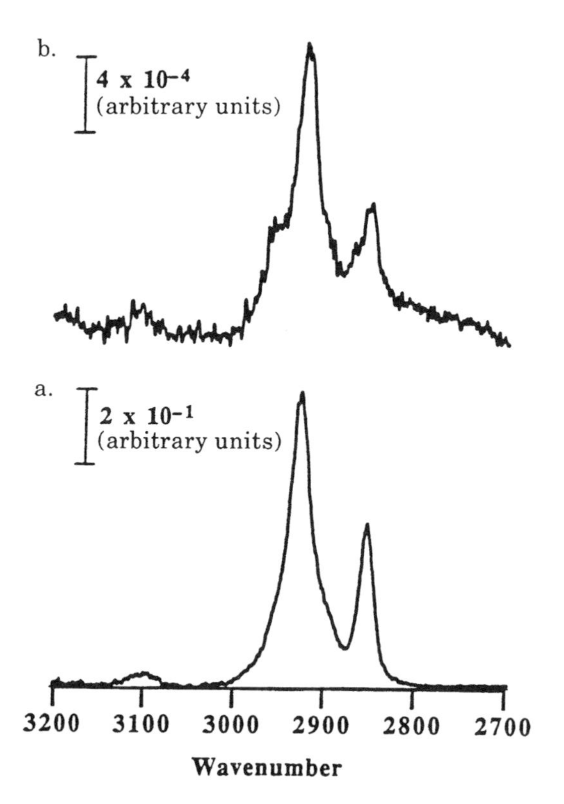

Fig. 3.22 Infrared specular reflectance spectrum between 3200 and 2700 cm^{-1} of (a) a KBr pellet of $FcC_{11}SH$ and (b) of a spontaneously adsorbed monolayer from $FcC_{11}SH$ at gold. The monolayer spectrum was collected at an incident angle of 82° using *p*-polarized light (adapted with permission from reference 43).

1445 and 1350 cm^{-1}. The presence of these modes, along with the methylene asymmetric and symmetric stretching modes between 2950 and 2800 cm^{-1}, confirms the formation of the $FcC_{11}SH$ monolayer.

A series of potential modulation spectra is shown in Figs. 3.23 and 3.24 for a monolayer from $FcC_{11}SH$ at gold in 1.0 M $HClO_4$(aq). In this mode a reflectance spectrum ($R_0(\bar{\nu})$) was first scanned at a base voltage of +0.20 V (vs. Ag/AgCl (saturated KCl)). The applied voltage was then stepped to a selected value at which another spectrum ($R(\bar{\nu})$) was scanned. The two single-beam spectra are then combined to yield a differential spectrum ($-\log(R(\bar{\nu})/R_0(\bar{\nu}))$. Several notable changes in the spectra are evident as the applied voltage becomes increasingly positive. The most notable changes in Fig. 3.23 are for the vibrational modes associated with the ester linkage between the ferrocenyl group and the polymethylene chains. Of the bands

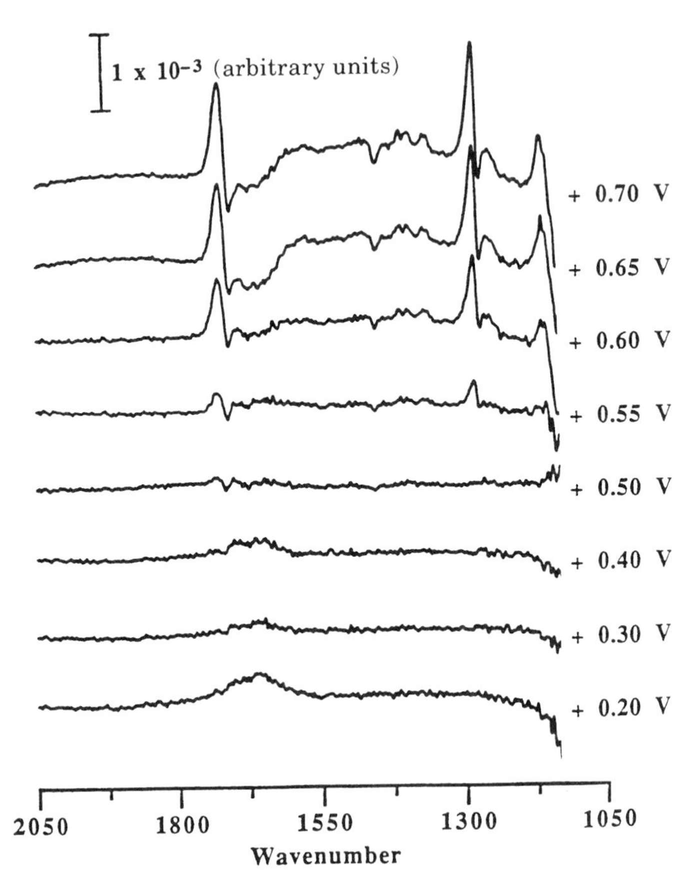

Fig. 3.23 *In situ* differential spectra of a monolayer from $FcC_{11}SH$ at gold in 1.0 M $HClO_4$ between 2050 and 1050 cm^{-1} as a function of applied voltage. The reference spectrum was taken at ±0.20 V (reproduced with permission from reference 43).

not obscured by features of the supporting electrolyte (i.e., the overlap of the C–O of the ferrocenyl group and the Cl–O mode of the supporting electrolyte at ~1140 cm^{-1}), derivative-shaped spectral features are observed. The shapes result from the shift of the positions of these vibrational modes to higher energy that results from the oxidation of the ferrocenyl group.

Changes in the high-energy region of the spectrum (Fig. 3.24) are also evident as a consequence of the oxidation of the ferrocenyl moiety. Six bands appear as the applied voltage becomes more positive. Importantly, the bands at 2929, 2872, and 2858 cm^{-1} correspond to the positions for disordered polymethylene chains. The appearance of these bands suggests

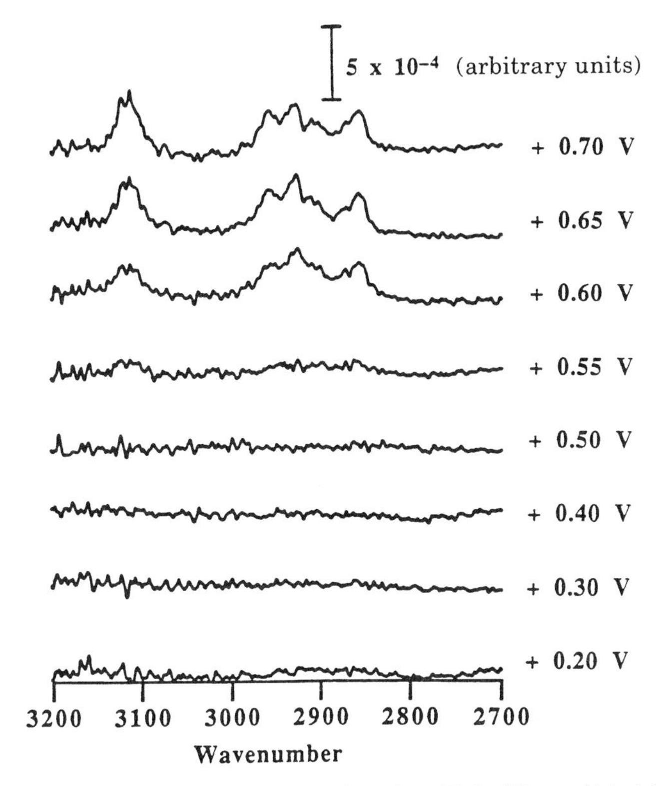

Fig. 3.24 *In situ* differential spectra of a monolayer from $FcC_{11}SH$ at gold in 1.0 M $HClO_4$ between 3200 and 2700 cm^{-1} as a function of applied voltage. The reference spectrum was taken at 0.20 V (reproduced with permission from reference 43).

that the oxidation of the ferrocenyl moiety indicates a change in the structure of the monolayer. However, control experiments using a series of bulk ferrocene esters indicate that such changes can also be induced on the α-methylene group by the oxidation of the ferrocenyl moiety. This finding, coupled with the low signal strength of the changes, argues against a structural change. Therefore, while we expect that changes in the structure of the monolayer should result from the redox transformation of the ferrocenyl group, the data in Figs. 3.21 and 3.22 are insufficient to support such a conclusion. Pending improvements in instrument performance may provide data that will allow a more definite interpretation. We note that similar conclusions have been reached for a structurally different monolayer system [44].

Aluminum Nitride Thin-Film Characterization

The examples discussed above have dealt with the use of IRS to characterize organic thin films. However, the use of this technique is continuously expanding, spanning into studies of thin films of inorganic materials. One illustration of this is recent work on the characterization of aluminum nitride (AlN) films. The chemical and thermal stability of AlN films, along with their electrical, optical, and acoustic properties, have generated considerable interest in the microelectronics research area. One example is the potential use of AlN as the transducer in high-performance piezoelectric mass sensors [52]. The efficient exploitation of this material, however, requires a thorough understanding of the effects of preparative variables on the resulting acoustical properties. Typically AlN films are formed by sputtering an aluminum target with a nitrogen plasma. Interestingly, recent studies indicate that the acoustical properties of AlN are improved if hydrogen is added as a component to the nitrogen plasma. To further diagnose the role of such processing, IRS has been used to probe the composition of AlN films fabricated under various conditions as well as to study the stability of the films by monitoring changes in the film composition as a function of exposure to different environments [53, 54, 55]. IRS is an attractive probe for this application because of the wealth of information that vibrational spectroscopy provides on the chemical and physical composition of the sample. In addition reflection spectroscopy is well suited for studying thin films, the form in which AlN is typically used.

In these studies AlN films with thicknesses ranging from 8 to 40 nm were deposited on both aluminum and gold mirrors. Although reflection spectra could be obtained using either substrate, gold has the advantage of being more chemically inert than aluminum and therefore preferable as a substrate. Reflection spectra were recorded using a Fourier transform interferometer with the impinging beam incident at $\sim 80°$. The immense value of the information provided by this technique is revealed by the species identified in these spectra. For films fabricated using a pure nitrogen plasma, the most prominent features in the spectra are a band at $910\,cm^{-1}$, a longitudinal optical (LO) mode of AlN; accompanied by a shoulder for a transverse optical (TO) mode on the low-energy side of the LO mode. The LO mode is associated with the crystalline form of AlN. Importantly, a small shift of the LO band maximum with increasing film thickness was detected, suggested an increase in the amorphous character of the film. This result indicates that IRS could provide a rapid and convenient method for evaluating film quality, which is important because the performance of AlN in many applications is enhanced by higher film crystallinity.

In addition to AlN, the spectroscoic data also revealed the presence of AlN_2 as a component of the plasma-deposited thin film. Studies in which the thickness of the films were varied between 8 and 40 nm and produced in a pure N_2 plasma indicated that the concentration of AlN_2 reached a maximum at 20 nm and dropped to a negligible level at 40 nm. In contrast, the absorbance feature for AlN_2 in films produced in a 25% H_2-enriched plasma increased linearly with film thickness. Spectral bands associated with Al–H and N–H were identified in the films prepared by both processes, along with bands for O–H and C–H vibrations. These latter findings suggest that the films are contaminated to a small degree during processing, demonstrating the potential for using IR reflection spectroscopy to monitor *in situ* the composition and purity of very thin films during fabrication.

The chemical information that IRS provides can be exploited to help understand the factors that influence the chemical stability of materials. Spectra of AlN films produced with pure N_2 and exposed for variable amounts of time to the laboratory ambient showed clear increases in the magnitudes of the N–H vibrations. This observation supports the view that moisture from the ambient reacts with AlN to form Al_2O_3 and NH_3, the latter of which is probably bound to the AlN surface via aluminum ions. Correlations with exposure time also indicated that the magnitude of the N–H signature increased for exposures up to $\sim$1.5 hr and then remained effectively constant. In contrast, there was not a significant decrease in the magnitudes of the AlN bands. This correlation argues that after a surface layer is hydrolyzed, the resulting Al_2O_3 film acts as a passivation layer toward further hydrolysis. This conclusion may account for the high stability of AlN toward further reactions with atmospheric moisture. Films produced with a hydrogen-enriched plasma and exposed to the atmosphere displayed an increase in the susceptibility to hydrolysis. In this case measurable decreases in the magnitude of the AlN vibrational bands, which were coupled with the appearance of features diagnostic of an immobilized NH_x species, were reported. The decrease in the magnitudes of the AlN bands indicated about a 15% depletion of the AlN film. This technique was also used to delineate the effects of exposing the AlN films to water. In this case hydrolysis occurred to equal extents with films produced with and without hydrogen added to the plasma.

This example of the application of external reflection infrared spectroscopy demonstrates how the technique can be used to address issues in microelectronics research area. The technique can be used to check the composition of thin films and to monitor for the presence of impurities. The lattice vibrations in IR spectra provide information on morphology of the films. Also IRS can be used to study the factors that influence the reactivity films to environmental influences.

3.5 CONCLUSIONS AND PROSPECTUS

We have discussed applications of infrared specular reflection spectroscopy as a probe of the composition and spatial orientation of materials immobilized on a wide array of substrates. Coupled with the high level of performance of infrared instrumentation, the ability to determine via classical electromagnetic theory the experimental conditions optimal for characterizations on surfaces with a range of reflectivities opens the way to tackling characterizations previously viewed as intractable. A particularly intriguing example of such an instance is the recent characterization of the structure of a single monolayer film at the air-water interface of a Langmuir trough [56]. Though not discussed as a specific application in Section 3.4, this example further demonstrates the insight afforded by consideration of the physical optics of a reflection measurement. As further improvements in instrumentation emerge, capabilities to follow surface reaction dynamics will undoubtedly appear. Such a capability will provide a much-needed access into the molecular reaction pathways of importance to catalysis and other important interfacial processes.

ACKNOWLEDGMENTS

The Ames Laboratory is operated for the U.S. Department of Energy by Iowa State University under Contract No. W-7405-eng-82. This work was supported by the Office of Basic Energy Sciences, Chemical Sciences Division.

REFERENCES

1. D. L. Allara, In *Characterization of Organic Thin Films,* A. Ulman, ed. Butterworth-Heinemann, Boston, 1995.
2. M. D. Porter, *Anal Chem.*, **60**, 1143A (1988).
3. H. Ishida, *Rubber Chem. Technol.*, **60**, 497 (1987).
4. A. M. Bradshaw, *Appl. Surf. Sci.*, **11/12**, 712 (1982).
5. W. G. Golden and D. D. Saperstein, *J. Electron. Spectrosc.*, **30**, 43 (1983).
6. F. M. Hoffman, *Surf. Sci. Rep.*, **3**, 107 (1983).
7. A. Ulman, *An Introduction to Ultra-Thin Organic Films From Langmuir-Blodgett to Self-Assembly.* Academic Press, San Diego, 1991.

8. D. R. Jung and A. W. Czanderna, *Crit. Rev. Solid State Mat. Sci.*, **19**, 1 (1994).

9. A. J. Bard, H. D. Abruña, C. E. Chidsey, L. R. Faulkner, S. W. Feldberg, K. Itaya, M. Majda, O. Melroy, R. W. Murray, M. D. Porter, M. P. Soriaga, and H. S. White, *J. Phys. Chem.*, **97**, 7147 (1993).

10. R. G. Greenler, *J. Chem. Phys.*, **44**, 310 (1966).

11. W. N. Hansen, *J. Opt. Soc. Am.*, **58**, 380 (1968).

12. W. N. Hansen, In *Advances in Electrochemistry and Electrochemical Engineering*, P. Delahay and C. W. Tobias, eds. Wiley, New York, 1973.

13. J. D. E. McIntyre, In *Advances in Electrochemistry and Electrochemical Engineering*, P. Delahay and C. W. Tobias, eds. Wiley, New York, 1973.

14. D. J. Griffiths, *Introduction to Electrodynamics*, Prentice-Hall, Englewood Cliffs, NJ, 1981.

15. J. P. Hawranek, P. Neelakantan, R. P. Young, and R. N. Jones, *Spectrochim. Acta. Part A*, **32**, 85 (1976).

16. H. A. Pearce and N. Sheppard, *Surf. Sci.*, **59**, 205 (1976).

17. J. A. Stratton, *Electromagnetic Theory*. McGraw-Hill, New York, 1941.

18. M. Born and E. Wolf, *Principles of Optics*, 5th ed. Pergamon Press, Oxford, 1975.

19. E. D. Palik, ed., *Handbook of Optical Constants of Solids*. Academic Press, Orlando, 1985.

20. F. A. Jenkins and H. E. White, *Fundamentals of Optics*, 3rd ed. McGraw-Hill, New York, 1957.

21. D. L. Allara, A. Baca, and C. A. Pryde, *Macromolecules*, **11**, 1215 (1978).

22. M. D. Porter, T. B. Bright, D. L. Allara, and T. Kuwana, *Anal. Chem.*, **58**, 2461 (1986).

23. D. L. Allara and R. G. Nuzzo, *Langmuir*, **1**, 52 (1985).

24. A. N. Parikh and D. L. Allara, *J. Chem. Phys.*, **96**, 927 (1992).

25. See Chapter 8 on Raman microspectroscopy.

26. M. M. Walczak, C. Chung, S. M. Stole, C. A. Widrig, and M. D. Porter, *J. Am. Chem. Soc.*, **113**, 2370 (1991).

27. C. J. Zhong and M. D. Porter, *J. Am. Chem. Soc.*, **116**, 11616 (1994).

28. R. V. Duevel and R. M. Corn, *Anal. Chem.*, **64**, 337 (1992).

29. C. Korzeniewski and S. Pons, *Prog. Anal. Spectrosc.*, **10**, 1 (1987).

30. S. M. Stole, D. D. Popenoe, and M. D. Porter, In *Electrochemical Interfaces: Modern Techniques for In Situ Interface Characterization*, H. D. Abruña, ed. VCH, New York, 1991.

31. B. J. Barner, M. J. Green, E. I. Saéz, and R. M. Corn, *Anal. Chem.* **63**, 55 (1991).

32. W. N. Richmond, P. W. Faguy, R. S. Jackson, and S. C. Weibel, *Anal. Chem.*, **68**, 621 (1996).

33. S. Wu, *Polymer Interface Adhesion*, Marcel Dekker, New York, 1982.

34. W. G. Golden, C. Snyder, and B. J. Smith, *J. Phys. Chem.*, **86**, 4675 (1982).

35. M. K. Debe, *J. Vac. Sci. Technol.*, **21**, 74 (1982).

36. J. F. Rabolt, F. C. Burns, N. E. Schlotter, and J. D. Swalen, *J. Chem. Phys.*, **78**, 946 (1983).

37. D. L. Allara and J. D. Swalen, *J. Phys. Chem.*, **86**, 2700 (1982).

38. J. Umemura, T. Kamata, T. Kawai, and T. Takenaka, *J. Phys. Chem.*, **94**, 62 (1990).

39. T. J. Lenk, V. M. Hallmark, J. F. Rabolt, L. Haussling, and H. Ringsdorf, *Macromolecules*, **26**, 1230 (1993).

40. K. Kelley, Y. Ishino, and H. Ishida, *Thin Solid Films*, **154**, 271 (1987).

41. M. Zhang and M. R. Anderson, *Langmuir*, **10**, 2807 (1994).

42. D. D. Popenoe, S. M. Stole, and M. D. Porter, *Appl. Spectrosc.*, **46**, 79 (1992).

43. D. D. Popenoe, R. S. Deinhammer, and M. D. Porter, *Langmuir*, **8**, 2521 (1992).

44. T. Sasaki, I. T. Bae, D. A. Scherson, B. G. Bravo, and M. P. Soriaga, *Langmuir*, **6**, 1234 (1990).

45. V. K. F. Chia, J. L. Stickney, M. P. Soriaga, S. D. Rosasco, G. N. Salaita, A. T. Hubbard, J. B. Benziger, and K.-W. Peter Pang, *J. Electroanal. Chem.*, **163**, 407 (1984).

46. B. E. Hayden and A. M. Bradshaw, *Surf. Sci.*, **125**, 787 (1983).

47. N. E. Schlotter, M. D. Porter, T. B. Bright, and D. L. Allara, *Chem. Phys. Lett.*, **132**, 93 (1986).

48. C. J. Zhong and M. D. Porter, *Anal. Chem.*, **67**, 709A (1995).

49. L. H. Dubois and R. G. Nuzzo, *Annu. Rev. Phys. Chem.*, **43**, 437 (1992).

50. R. G. Snyder, S. L. Hsu, and S. Krimm, *Spectrochim. Acta. Part A*, **34**, 946 (1978).

51. R. G. Snyder, H. L. Strauss, and C. A. Elliger, *J. Phys. Chem.*, **86**, 5145 (1982).

52. R. P. O'Toole, S. G. Burns, G. J. Bastiaans, and M. D. Porter, *Anal. Chem.*, **64**, 1289 (1992).

53. U. Mazur and A. C. Cleary, *J. Phys. Chem.*, **94**, 189 (1990).

54. U. Mazur, *Langmuir*, **6**, 1331 (1990).

55. X.-D. Wang, K. W. Hipps, and U. Mazur, *J. Phys. Chem.*, **96**, 8485 (1992).

56. M. L. Mitchell and R. A. Dluhy, *J. Am. Chem. Soc.*, **110**, 712 (1988).

4 Attenuated Total Reflection Spectroscopy

Francis M. Mirabella

4.1 INTRODUCTION

Internal reflection spectroscopy (IRS) became a popular spectroscopic technique beginning in the early 1960s. It has become more widely known by the name attenuated total reflection (ATR) spectroscopy. Before the development of ATR spectroscopy, the spectroscopic techniques available permitted analyses to be done by transmission through the bulk of the sample or external reflection from the surface of the sample. Transmission spectroscopy yields information about the bulk properties of the sample. External reflection spectroscopy yields information about the surface properties of the sample but requires that the sample surface be reflective for the radiation being used. ATR spectroscopy permits any surface to be brought

Modern Techniques in Applied Molecular Spectroscopy, Edited by Francis M. Mirabella.
Techniques in Analytical Chemistry Series.
ISBN 0-471-12359-5

in contact with a high index of refraction internal reflection element (IRE). However, since the radiation is trapped by total internal reflection inside the IRE and only interacts with the sample surface, it is not propagated through it.

The ATR technique has found abundant application for the analysis of a wide variety of sample types and for a wide range of spectral ranges. The surface of samples may be probed from as little as a few tens of nanometers to several micrometers by the ATR technique. The chemical composition, layer structure, diffusion, adsorption, chemical reaction monitoring, orientation, and physical state of surfaces are a few of the types of qualitative and quantitative analyses that can be done by ATR.

4.2 HISTORY

When radiation is passed into a medium with higher refractive index than the surroundings, it will be trapped inside if the angle of reflection at the surface of the medium exceeds a particular "critical angle" due to a phenomenon called total internal reflection (TIR). The TIR phenomenon is shown in Fig. 4.1. A standing wave is established at the reflecting interface, and under the proper conditions all the energy is reflected at the surface and propagated through the medium. This phenomenon has been known for a long time and was studied by Isaac Newton.

Newton made the important observation that the electromagnetic field extended beyond the reflecting interface. In fact, if the radiation is visible light, an object brought close to the outside surface of the reflecting interface, such as a knife-edge, will be illuminated. Since the light is totally reflected inside the medium, the electromagnetic field outside the medium must be nonpropagating and involves no energy loss for TIR. This electromagnetic field which extends beyond the reflecting interface is called an evanescent wave. This word is from the Latin root, *evanscere*, and means to vanish or pass away like a vapor. This nonpropagating evanescent wave decays in amplitude with distance from the reflecting interface. For total internal reflection no energy is withdrawn from this evanescent field, but energy can be extracted, and this is the basis of the ATR technique.

The evanescent wave decays in amplitude (intensity is the amplitude squared) with increasing distance from the reflecting interface. The decay in the amplitude of the evanescent wave is exponential and is shown in Fig. 4.2.

Taylor and coworkers of the University of Rochester performed what appear to be the first recorded experiments with ATR spectroscopy in the 1930s. These workers used glass prisms as IRE and passed visible light through them in order to measure the refractive index and absorption

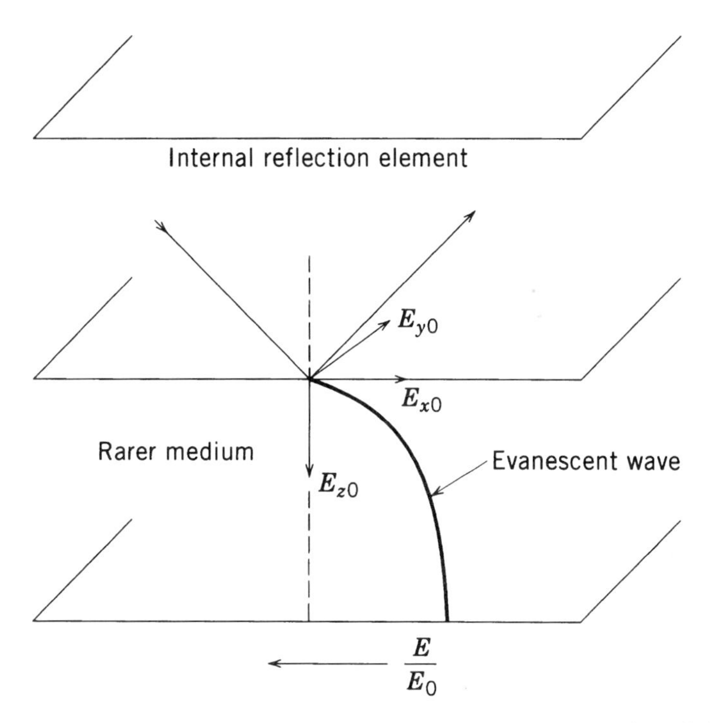

Fig. 4.1 Schematic diagram of ATR configuration. The radiation propagates through the IRE as a reflected wave and penetrates the sample as a nontransverse, exponentially decaying field (evanescent wave) with electric vector components in all spatial orientations.

spectra of potassium permanganate solutions. They described many of the properties and capabilities of ATR spectroscopy in their publications [1–3]. However, these seminal studies in ATR spectroscopy apparently went unnoticed during the next 30 years. In fact it was original work that later led to the development of ATR spectroscopy, while the relevance of the work of Taylor and coworkers was not noticed until after its development.

The modern development of ATR spectroscopy occurred in the late 1950s, due to the experiments of two independent workers in different parts of the world. N. J. Harrick was studying semiconductors at Philips Laboratories in New York, and J. Fahrenfort was conducting infrared absorption spectra studies of organic compounds at Shell Laboratories in Amsterdam.

Harrick found that infrared absorption could be obtained by the use of an IRE made of germanium in contact with a substance having a lower refractive index [4]. He later extended the technique and described many of its applications to the recording of absorption spectra [5–7].

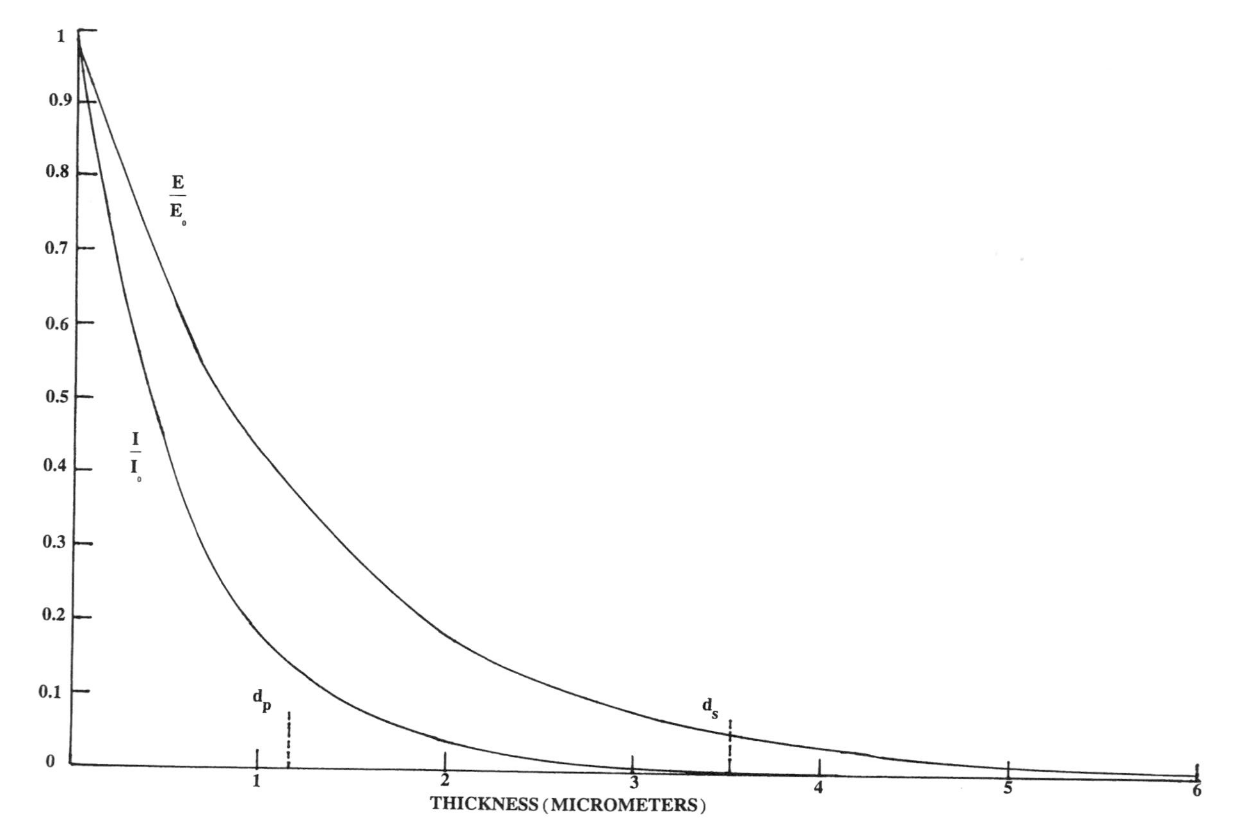

Fig. 4.2 Decay of the amplitude (E) and the intensity ($I = E^2$) of the evanescent wave with distance into the sample (thickness).

Fahrenfort was concerned with the need for recording infrared spectra of weakly or strongly absorbing or intractable materials. He found that external reflection spectroscopy had limited use for these purposes. He recognized the need of a new technique for recording absorption spectra of these types of substances. He developed the internal reflection technique so that a sample could be brought in contact with an IRE and a spectrum obtained without any sample preparation (e.g., of intractable organic compounds) and that worked for weak and strong absorbers [8].

The initial publications of Harrick and Fahrenfort describing the technique was followed by a period of relatively slow growth. The applications during this period were primarily toward the qualitative and semiquantitative analysis of a variety of sample types that did not pose particularly serious problems of sample contact to the IRE. Samples, such as sheets, films, and coatings on rigid surfaces, were routinely analyzed. The technique was not convenient for the analysis of liquids, pastes, and powders.

This situation persisted through the 1960s. But in the 1970s new sampling devices were introduced and more challenging problems were addressed, but it was not until the 1980s that a major upsurge in the application of ATR occurred. This upsurge in the applications of ATR followed significant redesign and optimization of the sampling systems, and it was further motivated by the widespread implementation of Fourier transform infrared spectroscopy. Several of the developments during this time period were the cylindrical IRE and the horizontally mounted IRE. The cylindrical IRE was the central component of newly designed liquid-handling systems, which became especially useful for process analysis, and the horizontally mounted IRE was the central component of solid-sampling systems, which became especially useful for high-productivity routine analysis. This trend of increasing application of ATR into new areas has continued in the 1990s. Several major new application areas are microscopic analysis and remote sensing devices, especially for process analysis.

The earlier generations of the ATR devices, namely in the 1960s and 1970s, had relatively complicated optics. The use of these devices required a rather high degree of expert technique from the practitioners in order to achieve good results. These devices were quite adequate, but the difficulty in the operation of these devices limited their usefulness to those who used them intensively. In the 1980s and 1990s the situation changed dramatically. Devices were designed for rapid emplacement into the spectrometer and with prealigned optics. These developments made operation of the devices extremely simple and broadened their use to a much wider group of practitioners.

The major applications of ATR are to be found in work using the mid-IR region of the spectrum. The initial applications were in the mid-IR, and subsequent growth of the technique was primarily the mid-IR. However, the use of ATR has been successfully extended to other spectral regions and spectroscopic techniques. Namely, near-IR, far-IR, UV, and visible spectral regions have been used. Raman spectroscopy has also been increasingly used.

4.3 PRINCIPLES AND THEORY

The theory of ATR spectroscopy has been derived in a both nonrigorous and rigorous fashion. The theory covers a large variety of cases. However, some very simple calculations from the nonrigorous theory can help to anticipate or explain the results one obtains from the large majority of cases in which ATR spectroscopy is employed.

A primary parameter that one needs to know is the critical angle (Θ_c). The angle of incidence, Θ, shown in Fig. 4.1, is measured from the normal, and can range between 0° and 90°. The critical angle is encountered as the angle of incidence is advanced toward the normal (0°). The critical angle is determined only by the ratio of the refractive index of the sample (n_2) and IRE (n_1) from

$$\Theta_c = \sin^{-1} n_{21}, \tag{4.1}$$

where $n_{21} = n_2/n_1$.

As the critical angle is approached from higher angles of incidence, the spectra appear normal until the critical angle is reached and crossed. The spectra then appear more and more distorted, as shown in Fig. 4.3. It can be observed in Fig. 4.3 that the noise level decreases and the spectra become more intense (i.e., the signal-to-noise ratio, S/N, increases), as the critical angle is approached from higher angles. The differences in these spectra are caused by the depth that the incident radiation penetrates into the sample. The incident radiation does not go into and pass out of the sample, as in the case of a reflection. Instead, the radiation penetrates into the sample as an exponentially decaying field or evanescent wave. The field amplitude decreases exponentially according to

$$E = E_0 \exp[-\gamma z], \tag{4.2}$$

where E is the amplitude in the sample at depth z and γ is a constant and E_0 is the amplitude at the sample surface ($z = 0$). The decay of the field amplitude and intensity ($I = E^2$) with depth into the rarer (i.e., lower refractive index medium; see Fig. 4.1) are shown in Fig. 4.2. Of course the

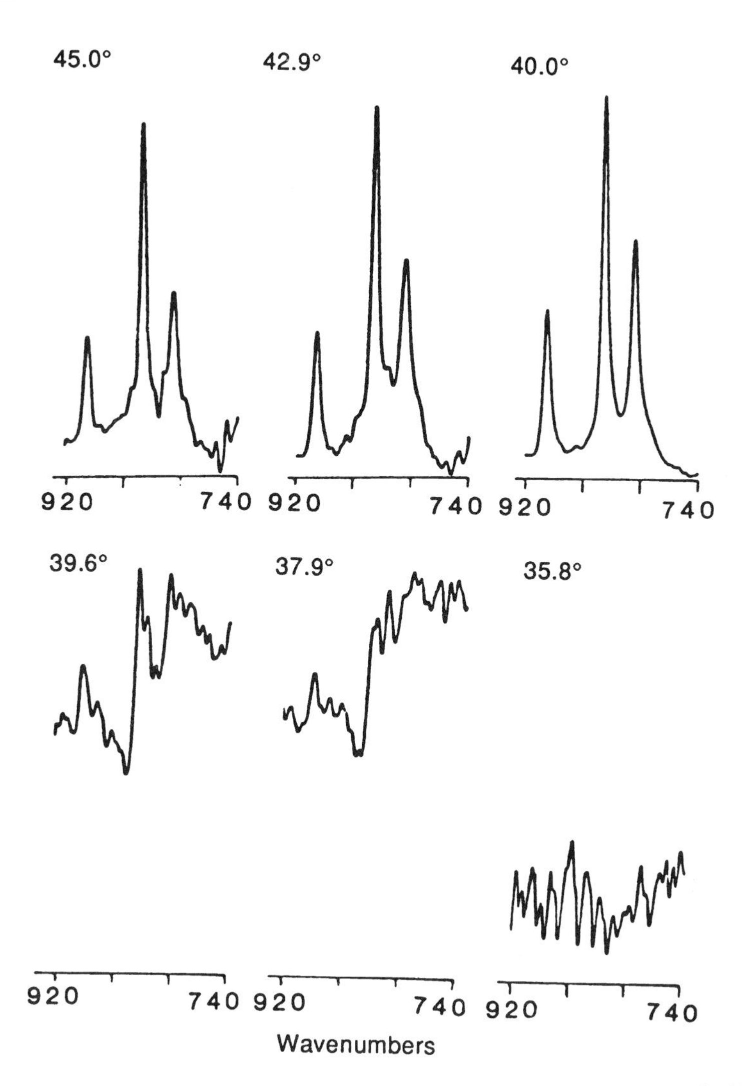

Fig. 4.3 ATR spectra of polypropylene on KRS-5 with variable angle of incidence (inside the IRE). $\theta = 45.0°$, 42.9°, 40.0°, 39.6°, 37.9°, and 35.8°.

evanescent wave only decays to zero intensity at infinite depth according to equation (4.2). However, this is not true in practice, and in any case the evanescent wave decreases to relatively negligible intensity at rather small depths.

The depth of penetration d_p is a measure of the depth into the surface of the sample that the radiation penetrates and that the resulting spectrum represents, and it is defined as

$$d_p = \frac{\lambda/n_1}{2\pi(\sin^2\Theta - (n_{21})^2)^{1/2}}, \tag{4.3}$$

where λ is the wavelength of the incident radiation, Θ is the angle of incidence, n_1 and n_2 are the refractive index of the IRE and sample, respectively, and $n_{21} = n_2/n_1$. Note that $\gamma = 1/d_p$ and that d_p is independent of polarization. In most practical studies d_p is used to estimate the depth that is probed in the ATR experiment, since it is the depth over which the majority of the spectral information comes. It should be noted that the spectral information is the average obtained for the exponentially decaying evanescent wave (see Fig. 4.2). Therefore the average can be seen to be composed of decreasing contributions from deeper layers in the sample as d_p is approached.

This parameter is sometimes mistakenly referred to as the "effective" penetration depth. This is not correct and confuses the penetration of the radiation into the surface layers of the sample with another parameter called the effective thickness, which will be discussed later. The depth of penetration is an arbitrary measure. At the depth of penetration there still remains about 13.5% of the intensity of the radiation at the surface of the sample. Therefore the spectrum obtained actually represents deeper layers of the sample than reflected by d_p. In other work it was shown that a more realistic depth sampled in the ATR mode can be estimated experimentally. In that work layers of polypropylene of known thickness were deposited on an IRE, and then an additional layer of Mylar was deposited over the polypropylene layers, as depicted in Fig. 4.4. The ratios of the absorbance of a band in the Mylar spectrum to a band in the polypropylene spectrum was plotted for varying thickness of the polypropylene layer, as shown in Fig. 4.5. When this ratio approaches zero for a particular thickness of the polypropylene layer,

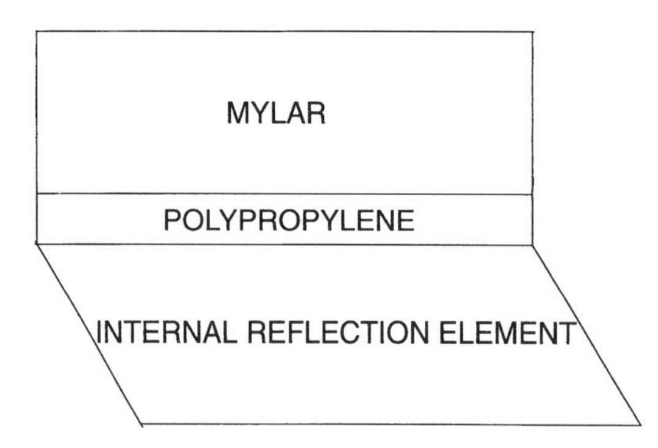

Fig. 4.4 Overlayer "sandwich" of Mylar on polypropylene on an IRE.

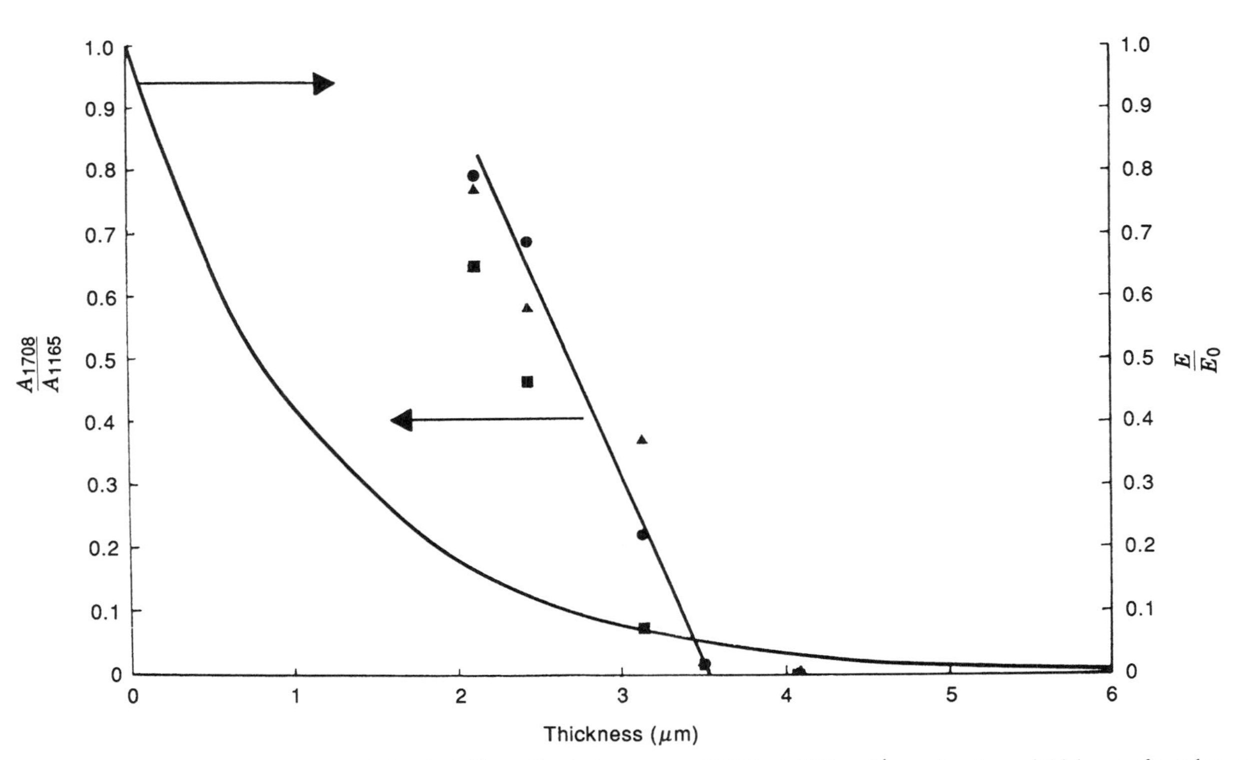

Fig. 4.5 Absorbance ratio A_{1708}/A_{1165} and electric field amplitude decay curve (E/E_0) at 1708 cm^{-1} as a function of thickness of a polypropylene layer with a Mylar overlayer (Fig. 4.4) on KRS-5 at 45° angle of incidence. The points refer to unpolarized (●), parallel-polarized (■), and perpendicular-polarized (▲) radiation. The IR band at 1165 cm^{-1} is due to polypropylene and that at 1708 cm^{-1} is due to Mylar only. The line is the linear least squares fit to the unpolarized radiation data.

TABLE 4.1 Spectroscopic Parameters for Polypropylene and Polystyrene and Sampling Depth

Parameter IRE	Polypropylene 49° KRS-5	Polystyrene 45° KRS-5
Angle of incidence	45°	45°
n_2	1.50	1.60
λ	5.85 μm	5.85 μm
λ_1	2.44 μm	2.44 μm
γ	0.852	0.608
d_p	1.17 μm	1.64 μm
$d_{e\parallel}$	3.40 μm	5.56 μm
$d_{e\perp}$	1.70 μm	2.78 μm
d_{eu}	2.55 μm	4.17 μm
d_s	3.53 μm	4.96 μm
$d_s/d_p = \gamma d_s$	3.02	3.02
E_s/E_0	0.05	0.05
$d_{s\parallel}$	3.45 μm	4.64 μm
$d_{s\perp}$	3.67 μm	4.84 μm

this indicates that the radiation no longer penetrates through the polypropylene underlayer. As shown in Table 4.1, in the results for polypropylene on a KRS-5 IRE (Fig. 4.5) and polystyrene on a KRS-5 IRE, d_s was about three times d_p for both materials. Therefore this thickness is defined as the "sampling" depth, d_s:

$$d_s = 3d_p. \tag{4.4}$$

At d_s the intensity of the radiation falls to 0.25% of that at the surface of the sample (see Fig. 4.2). Tables 4.2 to 4.7 present a useful selection of d_p and d_s values for a variety of IRE [germanium (Ge), thallium bromoiodide (KRS-5), zinc selenide (ZnSe), zinc sulfide (ZnS), cadmium telluride (CdTe), and silicon (Si)] and sample refractive index combinations. It can be seen that the materials with refractive index of 1.5 and 1.6, such as polypropylene and polystyrene, can be sampled to depths on the order of a fraction to several micrometers by proper selection of the IRE. Typically the most used IRE's are germanium for shallow penetration depths and KRS-5 and zinc selenide for deeper penetration depths.

The depth into the sample that the radiation penetrates depends directly on the wavelength of the radiation, as can be noted in equation (4.3). This is not the case for transmission spectroscopy. Therefore the ATR spectra appear different for ATR versus transmission spectra for the same material, as can be seen in Fig. 4.6. In Fig. 4.6 it may be noted that the bands in the ATR spectrum become more intense as the wavelength increases in comparison to the transmission spectrum. This wavelength dependence is due to the increase in d_p with increasing wavelength as seen in Tables 4.2 to 4.7.

It is often of interest to know what thickness of material is needed to produce a transmission spectrum of equal intensity to the corresponding ATR spectrum. This thickness is called the effective thickness, as was mentioned previously, and is designated by d_e. Of course, as just mentioned, this would depend on the spectral wavelength being considered. The thickness also depends on the state of polarization of the radiation being used. If the radiation is unpolarized, the effective thickness is calculated from

$$d_e = \frac{d_{ep} + d_{es}}{2}, \tag{4.5}$$

where d_{ep} and d_{es} are the parallel and perpendicular components of the effective thickness and

$$d_{ep} = \frac{n_{21}\lambda_1(2\sin^2\Theta - n_{21}^2)\cos\Theta}{\pi(1 - n_{21}^2)[(1 + n_{21}^2)\sin^2\Theta - n_{21}^2)](\sin^2\Theta - n_{21}^2)^{1/2}} \tag{4.6}$$

and

$$d_{es} = \frac{n_{21}\lambda_1\cos\Theta}{\pi(1 - n_{21}^2)(\sin^2\Theta - n_{21}^2)^{1/2}} \tag{4.7}$$

A useful set of effective thickness values is also given in Tables 4.2 to 4.7 for a variety of IRE, sample refractive indexes, and spectral wavelengths.

The number of reflections that the incident beam makes inside the IRE will be a factor in determining the intensity of the spectrum obtained. The number of reflections can be calculated from geometrical considerations. For example, if the sample covers both sides of a single-pass IRE, as in Fig. 4.7, then the number of reflections (N) that touch the sample is given by

$$N = \frac{l}{d}\cot\Theta, \tag{4.8}$$

where l and d are the length and thickness of the IRE, respectively, and Θ is the angle of incidence.

The ATR spectra have a "wavelength dependence," as was mentioned above. The ATR spectra are less intense at shorter wavelength and more intense at longer wavelength relative to transmission spectra of the identical materials. The severity of this "distortion" may be minimized by using a high index of refraction IRE, such as germanium, and by employing higher angles of incidence that are as far from the critical angle as is feasible.

TABLE 4.2 Depth of Penetration (d_p), Effective Thickness for Perpendicular (d_{es}) and Parallel (d_{ep}) Polarized Radiation, and Sampling Depth (d_s) for Germanium (Ge) IRE as a Function of Wavelength (λ) and Angle of Incidence (θ)

λ μm	θ degrees	d_p μm	d_{es} μm	d_{ep} μm	d_s μm	—	λ μm	θ degrees	d_p μm	d_{es} μm	d_{ep} μm	d_s μm
Refractive indexes: $n_1 = 4.00$, $n_2 = 1.50$												
2.5	24	0.632	0.504	1.993	1.895		5.0	48	0.310	0.181	0.357	0.930
2.5	28	0.352	0.271	0.735	1.057		5.0	52	0.287	0.154	0.299	0.861
2.5	32	0.266	0.197	0.461	0.797		5.0	56	0.269	0.131	0.252	0.807
2.5	36	0.220	0.155	0.337	0.659		5.0	60	0.255	0.111	0.211	0.765
2.5	40	0.191	0.127	0.264	0.572		10.0	24	2.526	2.014	7.970	7.578
2.5	44	0.170	0.107	0.215	0.510		10.0	28	1.409	1.086	2.942	4.226
2.5	48	0.155	0.091	0.178	0.465		10.0	32	1.063	0.787	1.843	3.188
2.5	52	0.144	0.077	0.150	0.431		10.0	36	0.879	0.621	1.348	2.637
2.5	56	0.135	0.066	0.126	0.404		10.0	40	0.762	0.510	1.057	2.286
2.5	60	0.127	0.056	0.106	0.382		10.0	44	0.680	0.427	0.859	2.041
5.0	24	1.263	1.007	3.985	3.789		10.0	48	0.620	0.362	0.713	1.860
5.0	28	0.704	0.543	1.471	2.113		10.0	52	0.574	0.308	0.598	1.722
5.0	32	0.531	0.393	0.921	1.594		10.0	56	0.538	0.263	0.504	1.614
5.0	36	0.440	0.310	0.674	1.319		10.0	60	0.510	0.222	0.423	1.529
5.0	40	0.381	0.255	0.528	1.143							
5.0	44	0.340	0.214	0.430	1.021							

Refractive indexes: $n_1 = 4.00$, $n_2 = 1.60$											
2.5	24	1.349	1.174	6.287	4.048	10.0	24	5.397	4.696	—	—
2.5	28	0.405	0.340	0.999	1.214	10.0	28	1.619	1.361	3.996	4.857
2.5	32	0.286	0.231	0.560	0.859	10.0	32	1.145	0.925	2.240	3.434
2.5	36	0.231	0.178	0.392	0.693	10.0	36	0.924	0.712	1.570	2.772
2.5	40	0.198	0.144	0.301	0.593	10.0	40	0.791	0.577	1.204	2.372
2.5	44	0.175	0.120	0.242	0.525	10.0	44	0.701	0.480	0.967	2.102
2.5	48	0.159	0.101	0.199	0.476	10.0	48	0.635	0.405	0.796	1.906
2.5	52	0.147	0.086	0.166	0.440	10.0	52	0.586	0.344	0.664	1.758
2.5	56	0.137	0.073	0.139	0.411	10.0	56	0.548	0.292	0.556	1.644
2.5	60	0.130	0.062	0.116	0.389	10.0	60	0.518	0.247	0.466	1.554
5.0	24	2.699	2.348	12.58	8.096						
5.0	28	0.809	0.681	1.998	2.428						
5.0	32	0.572	0.462	1.120	1.717						
5.0	36	0.462	0.356	0.785	1.386						
5.0	40	0.395	0.288	0.602	1.186						
5.0	44	0.350	0.240	0.483	1.051						
5.0	48	0.318	0.202	0.398	0.953						
5.0	52	0.293	0.172	0.332	0.879						
5.0	56	0.274	0.146	0.278	0.822						
5.0	60	0.259	0.123	0.233	0.777						

TABLE 4.3 Depth of Penetration (d_p), Effective Thickness for Perpendicular (d_{es}) and Parallel (d_{ep}) Polarized Radiation, and Sampling depth (d_s) for Thallium Bromoiodide (KRS-5) IRE as a Function of Wavelength (λ) and Angle of Incidence (θ)

λ μm	θ degrees	d_p μm	d_{es} μm	d_{ep} μm	d_s μm	λ μm	θ degrees	d_p μm	d_{es} μm	d_{ep} μm	d_s μm
Refractive indexes: $n_1 = 2.35$, $n_2 = 1.60$											
2.5	40	2.233	3.684	8.866	6.698	10.0	40	25.500	14.737	35.464	76.500
2.5	44	0.618	0.957	1.965	1.853	10.0	44	2.471	3.829	7.859	7.413
2.5	48	0.445	0.641	1.209	1.335	10.0	48	1.780	2.565	4.835	5.339
2.5	52	0.366	0.486	0.869	1.099	10.0	52	1.466	1.944	3.477	4.397
2.5	56	0.320	0.386	0.666	0.960	10.0	56	1.280	1.542	2.664	3.841
2.5	60	0.289	0.312	0.525	0.868	10.0	60	1.157	1.246	2.101	3.471
5.0	40	4.465	7.369	17.732	13.395	15.0	40	25.500	22.106	53.196	76.500
5.0	44	1.235	1.915	3.929	3.706	15.0	44	3.706	5.744	11.788	11.119
5.0	48	0.890	1.283	2.418	2.669	15.0	48	2.669	3.848	7.253	8.008
5.0	52	0.733	0.972	1.739	2.198	15.0	52	2.198	2.916	5.216	6.595
5.0	56	0.640	0.771	1.332	1.920	15.0	56	1.920	2.313	3.996	5.761
5.0	60	0.579	0.623	1.051	1.736	15.0	60	1.736	1.870	3.152	5.207
Refractive indexes: $n_1 = 2.35$, $n_2 = 1.50$											
2.5	44	1.229	2.243	4.636	3.686	10.0	44	4.914	8.973	18.545	14.743
2.5	48	0.568	0.966	1.795	1.705	10.0	48	2.274	3.862	7.182	6.822
2.5	52	0.427	0.667	1.166	1.280	10.0	52	1.707	2.668	4.664	5.121
2.5	56	0.358	0.508	0.853	1.074	10.0	56	1.432	2.032	3.414	4.295
2.5	60	0.316	0.402	0.656	0.949	10.0	60	1.265	1.606	2.625	3.796
5.0	44	2.457	4.487	9.273	7.372	15.0	44	25.500	13.460	27.818	76.500
5.0	48	1.137	1.931	3.591	3.411	15.0	48	3.411	5.793	10.773	10.233
5.0	52	0.854	1.334	2.332	2.561	15.0	52	2.561	4.002	6.995	7.682
5.0	56	0.716	1.016	1.707	2.148	15.0	56	2.148	3.049	5.121	6.443
5.0	60	0.633	0.803	1.313	1.898	15.0	60	1.898	2.409	3.938	5.694

TABLE 4.4 Depth of Penetration (d_p) and Effective Thickness for Perpendicular (d_{es}) and Parallel (d_{ep}) Polarized Radiation for Zinc Selenide (ZnSe) IRE as a Function of Wavelength (λ) and Angle of Incidence (θ)

λ μm	θ degrees	d_p μm	d_{es} μm	d_{ep} μm	λ μm	θ degrees	d_p μm	d_{es} μm	d_{ep} μm
Refractive indexes: $n_1 = 2.42$, $n_2 = 1.50$									
2.5	40	0.966	1.489	3.508	10.0	40	3.863	5.958	14.032
2.5	44	0.524	0.759	1.554	10.0	44	2.097	3.037	6.217
2.5	48	0.401	0.540	1.023	10.0	48	1.604	2.161	4.094
2.5	52	0.338	0.419	0.756	10.0	52	1.352	1.675	3.023
2.5	56	0.299	0.336	0.587	10.0	56	1.195	1.345	2.348
2.5	60	0.272	0.274	0.467	10.0	60	1.087	1.094	1.867
5.0	40	1.932	2.979	7.016	15.0	40	5.795	8.936	21.048
5.0	44	1.049	1.518	3.108	15.0	44	3.146	4.555	9.325
5.0	48	0.802	1.080	2.047	15.0	48	2.406	3.241	6.140
5.0	52	0.676	0.838	1.511	15.0	52	2.027	2.513	4.534
5.0	56	0.597	0.672	1.174	15.0	56	1.792	2.017	3.522
5.0	60	0.544	0.547	0.934	15.0	60	1.631	1.642	2.801

TABLE 4.5 Depth of Penetration (d_p) and Effective Thickness for Perpendicular (d_{es}) and Parallel (d_{ep}) Polarized Radiation for Zinc Sulfide (ZnS) IRE as a Function of Wavelength (λ) and Angle of Incidence (θ)

λ μm	θ degrees	d_p μm	d_{es} μm	d_{ep} μm	λ μm	θ degrees	d_p μm	d_{es} μm	d_{ep} μm
Refractive indexes: $n_1 = 2.22$, $n_2 = 1.50$									
2.5	44	1.111	1.988	4.104	5.0	56	0.746	1.038	1.749
2.5	48	0.579	0.964	1.795	5.0	60	0.662	0.823	1.350
2.5	52	0.442	0.677	1.186	10.0	44	4.445	7.951	16.416
2.5	56	0.373	0.519	0.875	10.0	48	2.317	3.855	7.182
2.5	60	0.331	0.411	0.675	10.0	52	1.768	2.707	4.746
5.0	44	2.223	3.975	8.208	10.0	56	1.492	2.075	3.499
5.0	48	1.159	1.928	3.591	10.0	60	1.323	1.645	2.700
5.0	52	0.884	1.353	2.373					

Since the ATR sectra are "distorted" relative to the corresponding transmission spectra (see Fig. 4.6), computer software has been developed to "correct" the ATR spectra so that they appear more like the transmission spectra. An example of an ATR spectrum that has been corrected to appear like a transmission spectrum is shown in Fig. 4.8. The correction software

TABLE 4.6 Depth of Penetration (d_p) and Effective Thickness for Perpendicular (d_{es}) and Parallel (d_{ep}) Polarized Radiation for Cadmium Telluride (CdTe) IRE as a Function of Wavelength (λ) and Angle of Incidence (θ)

λ μm	θ degrees	d_p μm	d_{es} μm	d_{ep} μm	λ μm	θ degrees	d_p μm	d_{es} μm	d_{ep} μm
Refractive indexes: $n_1 = 2.65$, $n_2 = 1.50$									
2.5	36	0.948	1.277	3.486	10.0	36	3.791	5.110	13.945
2.5	40	0.493	0.629	1.413	10.0	40	1.972	2.516	5.654
2.5	44	0.373	0.447	0.909	10.0	44	1.491	1.787	3.637
2.5	48	0.312	0.348	0.667	10.0	48	1.247	1.390	2.667
2.5	52	0.274	0.281	0.518	10.0	52	1.095	1.123	2.073
2.5	56	0.248	0.231	0.415	10.0	56	0.992	0.924	1.658
2.5	60	0.229	0.191	0.336	10.0	60	0.916	0.763	1.344
5.0	36	1.896	2.555	6.972	15.0	36	5.687	7.664	20.917
5.0	40	0.986	1.258	2.827	15.0	40	2.958	3.774	8.481
5.0	44	0.746	0.894	1.819	15.0	44	2.237	2.681	5.456
5.0	48	0.624	0.695	1.333	15.0	48	1.871	2.085	4.000
5.0	52	0.548	0.562	1.036	15.0	52	1.643	1.685	3.109
5.0	56	0.496	0.462	0.829	15.0	56	1.487	1.385	2.488
5.0	60	0.458	0.382	0.672	15.0	60	1.374	1.145	2.016

usually accomplishes the correction by simply multiplying each point of absorbance in the spectrum by a factor, such as λ_{ref}/λ. If, for example, $\lambda_{ref} = 10\,\mu m$, then at wavelengths less than $10\,\mu m$ this fraction will be greater than one, and at wavelengths greater than $10\,\mu m$ it will be less than one. Therefore multiplying the spectrum absorbances by this ratio will have the effect of decreasing the intensities toward longer wavelengths and increasing the intensities toward shorter wavelengths, with $10\,\mu m$ as the pivot point. This will "correct" the intensity of the ATR spectrum to be more like that of a transmission-spectrum, as demonstrated in Fig. 4.8.

4.4 PRACTICAL CONSIDERATIONS

The sample material must be brought into contact with the internal reflection element in order to use ATR spectroscopy. This is an additional and somewhat unique problem associated with ATR that is not encountered with other spectroscopic techniques. The operation of contacting the sample to the IRE falls into several classes: very easy, difficult but not apparently so, or obviously very difficult. The readily available, appropriate accessories make the analysis of films, solids that can be cast as a film on the IRE, greases, liquids, some powders, and so on, typically easy to place in contact

TABLE 4.7 Depth of Penetration (d_p), Effective Thickness for Perpendicular (d_{es}) and Parallel (d_{ep}) Polarized Radiation, and Sampling Depth (d_s) for Silicon (Si) IRE as a Function of Wavelength (λ) and Angle of Incidence (θ)

λ μm	θ degrees	d_p μm	d_{es} μm	d_{ep} μm	d_s μm	—	λ μm	θ degrees	d_p μm	d_{es} μm	d_{ep} μm	d_s μm
Refractive indexes: $n_1 = 3.42$, $n_2 = 1.50$												
2.5	28.0	0.695	0.666	2.350	2.084		10.0	28	2.779	2.665	9.401	8.338
2.5	32.0	0.391	0.360	0.934	1.174		10.0	32	1.565	1.441	3.736	4.694
2.5	36.0	0.297	0.261	0.593	0.892		10.0	36	1.189	1.045	2.373	3.568
2.5	40.0	0.248	0.206	0.435	0.743		10.0	40	0.990	0.824	1.740	2.971
2.5	44.0	0.216	0.169	0.340	0.648		10.0	44	0.864	0.675	1.362	2.592
2.5	48.0	0.194	0.141	0.276	0.582		10.0	48	0.776	0.564	1.103	2.327
2.5	52.0	0.178	0.119	0.228	0.533		10.0	52	0.711	0.475	0.910	2.133
2.5	56.0	0.165	0.100	0.189	0.496		10.0	56	0.661	0.402	0.757	1.984
2.5	60.0	0.156	0.085	0.158	0.467		10.0	60	0.623	0.338	0.630	1.870
5.0	28.0	1.390	1.333	4.701	4.169		15.0	28	4.169	3.998	14.102	12.507
5.0	32.0	0.782	0.721	1.868	2.347		15.0	32	2.347	2.162	5.604	7.041
5.0	36.0	0.595	0.522	1.186	1.784		15.0	36	1.784	1.567	3.559	5.352
5.0	40.0	0.495	0.412	0.870	1.486		15.0	40	1.486	1.236	3.609	4.457
5.0	44.0	2.315	0.337	0.681	6.945		15.0	44	1.296	1.012	2.043	3.887
5.0	48.0	0.388	0.282	0.552	1.164		15.0	48	1.164	0.846	1.655	3.491
5.0	52.0	0.355	0.238	0.455	1.066		15.0	52	1.066	0.713	1.365	3.199
5.0	56.0	0.331	0.201	0.379	0.992		15.0	56	0.992	0.603	1.136	2.977
5.0	60.0	0.312	0.169	0.315	0.935		15.0	60	0.935	0.508	0.946	2.804

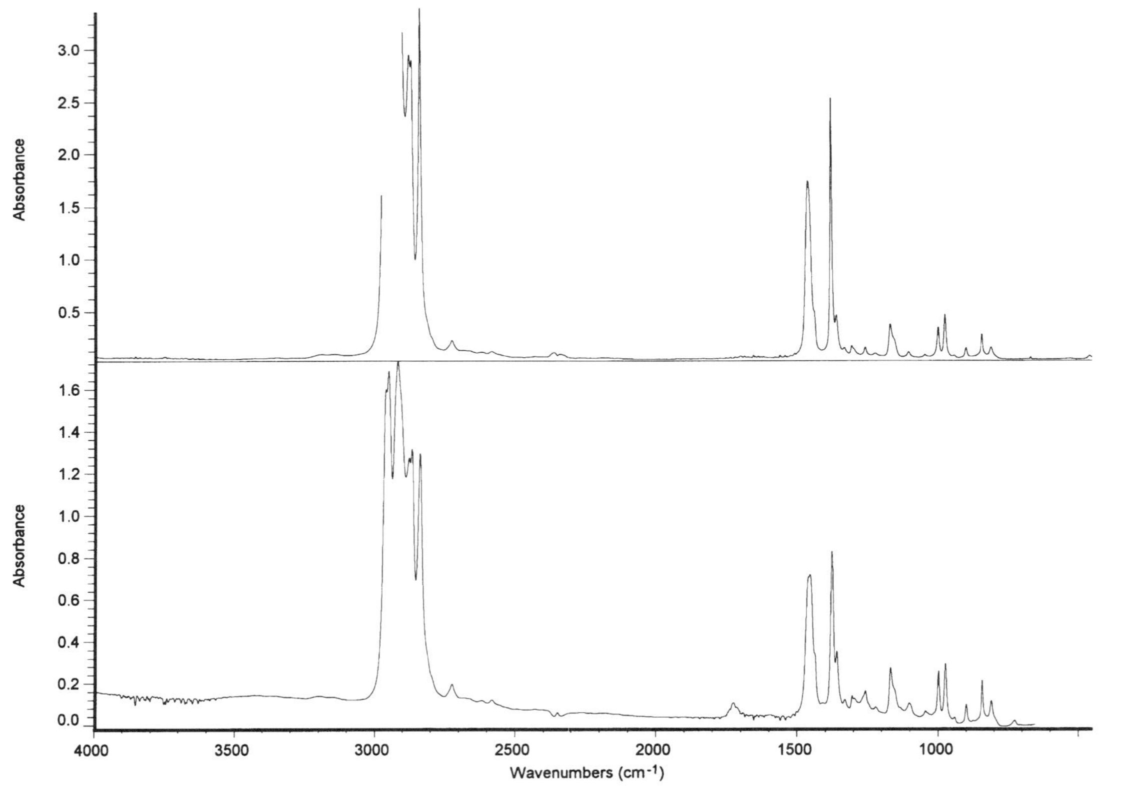

Fig. 4.6 Effect of wavelength dependence on an ATR spectrum (lower spectrum) compared to a transmission spectrum (upper spectrum) of polypropylene.

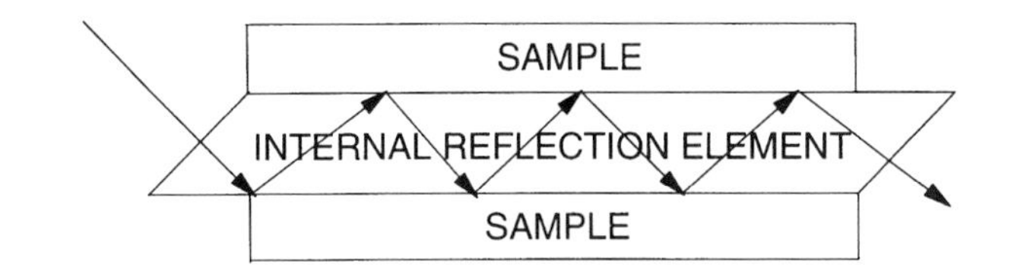

Fig. 4.7 Internal reflections inside an IRE.

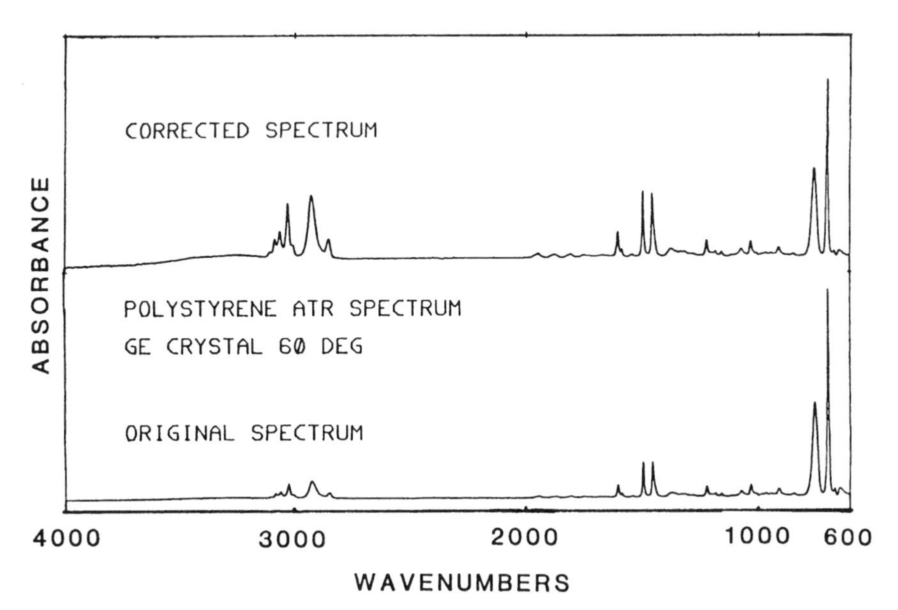

Fig. 4.8 ATR spectra of polystyrene: The uncorrected spectrum is the lower spectrum, and the corrected spectrum is the upper spectrum. Reprinted with permission of BioRad, Digilab Division, Cambridge, MA.

with the IRE and yield good spectra. Some films, powders, or other solids may appear to be in contact with the IRE but may not make even good physical contact. This often results in poor spectra, which are unexpectedly obtained. Materials on metal surfaces (especially metal surfaces that are not flat), coarse powders, rough surfaces, and so on, often cannot readily be brought into good contact with the IRE and may pose great difficulty in obtaining spectra.

The equipment that has been available since the early years of ATR spectroscopy consisted of various geometry IRE types, as shown in Fig. 4.9. The IRE in Fig. 4.9 have a face, called the entrance aperture, through which the radiation enters the IRE and a face, called the exit aperture, through which the radiation emerges from the IRE. The angles cut for the entrance

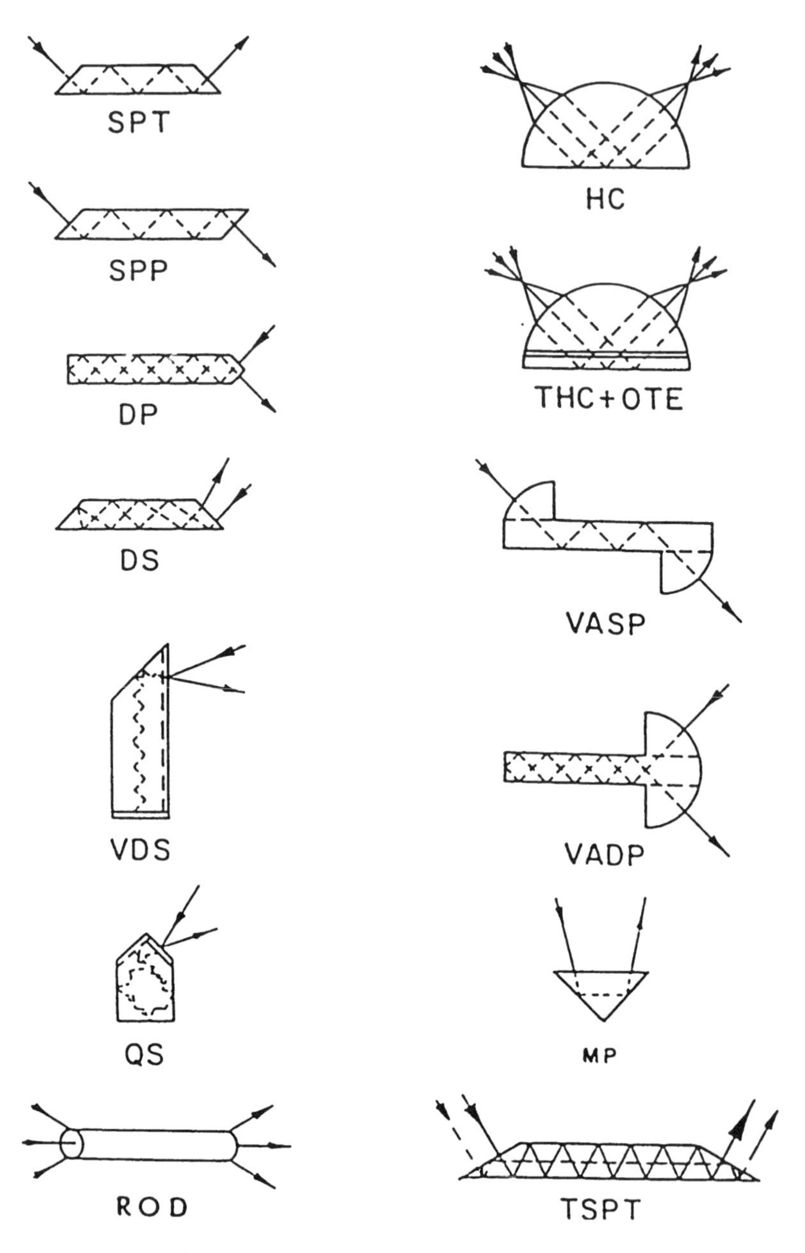

Fig. 4.9 Various configurations of internal reflection elements. Acronyms are according to Harrick (Harrick Scientific Corp., Ossining, NY) and indicate the path of the rays in the IREs.

Fig. 4.10 Optical accessory for ATR spectroscopy. The mirrors guide the radiation from the source through the IRE and onto the detector. The angle of incidence is adjustable. Photo courtesy of Pike Technologies, Madison, WI.

and exit apertures are normally the same and are fixed angles for the particular IRE. An example of the optical accessories used to guide the radiation through these IRE's is shown in Fig. 4.10. The use of these "older-style" accessories typically requires some "art" on the part of the practioner. The alignment and optimization of the signal through the accessories, and the quality and intensity of the spectra obtained depend considerably on the skill of the operator and involve the attendant tedious adjustment of the devices.

New accessories have improved this situation to a large extent and obviated the need for much user intervention to obtain high-quality spectra. As a result the number of sample types that pose great difficulty in obtaining ATR spectra have been minimized. The advantages of the current generation of accessory devices include optimized energy throughput, permanent precision alignment, large sampling area, purgable optics compartment, and rugged construction. Probably the most popular and useful innovations are the devices with horizontally mounted internal reflection elements. The

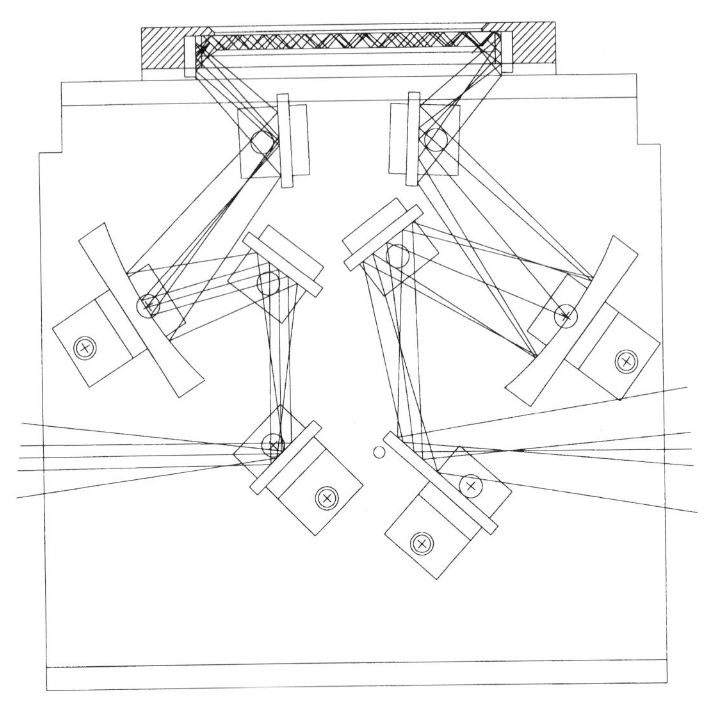

Fig. 4.11 Schematic of the ray of a horizontal ATR accessory. Reprinted with permission of Pike Technologies, Madison, WI.

basic geometry of this type of accessory for ATR spectroscopy is shown in Fig. 4.11. This type of accessory permits a wide variety of sample types to be analyzed by simply placing the sample onto the top of the IRE and pressing the sample down onto the IRE with a top-mounted clamp. This accessory usually employs two types of configurations for the IRE. These are shown in Fig. 4.12.

One geometry is the "flat-plate" IRE that is mounted flush to the surface of the accessory, as shown in Fig. 4.12*b*. This geometry is suitable for sample types, which may be thick or thin films, opaque solids, surface coatings, films on opaque substrates, soft solids, pastes, powders, liquids, and so on. A clamping plate is available for the "flat-plate" IRE (see Fig. 4.11) which permits clamping of samples against the IRE surface. The other geometry is the "trough-plate" IRE configuration in which the IRE is recessed from the surface of the accessory so that there is an empty volume over the surface of the IRE, as shown in Fig. 4.12*a*. This geometry is especially suitable for liquids, pastes, and gels which can be poured into the cavity above the IRE. Many other devices have become available for sampling applications of these types.

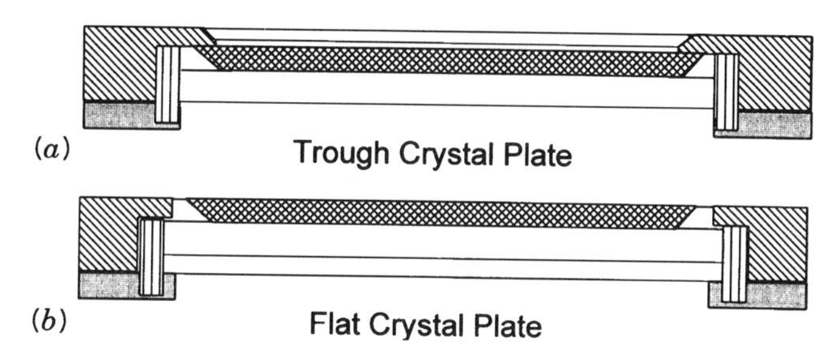

Fig. 4.12 Two types of horizontal ATR accessories: (*a*) Trough-plate crystal; (*b*) flat-plate crystal. Reprinted with permission of Pike Technologies, Madison, WI.

The basic characteristics of these devices are the same as those mentioed above, however, additional capabilities such as temperature control, calibrated pressure plates, polarization of radiation, static or flow-through liquid cells, and other modifications are available. The angle of incidence inside the IRE is fixed in these accessories. However, the angle of incidence and/or type of IRE can normally be easily changed by replacing it with another IRE in a "locating" plate that fixes the IRE into the proper alignment. The major applications of these devices are for collecting survey spectra and qualitative data, but quantitative analysis is achievable in many cases with these accessories.

Another accessory that is used for liquid sample applications is based on a cylindrical internal reflection element. The cylindrical IRE can be sealed into a cell in such a way that static measurements can be made on liquids in an "open-boat" cell or in such a way that dynamic measurements on liquid streams can be made in a sealed cell. Specialized devices, based on the cylindrical IRE, are available from various manufacturers, especially for applications in soda-based beverage, fruit juice, beer and liquors, pharma-

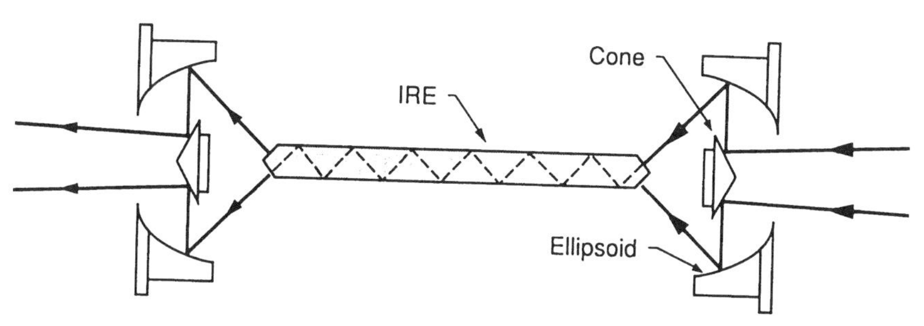

Fig. 4.13 Geometry of a liquid cell ATR accessory with cylindrical rod ATR crystal.

ceutical industries, and various others. The optical arrangement of this type of cell for liquid measurements is shown in Fig. 4.13.

Devices employed for strictly quantitative applications are often the type that permit variation of the angle of incidence. There are basically two types of devices for these applications. One type employs a multiple reflection IRE, which is cut to a specific entrance aperture angle and permits variation of the angle of incidence at the entrance aperture of the IRE. The geometry of this type of device is shown in Fig. 4.14. This type of device employs nonnormal incidence with the entrance aperture of the IRE, as shown in Fig. 4.15, whenever the angle is set off the normal. The true angle of incidence inside the IRE must be calculated in these cases using Snell's law because refraction occurs at the entrance aperture of the IRE. The true angle according to Snell's law is calculated from

$$n_a \sin\theta = n_b \sin\phi. \tag{4.9}$$

Here n_a is the refractive index of the medium outside the IRE (usually air), θ is the angle of incidence, relative to the normal, n_b is the refractive index of the IRE, and ϕ is the angle of reflection inside the IRE, relative to the normal. For example, if an IRE is cut with a 45° entrance aperture, when the radiation is input along the normal to the entrance aperture a 45° reflection inside the IRE is obtained, as shown in Fig. 4.15. If the angle is increased by 10° to the normal, as shown in Fig. 4.15, the angle of reflection is not 55°. Using equation (4.9), we can calculate ϕ from $\theta = 10°$ and $n_a = 1.0$ and $n_b = 2.4$. This gives $\phi = 4.1°$. Therefore the angle of reflection inside the IRE is $45 + 4.1 = 49.1°$. Increasingly the angle outside the IRE by 10°, only increases the angle of reflection inside the IRE by 4.1°, because the

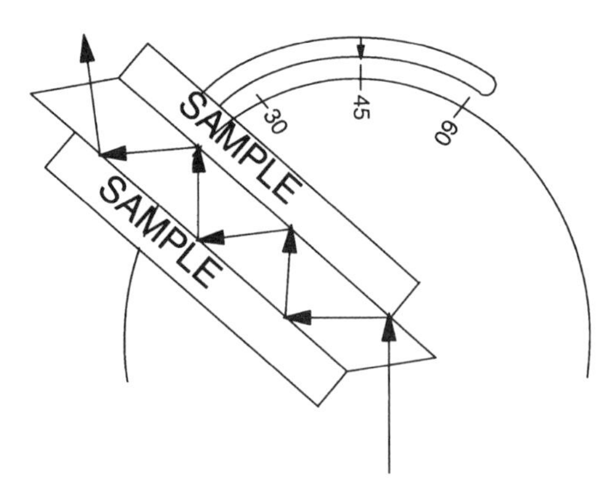

Fig. 4.14 Schematic diagram of ATR accessory permitting variation of the angle of incidence.

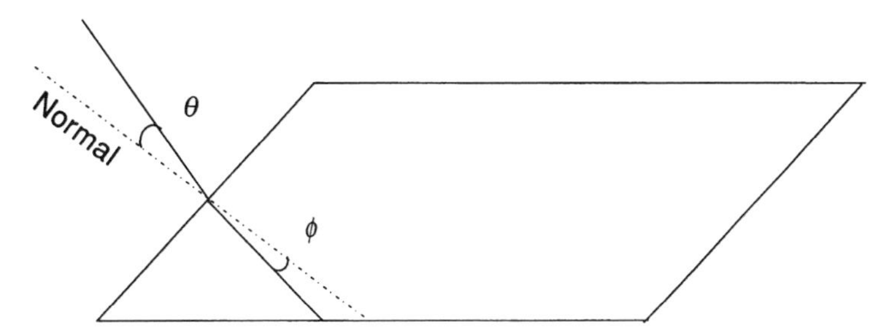

Fig. 4.15 Schematic diagram showing the effect of nonnormal incidence on the IRE entrance face. The ray is "pulled" toward the normal inside the IRE, and the actual angle of incidence inside the IRE is calculated by Snell's law.

higher index of refraction of the IRE material "pulls" the beam closer to the normal. This type of accessory obviously only permits relatively small angular changes to be obtained. However, sensitivity can be high because of the multiple reflections with the sample.

The other type of device employs a hemispherical IRE, as shown in Fig. 4.16. This type of device permits continuously variable angles, with no refraction inside the IRE. However, since only one reflection off the sample is obtained, the spectral intensity can be low. Therefore angular selection is

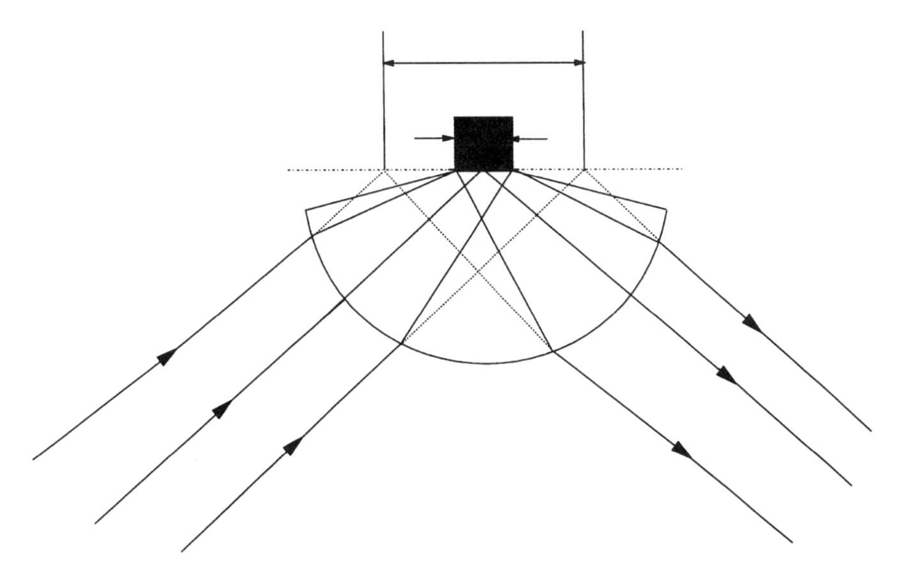

Fig. 4.16 Optical ray diagram for hemispherical IRE.

advantageous and can be conveniently controlled, whereas spectral intensity can be low because of the single reflection.

The rectangular shape of the entrance aperture of most IREs was due to the rectangular shaped beams in dispersive infrared spectrometers in the 1950s and 1960s. The rectangular shape of the beams in the dispersive instruments "fit" well into the rectangular apertures of the typical IREs. The advent of FT-IR spectrometers with circularly shaped beams has changed this situation. The circular beams are typically of small diameter so that they fit inside the rectangular apertures if the beam is at its focus at the plane of the aperture. Alternatively, optics have been employed to "ovalize" the FTIR beams, in order for them to better match the rectangular IRE apertures. Also IREs with circular geometry have been fabricated to match the circular beam geometry.

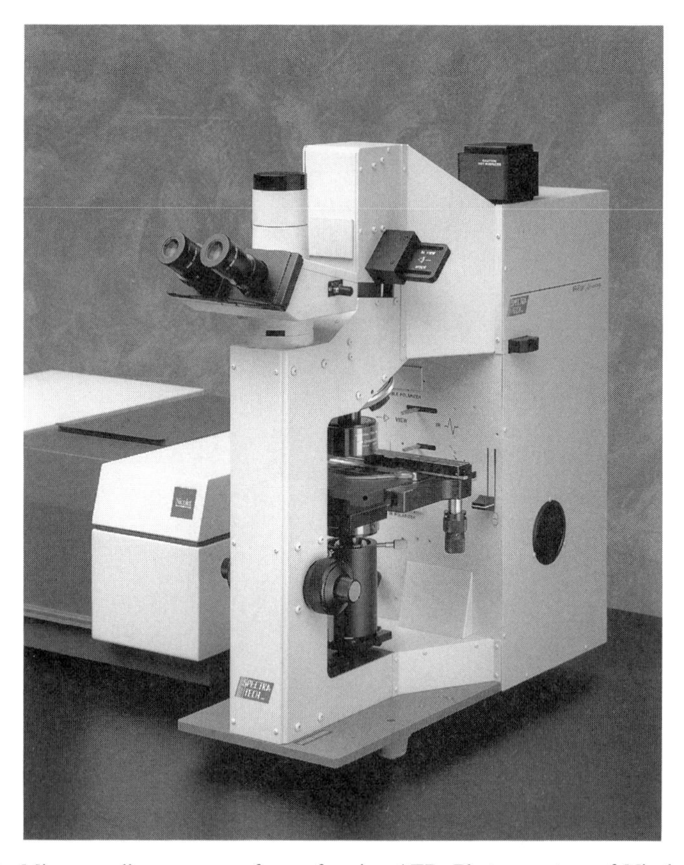

Fig. 4.17 Microsampling accessory for performing ATR. Photo courtesy of Nicolet Instrument Corp., Madison, WI.

Microsampling applications have been conveniently addressed with ATR accessories. The hemispherical IRE is typically employed in a device that permits microscopic viewing of a small area of a sample, followed by contacting of a small area on the flat face of the hemispherical IRE to the desired sample area. This type of accessory is shown in Fig. 4.17. Fig. 4.16 shows the hemispherical IRE used in the device shown in Fig. 4.17. The microscopic ATR accessories usually employ a pressure transducer, which permits actual pressure readings to be obtained, or an alarm system, which can prevent breakage of the IRE by over pressuring the contact to the sample. Microsampling accessories, such as that shown in Fig. 4.17, include provisions for viewing and contacting the area of interest on a specimen surface with "cross-hair" contacting for accuracy, photographic and/or TV monitoring capability, high-sensitivity detectors, and imaging software for data handling.

4.5 ADVANTAGES AND DISADVANTAGES

Advantages

The fact that the incident radiation does not have to pass through the sample is often a great advantage of ATR. Samples that are very thick, opaque due to being filled or colored, have a very high absorption coefficient, and so forth, are not tractable in transmission but can often be easily analyzed by ATR. Good spectra of these types of samples may be obtained by ATR spectroscopy.

A problem encountered in transmission spectroscopy with thin films with parallel surfaces is the superposition of interference fringes on the spectrum. This is easily eliminated with ATR spectroscopy because the radiation does not pass through the sample. This is demonstrated in Fig. 4.18 in which spectra of a polypropylene film obtained by transmission and ATR spectroscopy, respectively, are shown. The interference fringes become larger in amplitude as wavelength across the transmission spectrum increases. This makes the spectrum more difficult to interpret and causes large errors in peak measurement. The ATR spectrum has no interference fringes.

The ATR technique is sensitive to the surface of the sample up to depths into the sample, indicated in Tables 4.2 to 4.7. Normally the depth ranges from a few tenths to a few micrometers into the sample surface. This is ideal for many studies in which components are concentrated near the surface of the sample.

The polarization of the radiation can be varied in order to perform polarization studies, for example, on systems that contain some degree of orientation induced by mechanical deformation, electrical poling, chemical

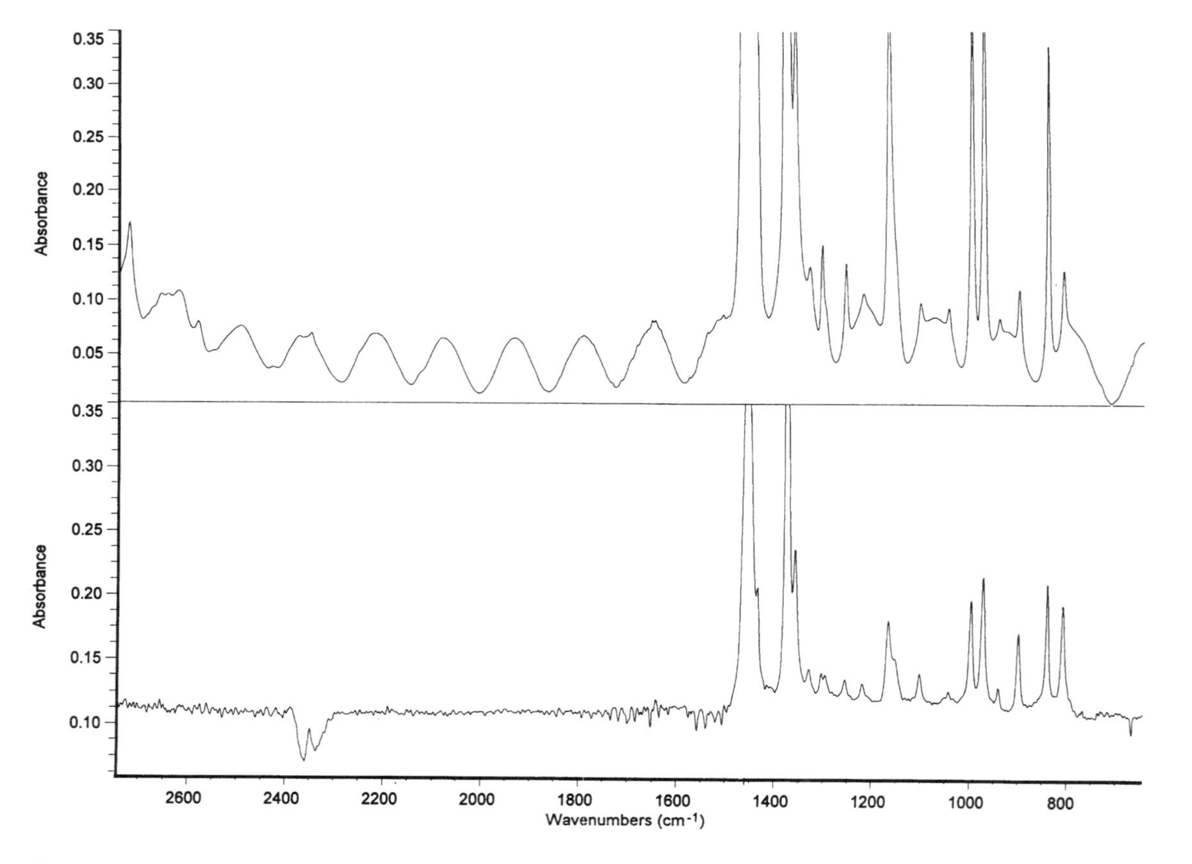

Fig. 4.18 Interference fringes in a transmission (upper) spectrum compared to an ATR (lower) spectrum of polypropylene.

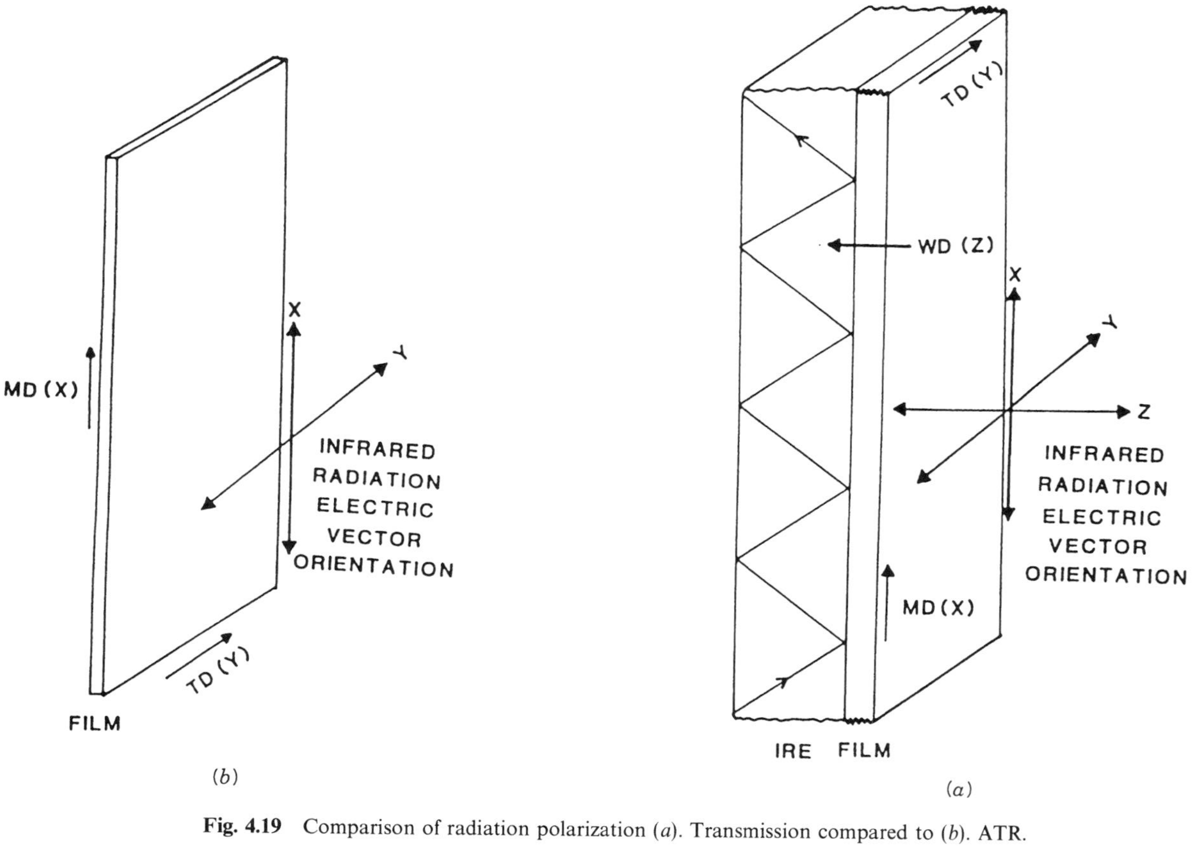

Fig. 4.19 Comparison of radiation polarization (*a*). Transmission compared to (*b*). ATR.

poling, and so on. In transmission spectroscopy the radiation can be polarized in a plane perpendicular to the beam propagation direction; this is called plane polarization and is shown in Fig. 4.19. In ATR spectroscopy the radiation can be polarized in three dimensions, which is in all spatial orientations as shown also in Fig. 4.19. Therefore ATR spectroscopy can provide information of preferential orientation, or anisotropy, in all three dimensions, while transmission spectroscopy is limited to two dimensions in any measurement.

Although physical contact of the sample with the IRE is necessary to obtain a spectrum of the sample, perfect, or so-called optical contact, is not necessary. In fact imperfect contact of varying degrees will typically result in good spectra. It should also be remembered that the evanescent wave can bridge an air gap of up to several micrometers.

Disadvantages

The fact that the sample is contacted with the IRE is a potential disadvantage of ATR spectroscopy. The IRE can be contaminated with material from a previous sample, cleaning media used to clean the IRE between samples, the atmosphere, the operator's fingers, clothes, hair, and so on. These many sources of contamination can confound the analyses done by ATR. A good practice is to obtain a spectrum of the free-standing IRE so that any contamination will be observed. Some of the common contaminants to watch out for are (1) hand lotion, (2) fingerprints, (3) silicone oil and (4) polytetrafluoroethylene C-clamp pads (caution is required to avoid contacting any pads on the C-clamp to the IRE because this will contaminate the sample spectrum). These contaminants are shown in the spectra in Fig. 4.20.

Simple cleaning procedures include dipping the IRE in an appropriate solvent, followed by gentle wiping with a nonabrasive optical lens tissue. More rigorous cleaning involves washing in a series of solvents, rinsing thoroughly, and drying in a solvent vapor dryer. The most rigorous cleaning is done by radio frequency glow discharge and vacuum baking methods, which are used infrequently and only for the most demanding cleaning operations.

The apparent "advantage" that ATR spectroscopy is surface sensitive can, in some cases, turn out to be a serious disadvantage. For example, since only the surface is probed, if there is a coating on the surface of a sample, this coating may be the only apparent component evidenced in the ATR spectrum. If such a coating is of unknown origin, it may actually hide the presence of the bulk material of interest. This problem may be more prevalent when a microsampling accessory (Fig. 4.17) is used. This type of device has been found to be convenient for obtaining numerous survey spectra of various materials, which is apparently the way many analytical

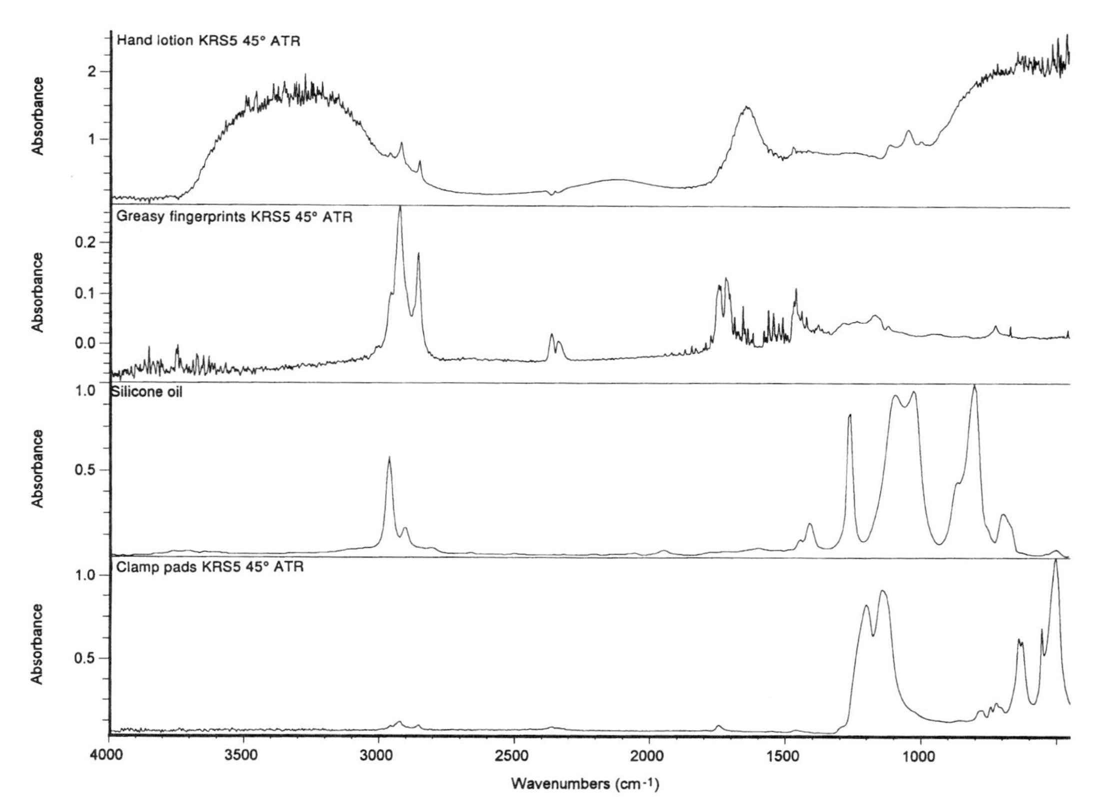

Fig. 4.20 Common contaminants in ATR spectroscopy. From the top down: (1) hand lotion, (2) fingerprints, (3) silicone oil, and (4) polytetrafluoroethylene C-clamp pads.

laboratories use it. However, if a surface coating is present, the underlying material of interest may not be probed. Any surface coatings that exist variably over a surface can be isolated to only a small portion of the sample surface, and as that area happens to be contacted by the microsampling IRE, the sample becomes misidentified.

Although only physical contact between the sample and IRE is necessary, particular sample types can present serious difficulties in obtaining sufficient contact to the IRE. This might be the case with very hard materials, rough surfaces, large and nonflat surfaces, for example. Samples that are required to be maintained in a particular atmosphere (water-free, oxygen-free, etc.) may be difficult to handle in ATR spectroscopy if special enclosures are required to hold the sample in the desired atmosphere. Then a specially mounted IRE is needed to contact the sample in the enclosure.

One of the apparent advantages of ATR spectroscopy is that is has sensitivity to the near-surface layers of the sample. It is not a true surface-sensitive method, such as the electron or ion spectroscopies, which are sensitive to only the first several nanometers of a surface. This is particularly true in the mid-IR in which ATR spectroscopy is sensitive to at least the first few hundred or thousand nanometers of a surface. This can pose a disadvantage in cases where the very near-surface layers are to be probed. As spectra are collected in the mid-IR, the sensitivity may be to layers much deeper than desired, and the data may be more representative of the bulk rather than near-surface layers.

The horizontal and microsampling ATR accessories have proved so convenient that many laboratories are known to leave these accessories in place on the spectrometers in order to collect survey spectra of numerous samples. There are many advantages to this practice, such as little or no sample preparation, easy sample mounting, and the instrument purge is not broken, that save time. However, some potential disadvantages should be borne in mind in order to avoid misleading results from this practice. The spectra obtained by ATR spectroscopy exhibit more or less of a "wavelength-dependent" character depending on the conditions employed. These spectra can be found to be more difficult to use for chemical identification if the transmission spectra are used as a comparison, for example, in a computer library searching function. Correction to a more "transmissionlike" spectrum or use of a high index of refraction IRE and/or high angles of incidence can alleviate such problems. However, employing conditions to obtain more "transmissionlike" spectra will also yield shallower penetration into the sample. Whatever conditions are used, the relative shallowness of penetration into the sample that will obtain in ATR spectroscopy will be a potential disadvantage, since any surface coating on the sample can lead to misidentification. Also contamination of the IRE by sequential contacting of samples can lead to sample misidentification or confounding of sample composition. Therefore there can be some disadvantages to using ATR for repetitive sample analyses unless proper precautions are taken and subsequent data interpretation is done with an understanding of the characteristics of ATR spectroscopy.

4.6 SPECTRAL REGIONS AND SPECTRAL TECHNIQUES EMPLOYING ATR

The overwhelming majority of ATR spectroscopy studies reported in the scientific literature were done in the mid-IR region (4000–400 cm^{-1}) of the spectrum. This is undoubtedly a reflection of the applications of ATR in industrial, government, and academic laboratories in unpublished work. This further mirrors the predominant usage of the mid-IR region of the spectrum among the other techniques of molecular spectroscopy. The preference for the usage of the mid-IR region is a consequence of the distinctive spectra obtained, since fundamental vibrations, such as stretching, bending, and rocking of bonds, occur in this spectral region. Although the mid-IR region is predominantly employed, increasing use of other spectral regions and techniques is occurring in ATR applications.

The increasing popularity of near-infrared (NIR) spectroscopy has spurred on applications of ATR in this spectral region. This region covers 12,000–4000 cm^{-1}. Overtone bands and combination bands occur in the NIR region, which are difficult to assign to specific chemical groups, and this is a disadvantage of NIR relative to mid-IR. A further disadvantage of the NIR spectral region is that bands are about one order of magnitude less intense in the NIR relative to the mid-IR. Therefore only a few papers have been published of ATR studies in the NIR region. The NIR region poses a potential advantage for its use in ATR spectroscopy, since the wavelength is on average about 1 μm in this spectral region. Therefore, assuming that the average wavelength is about 10 μm in the mid-IR, the depth probed in the NIR with ATR spectroscopy is about an order of magnitude smaller. This might permit analysis of a surface at shallower depths of penetration with NIR, relative to the mid-IR.

The ultraviolet/visible (UV/vis) spectal region has been steadily employed in a variety of spectroscopic work, but ATR studies in the UV/vis region are about as few as in the NIR region. This region covers 28,570–12,000 cm^{-1} in the visible, 50,000–28,570 cm^{-1} in the quartz UV, and 100,000–50,000 cm^{-1} in the vacuum UV. Electronic absorption and emission spectra in these regions are useful in trace analyses of biological and medical studies.

Reports of ATR studies in the far-IR region appear to be extremely rare in the literature. This region extends from 400–30 cm^{-1}. One potential disadvantage in the far-IR is that the wavelengths are very large in this region, and the depth of penetration so large, that ATR spectroscopy cannot be distinguished from transmission spectroscopy.

4.7 APPLICATIONS OF ATR SPECTROSCOPY

Surface Chemistry

The most obvious applications of ATR spectroscopy are for analyzing surface chemistry of many types of systems. These surface chemistry studies are the primary types of studies available in the literature. There are many systems in which the chemical changes occur or are caused to occur by purposeful chemical reactions in the near-surface layers of the system. Transmission spectroscopy is often deficient for characterizing such systems because the near-surface layers contribute such a small fraction to the total spectrum; namely the bulk predominates. Therefore any changes in these layers are not apparent in the transmission spectra.

In order to alleviate this problem, transmission spectroscopy is often done on thin layers cut (e.g., by microtomy) or material solvent-extracted, scraped, sanded, and so forth, from the near-surface layers of a sample. These methods are often tedious, irreproducible, and of course destructive. These types of applications are particularly well-suited to ATR spectroscopy. Chemical changes in the near-surface layers can be characterized, both qualitatively and quantitatively, conveniently and with great sensitivity, by ATR.

One of the most popular applications of ATR spectroscopy for surface chemistry studies is the qualitative characterization of chemical changes that occur in near-surface layers of polymers due to some intentional or unintentional treatment. In a study of the photodegradation of epoxy resins by UV radiation emitting in the broad range of 254–600 nm, the chemical changes near the surface of the polymer were investigated by ATR [9]. A KRS-5 IRE with angle of incidence of 45° was used in the mid-IR in this ATR study. The changes in the near-surface layers can be observed in Fig. 4.21. Fig. 4.21 shows the emergence of new infrared bands at 1721 and 1676 cm^{-1} which were assigned to aldehydes, esters, and ketones (1721 cm^{-1}) and conjugated ketones, quinones, and semiquinones (1676 cm^{-1}). No calculations were done to determine the depth of penetration of the IR radiation into the polymer in the ATR experiment. For KRS-5 at 45° for an epoxy having a refractive index of roughly 1.5, according to Table 4.3 the d_p is about 1 μm for a wavelength of 5.0 μm (2000 cm^{-1}). The "sampling" depth for this system would, then, be on the order of about 3 μm. By cutting off a surface layer of 100 μm and reanalyzing the new surface, it was found that carbonyl groups were still evident in the

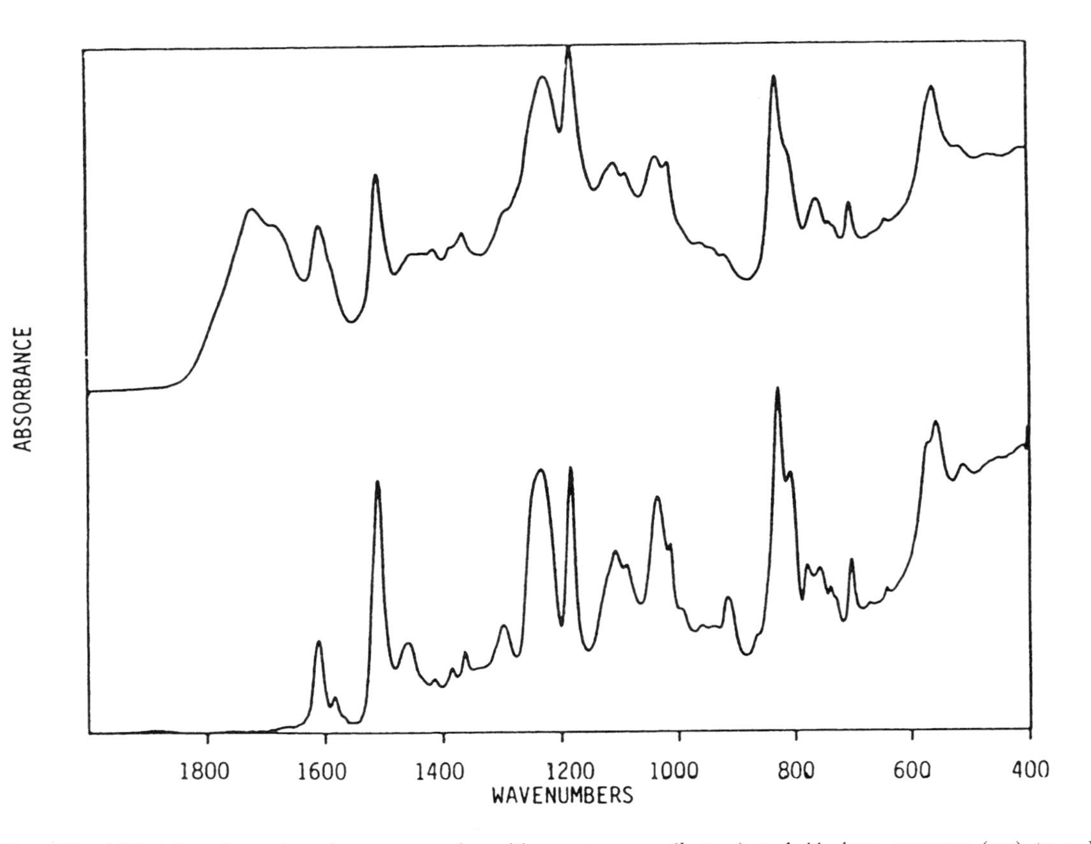

Fig. 4.21 ATR infrared spectra of epoxy samples with no exposure (*bottom*) and 11 days exposure (*top*) to a UV-A lamp. Reproduced with permission from Ref. 9.

ATR spectrum [9]. Therefore, the chemical changes from the UV treatment affected layers much deeper than those probed by the ATR spectrum.

The surface of polymers is often not of the required chemical composition to achieve a particular surface-related property, such as adhesion to another surface. This situation is typically overcome by pretreating the polymer surface to change the chemical composition as required. The capabilities of ATR spectroscopy are particularly well-suited to the qualitative monitoring of these chemical modification processes. Inagaki et al. [10] followed the chlorination of a polypropylene ($\eta_2 = 1.50$) surface with a $CHCl_3$ plasma. The ATR spectra were collected with a KRS-5 IRE at a 45° angle of incidence in the mid-IR. Although it was not mentioned in this publication, the depths probed at particular wavelengths were estimated by calculating d_p. Without referring to d_p or where they came from the depths probed were given as 0.66–5.0 μm for the respective wavenumber range of 3000–400 cm^{-1}. By use of equation (4.3), or the data in Table 4.3, this range of d_p's is obtained over the cited wavenumber range. The changes in the mid-IR ATR spectra upon plasma treatment are shown in Fig. 4.22. The $CHCl_3$ plasma caused new absorption bands to appear in the 1800–1600 and 800–600 cm^{-1} regions. The absorption bands in the 800–600 cm^{-1} region were assigned to C–Cl, C–Cl_2, and Cl–Cl_3 groups, while those in the 1800–1600 cm^{-1} region were speculated to be nonconjugated and

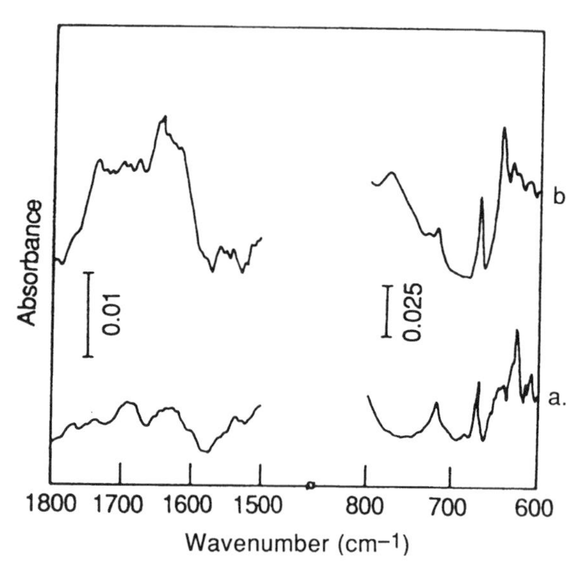

Fig. 4.22 ATR infrared spectrum of $CHCl_3$ plasma-treated polypropylene: (*a*), Untreated and (*b*), treated with $CHCl_3$ plasma for five minutes. Reproduced with permission from Ref. 10.

conjugated C=C groups. No further discussion was made about the depth of penetration of the radiation in the ATR measurements relative to the depth to which the $CHCl_3$ plasma treatment ingressed into the polymer surface.

Many studies have been concerned with designing the ATR experiment in such a way as to limit the penetration of the probing radiation into the surface to a very small distance. This has been especially in response to the need for a technique that is sensitive to surface layers only a few tens of a nanometer in thickness. Although vacuum techniques, such as X-ray photoelectron spectroscopy (XPS), are very sensitive to the outermost surface layers of only a few nanometers in thickness, these techniques may not succeed for layers of a few tens to a few hundred nanometers in thickness. The use of a high index of refraction IRE, coupled with high angles of incidence, is the basic requirement to achieve small penetration depths in ATR spectroscopy. Garton et al. [11] explored the lower limits of penetration that can be obtained for ATR in the mid-IR. These workers were interested in developing a method that would "bridge the gap" between the depth probed in the electron spectroscopies and typical ATR methods. In order to limit the penetration in the ATR, they used a germanium IRE at 60°. Consulting equation (4.3) or Table 4.2, shows that at about 1700 cm^{-1} for germanium at 60°, the d_p is only 0.3 μm (300 nm), which is what was claimed by these authors. The spectra of a polypropylene surface before and after treatment with a plasma are shown in Fig. 4.23. The appearance of a carbonyl at about 1720 cm^{-1}, as a result of oxidation by the plasma, can be seen in the figure. This chemical modification on the polypropylene surface leads to improved adhesion or printability. It was pointed out by Garton et al. that the carbonyl absorption at about 1720 cm^{-1} has the intensity of only about 0.003 absorbance units, which is partially due to the very shallow penetration of the IR radiation into the surface. This work showed that ATR spectroscopy is useful for probing shallow depths with appropriate experimental design, even in the mid-IR spectral region.

An alternate approach to probing shallow depths is the so-called "barrier film" technique. This technique was alluded to previously in relation to determining the practical depth that is penetrated in ATR spectroscopy (see Fig. 4.5 and associated text). The approach in this technique is to interpose a "barrier layer" between the IRE and the sample. The barrier layer is controlled to be of sufficient thickness that the evanescent wave has just enough intensity beyond it to probe the sample surface only to a very shallow penetration depth. This setup is shown schematically in Fig. 4.24. The evanescent wave decays exponentially from the surface of the IRE as it extends through the barrier layer (base-layer) and only a small fraction of it's intensity remains in the sample (overlayer). This is shown schematically in Fig. 4.25.

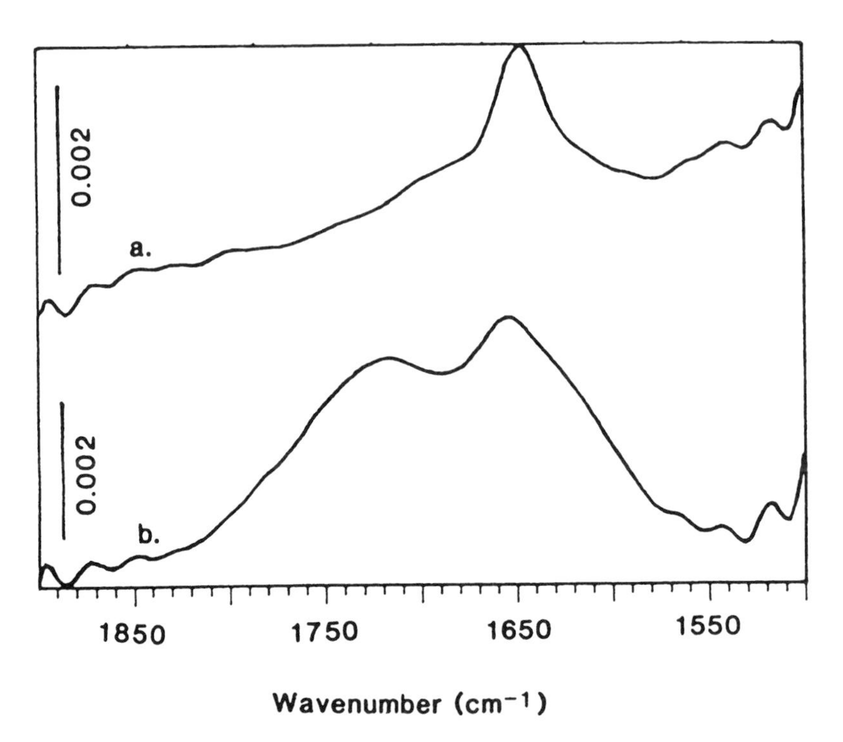

Fig. 4.23 ATR infrared spectrum of polypropylene (*a*) before and (*b*) after plasma treatment. Reproduced with permission from Ref. 11.

The barrier-film techique can be used empirically in a very simple way to probe the very shallow near-surface layers of a sample. Initially a very thick base-layer is placed on the IRE with the sample overlayer placed on top of this layer. The base-layer is then sequentially reduced in thickness in a series of experiments, while spectra are being collected. At particular thicknesses of the base-layer, absorption bands of the sample overlayer will just appear in the spectrum (which initially will be only composed of bands of the base-layer). Choices of base-layer thickness can then be made to probe some

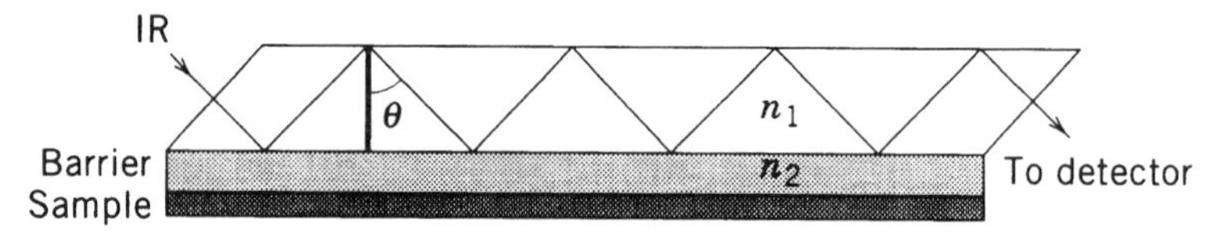

Fig. 4.24 Arrangement of the IRE, "barrier-layer," and sample for ATR depth-profiling studies.

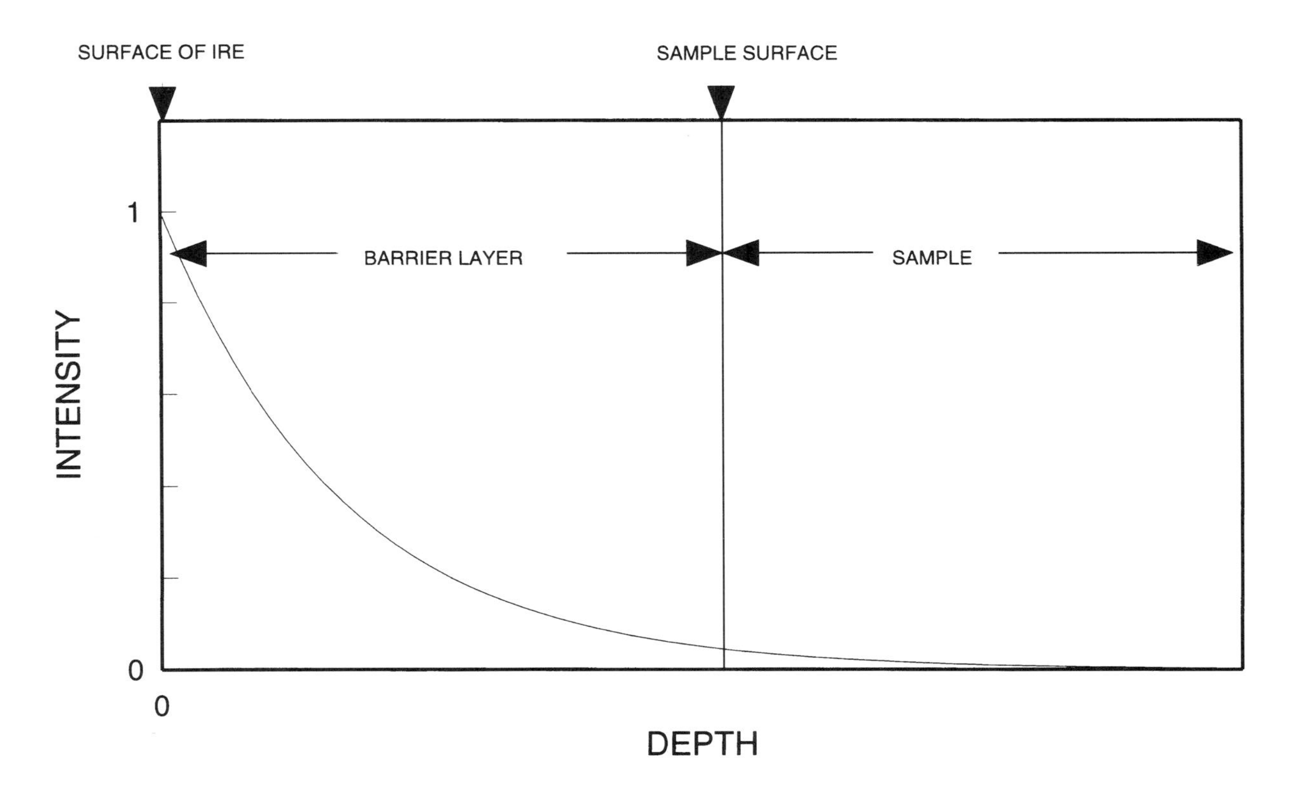

Fig. 4.25 Exponential decay of evanescent-wave intensity in the "barrier-layer" technique showing that only a small fraction of the intensity of the wave remains in the sample.

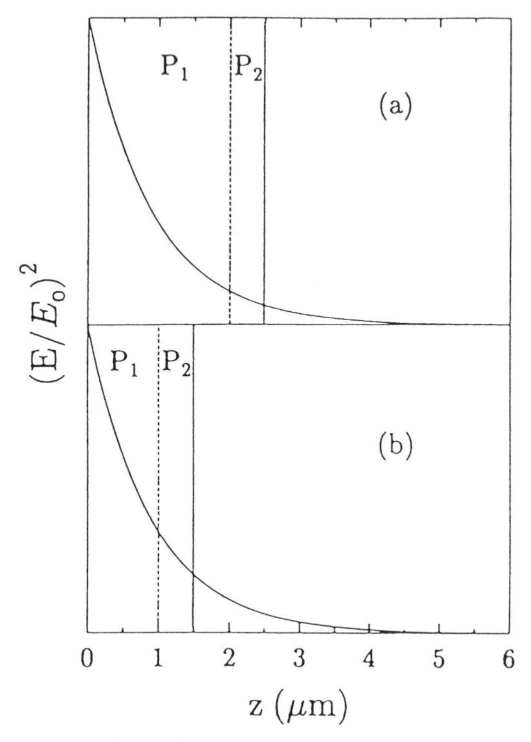

Fig. 4.26 The approximate electric field intensity distribution within the overlayer film (P_2) of a laminate system (*a*) for a thick base-layer (P_1) and (*b*) for a much thinner base-layer (P_1). Reproduced with permission from Ref. 12.

relatively shallow surface layers of the sample overlayer. This configuration can be used with any combination of IRE, base-layer and overlayer, and the base-layer thickness adjusted to control the probing depth into the near-surface layers of a sample material. Since this procedure can be done empirically, no calculations are necessary. However, approximate calculations can speed up the process.

Pereira and Yarwood [12] tested this approach as a solution to the "depth-profiling" problem in ATR spectroscopy. Fig. 4.26 shows the intensity decay of the infrared radiation into the sample overlayer for a thick and thin base-layer, respectively. The intensity decay curve may be calculated from equation (4.2), also noting that $I = E^2$. These workers tested this model by comparing the integrated absorbance (area under the peak) for the 1093-cm^{-1} band of a poly (vinyl alcohol) (PVOH) overlayer (P_2) and a poly (methyl methacrylate) (PMMA) base-layer (P_1) on a ZnSe IRE. The intensity decay curve and 1093 cm^{-1} band areas (normalized in each case to the band area for pure PVOH placed directly on the IRE) are plotted in

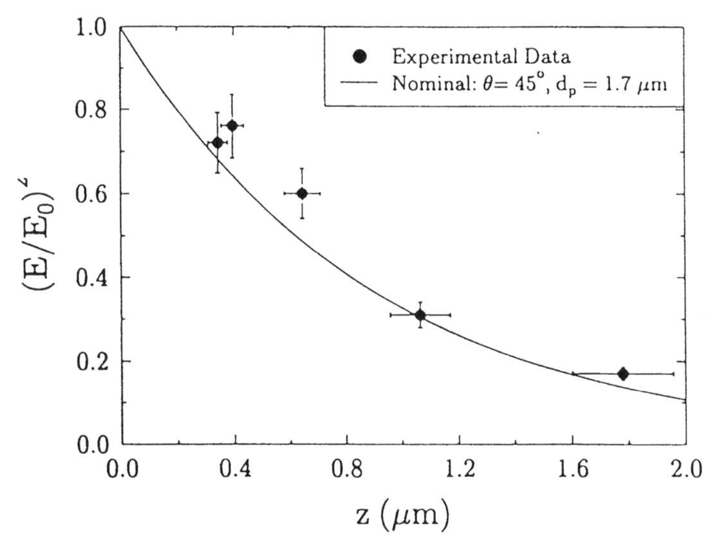

Fig. 4.27 Comparison of the observed poly(vinyl alcohol) band intensities with the computed electric field intensity distribution as a function of the base-layer thickness. Nominal $\theta = 45°$. Reproduced with permission from Ref. 12.

Fig. 4.27. The points over the decay curve in Fig. 4.27 are plotted at depths (z) determined by the thickness of the base-layer, namely $z = P_1$. The normalized band areas may be seen to decrease in good agreement with the intensity decay curve in Fig. 4.27. This indicates that more shallow depths into the sample overlayer are probed as the base-layer thickness is increased. The thickness (t) of the sample surface layers probed may be approximated by $t = d_p - P_1$. This approach can be employed to probe reasonably well-controlled and relatively shallow depths into the near-surface layers of a sample material. Of course, using relatively thin barrier layers makes control of deeper penetration possible. In the case described above, a ZnSe IRE, which has relatively low refractive index, was used. If a higher refractive index IRE such as germanium is employed, which has relatively small total penetration, it would be feasible to obtain extremely small penetration depths with relatively thin barrier layers. Studies of this type do not appear to be available in the literature.

The foregoing discussion centered on methodologies to generate some approximate information on the profile of species concentration in near-surface layers of a sample. The determination of a quantitative concentration depth-profile is of greater interest. The depth-profile (i.e., the chemical or physical information as a function of depth into a surface) is used in characterizing changes that occur in the near-surface layers of systems of interest. The fact that the evanescent wave exponentially decays with increasing distance into the sample surface causes difficulty in extracting depth-profiling information from ATR spectra. The further fact that the actual concentration depth-profile is unknown and may be linear, stepwise, exponential, and so on, makes the depth-profiling problems extremely complex. This problem has remained largely unsolved for a long time.

The problem of extracting quantitative depth-profile information has been approached in a variety of ways. The primary mode of attacking the problem has been to use multiple IRE crystals and multiple angles of incidence to cover a wide range of depths into the sample surface. The depth-profile function is then extracted from the large data set by the use of a complex mathematical model. The mathematical model must be developed with the aid of simplifying assumptions. The resulting equations must be solved numerically. Some useful results and interesting examples have been published [13, 14, 15]. However, the methods are not straight forward. Further detailed discussion is beyond the scope of this text but may be found in the above references.

Raman spectroscopy in the total internal reflection (TIR) mode presents a particularly ideal case for probing extremely thin near-surface layers. This is due to the fact that a monochromatic radiation source is used with wavelength typically in the visible spectral region. Since the wavelength is short, penetration is very shallow particularly relative to the mid-IR. Further the wavelength is constant, so the penetration is likewise constant across the whole spectrum and there is no wavelength dependence of the spectral intensities, as observed in the infrared. A schematic representation of the experimental Raman TIR setup is shown in Fig. 4.28. This shows that the laser source is totally reflected at the IRE/sample interface and that the Raman spectrum arises due to scattering of the laser radiation normal to this interface. Therefore this is total internal reflection, since absorption is not the mechanism that gives rise to the Raman spectrum. The typical problems of Raman spectroscopy, primarily weak signals and fluorescence of the sample, also appear to inhibit some applications of Raman TIR. Additional problems associated with Raman TIR are background scattering from the IRE and intrinsic Raman peaks arising from the IRE.

Several studies have been done that demonstrated the capability of Raman TIR for probing very shallow near-surface layers of polymers. Iwamoto et al. [16] demonstrated that this technique can yield good spectra of thin layers of polystyrene. These workers were successful in obtaining Raman TIR spectra of thin layers of polystyrene on thick backing layers of polycarbonate on a sapphire IRE. Spectra of polystyrene for layers of from 6 nm to 1 μm thickness were obtained. The absorption bands for polystyrene were clearly apparent, even down to 6-nm thick films.

The other extreme in the area of depth-profiling studies refers to the determination of the maximum depth probed in the ATR experiment. It was

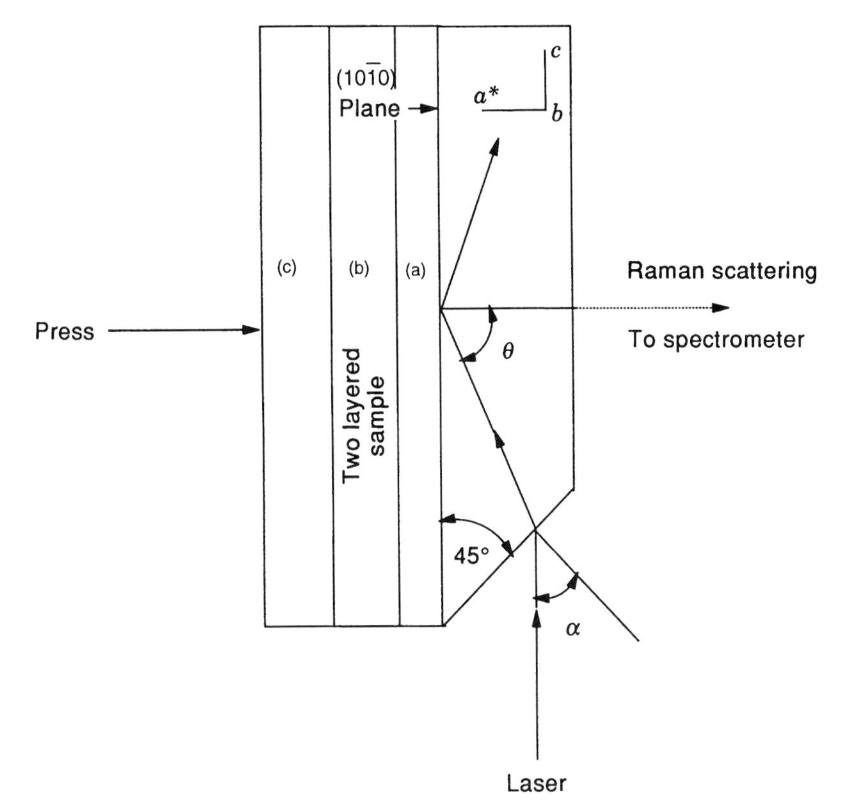

Fig. 4.28 Schematic diagram of the measurement optics for total internal reflection Raman spectroscopy and the relation of the polarization of the exciting laser beam to the crystal axis of the sapphire IRE. (*a*) The surface layer, (*b*) the base-layer, and (*c*) the silicone rubber clamp pad serve to distribute the pressure uniformly to ensure intimate contact at the interface.

demonstrated previously (see Fig. 4.5, equation (4.4), and related text) that the maximum depth sampled can be reasonably approximated by the so-called sampling depth, d_s, where $d_s = 3\,d_p$. This approach was used by Gardella et al., in studies of biomedical polymers used for heart valves and interaortic balloon pumps [17].

Sampling depths, d_s, were calculated according to equation (4.4) and were found to correspond to the actual depth that was probed to a much better degree than d_p. Therefore the depth probed was much greater than d_p. These workers used d_s as a measure of the depth that was probed and found that a consistent picture of the compositions as a function of depth could be constructed by this ATR information, along with data from transmission infrared spectroscopy and X-ray photoelectron spectroscopy. Conclusions could then be drawn about the bulk composition and the near-surface composition and bonding in biomedical polymers used for blood compatible *in vivo* devices.

For quantitative applications of ATR spectroscopy, Beer's law can be used in calibration plots provided that absorptions are kept low. This is really no different from transmission spectroscopy in which curvature of the Beer's law plots are typically observed if absorptions become too high. Mirabella [18] developed ATR methods to determine the vinyl acetate (VA) content of poly(ethylene-vinyl acetate) (EVA) copolymers. It was shown that quantitative calibration can be achieved if two absorption bands in the spectrum are ratioed. This is necessary in ATR spectroscopy because of imperfect, and therefore uncertain, surface contact to the IRE. The variation in contact of the sample surface to the IRE surface will cause the spectral intensity to vary greatly for identical samples even if a torque wrench is used to obtain identical C-clamp pressure [18]. For this reason band ratioing techniques are generally used for all quantitative ATR work.

A calibration line for the determination of the vinyl acetate (VA) content in poly(ethylene-vinyl acetate) (EVA) copolymers, based on the absorbance

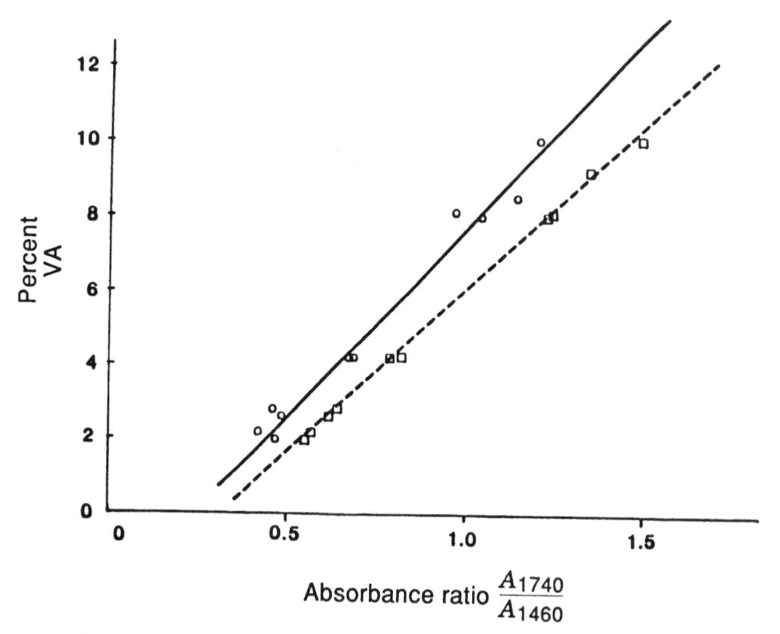

Fig. 4.29 Vinyl acetate concentration (w%) in poly(ethylene-vinyl acetate) copolymers versus the absorbance ratio A_{1740}/A_{1460} for the ATR (○) and the transmission (□) techniques. The lines are the linear least squares fits to the data.

ratio of the 1740- and 1460-cm^{-1} bands, is given for ATR and compared to the corresponding calibration line for transmission in Fig. 4.29. As can be seen, the lines are parallel but shifted apart. This is due to the wavelength dependence of ATR, which causes the absorbance at 1740 cm^{-1} to be increased in intensity relative to the 1460-cm^{-1} absorbance. Then the A_{1740}/A_{1460} ratio increases and shifts the ATR calibration line away from the transmission calibration line. However, the ATR calibration line is observed to provide a good means for determining the percent VA between about 2 and 10% VA. Fig. 4.30 shows ATR and transmission calibration lines for the methyl content per 1000 carbon atoms (Me/1000C) in polyethylene based on the ratio of the 1379- and 1368-cm^{-1} absorbances. In this case it is observed in the figure that the ATR and transmission calibration lines are the same, within experimental error. This is because the absorbances are so close together that the wavelength dependence in the ATR is negligible and has no effect on the absorbance ration (A_{1379}/A_{1368}). Again, the ATR line provides a useful means for determining the methyl content in polyethylenes.

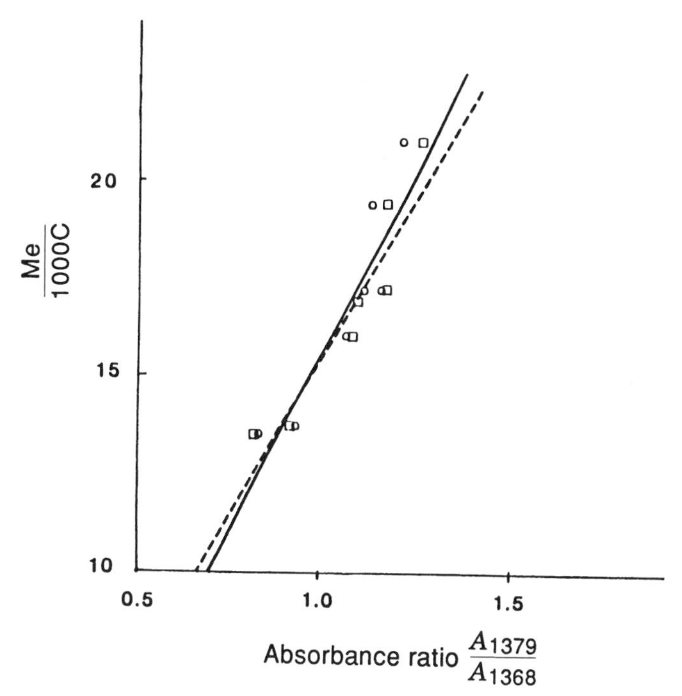

Fig. 4.30 Methyl concentration (methyls/1000C atoms) in polyethylenes versus the absorbance ratio A_{1379}/A_{1368} for the ATR (○) and the transmission (□) techniques. The lines are the linear least squares fits to the data.

Figures 4.29 and 4.30 employ the ATR calibrations to probe the surface chemistry of polymers, but the bulk chemistry can be probed with the transmission calibrations, and then the surface and bulk compared. Figures 4.29 and 4.30 show reasonable agreement with Beer' law, since the calibrations are fairly linear over the calibration ranges investigated. This is expected for relatively weak absorbers and in the concentration range employed.

There are several advantages to using band-ratioing techniques for ATR calibration development. Although sample contact problems can seriously affect the quality of ATR spectra, it has been demonstrated that identical sample contact is not required in quantitative IRS work when using band-ratioing techniques [19]. This is shown in Table 4.8 for a poly (ethylene-vinyl acetate) copolymer sample analyzed using an established IRS method, but for a wide range of contact levels. The calibration curve for IRS in Fig. 4.29 was used for these measurements. It can be observed from the absorbance values and A_p/A_s values (this ratio approaches 2 as optical, i.e., perfect, contact is approached) that the level of contact increases with increasing torque on the C-clamp used to hold the sample against the IRE. Also the level of contact increases as the films are melted and brought into more intimate contact with the IRE (designated "melt" in Table 4.8).

TABLE 4.8 ATR Data on 4.4 wt% Vinyl Acetate Poly(ethylene-vinyl acetate) Employing a Variety of Conditions

Sample	Clamping Torque (in.-lb)	Type of Contact	Polarization	Absorbance (cm^{-1})		$A_{\parallel}/A_{\perp}$ ($1740 cm^{-1}$)	$A_{\parallel}/A_{\perp}$ ($1460 cm^{-1}$)	IRS Calculated (%VA)
				A_{1740}	A_{1460}			
EVA-2	5	Solid	No	1.0530	1.5596			4.41
	5	Solid	$\parallel$	1.2939	1.8402	1.4813	1.3321	4.69
	5	Solid	$\perp$	0.8735	1.3814			3.97
	10	Solid	No	1.2034	1.7335			4.60
	10	Solid	$\parallel$	1.5091	2.2310	1.4455	1.3892	4.42
	10	Solid	$\perp$	1.0440	1.6059			4.16
	15	Solid	No	1.3218	1.7970			5.02
	15	Solid	$\parallel$	1.4866	2.3002	1.1306	1.3563	4.11
	15	Solid	$\perp$	1.3148	1.6974			5.41
	15	Melt	No	1.3514	1.9029			4.76
	15	Melt	$\parallel$	2.1178	3.1410	1.7845	1.6973	4.40
	15	Melt	$\perp$	1.1868	1.8506			4.07
							Mean $\pm$ S.D. 4.42 $\pm$ 0.29	

Source: Ref. [19].

[a]The percentage VA was measured by a pyrolysis technique.

However, it can be noted in Table 4.8 that the percentage of vinyl acetate calculated over this range of levels of contact is unaffected by the changing quality of contact. This result was obtained over the entire range of goodness of contact, except in the case of very poor contact, for which very low signal-to-noise ratios and very poor spectra were obtained. In the case of very good contact the spectra become so intense that deviations from the linear calibration in Fig. 4.29 were observed.

It was further shown theoretically in this study that a ATR calibration obtained by band ratioing for a particular IRE and angle of incidence is equally valid for any other IRE and angle of incidence [19]. This was demonstrated experimentally for the same poly(ethylene-vinyl acetate) copolymer sample as in Table 4.8. Table 4.9 presents measured values of the percentage of vinyl acetate for this sample obtained on a variety of IRE and angles of incidence combinations with unpolarized radiation. The values in Table 4.9 were calculated with the calibration in Fig. 4.29, which was obtained with KRS-5 at 45°. The one caveat is that a large range of sampling depths are covered as the IRE and angle of incidence are changed. The variation of the depth sampled may have an effect on the composition values obtained depending on the presence of any concentration depth-profile in the sample being considered.

However, Table 4.9 clearly shows that the percent VA values obtained at all conditions were similar to those on the same sample presented in Table 4.8, although a larger experimental error range was observed. The larger experimental error can be attributed to the lower signal-to-noise ratio obtained under some of the conditions in Table 4.9 compared to the

TABLE 4.9 Measurements of % Vinyl Acetate in Poly(ethylene-vinyl acetate) for a 4.4 wt% Vinyl Acetate Sample Employing a Variety of ATR Conditions and Using Unpolarized Radiation

Sample	IRE	Angle of Incidence	Torque (in.-lb)	Absorbance (cm^{-1})		Percent VA[a]	Mean ± Standard Deviation
				A_{1740}	A_{1460}		
EVA-2	KRS-5	60°	5	0.2220	0.3468	4.05	
	Ge	30°	5	0.4832	0.6689	4.89	
	Ge	45°	5	0.1317	0.2043	4.10	
4.4%	Ge	60°	5	0.0543	0.0788	4.55	4.33 ± 0.37
	KRS-5	60°	15	0.2458	0.3852	4.03	
	Ge	30°	15	0.5898	0.8226	4.83	
	Ge	45°	15	0.1549	0.2407	4.09	
	Ge	60°	15	0.0537	0.0835	4.08	

Source: Ref. [19].

[a]Calculated from Fig. 4.29.

conditions used in Table 4.8 (i.e., the effective thicknesses obtained under the various conditions in Table 4.9 were generally much smaller than those obtained under the conditions used in Table 4.8, since larger angles of incidence or IREs with higher *n* were used). The increase in absorbance with the higher clamping torque can also be observed in Table 4.9. However, this has essentially no effect on the percent VA values.

Thus the use of band-ratioing methods in ATR permits great latitude in conditions chosen for quantitative measurements. The level of contact of the sample to the IRE is not an impediment to using the calibration curves generated by band-ratioing methods at another level of contact. However, a reasonable level of contact is prudent in order to ensure a high signal-to-noise ratio and good spectra. The type and angle of incidence of the IRE is not an impediment to using the calibration curve generated by band-ratioing methods for another IRE and angle of incidence. However, variations may be caused by large changes in signal-to-noise ratios and for changes in penetration depth into a sample, which may have a depth dependence of composition.

ATR spectroscopy is a powerful technique for kinetic studies involving surfaces or interfaces. Specific studies of this type that have been done and published rather extensively are diffusion of small molecules in polymers, interdiffusion of polymers, and adsorption of small-molecules and polymers from solution onto specific surfaces. The determination of diffusion coefficients of small-molecules and of polymers by ATR spectroscopy has been thoroughly reviewed by Van Alsten [20]. Many ATR spectroscopy studies of adsorption kinetics from solution onto surfaces have been published; an example of the determination of the adsorption kinetics of a polymer has been published by Couzis and Gulari [21].

For studies of small-molecule penetrants into polymers, a polymer film is usually cast directly onto the IRE surface to ensure excellent contact. The film is controlled so that its thickness encompasses a predetermined extent of the ATR evanescent wave penetration. The supported polymer film is then contacted with the liquid or gas penetrant in an appropriate cell and the experimental measurements commenced. Spectra are collected as a function of time to monitor the sorption of the penetrant until saturation is reached. The increasing appearance of penetrant molecules in the ATR detection volume is thus monitored. Diffusion coefficients are typically calculated with a mathematical treatment based on Fick's Law and taking into account the exponential decay of the ATR evanescent wave [20]. The diffusion of heavy water into polyimides and urea into silicone elastomer are examples of systems reported in the review by Van Alsten [20].

The experimental configuration for the measurement of polymer interdiffusion is determined by a consideration of the polymers involved. In contrast to small molecules, polymers diffuse much more slowly. The faster diffusing polymer is often cast as a thin film onto the IRE at a thickness that encompasses a large fraction of the evanescent wave. The other polymer is cast on top of this polymer film as a thick-film. The loss of the thin-film diffusant from the ATR detection volume is then monitored. The diffusion coefficients may be calculated from the appropriate theory for polymers with the correction applied to the data for the decay of the evanescent wave [20]. The interdiffusion of low molecular weight (which will be fast diffusing) deuterated polystyrene into high molecular weight isotactic polystyrene matrices was one of many systems reported in the review by Van Alsten [20].

The experimental arrangement for the determination of the adsorption kinetics of small molecules or polymers from solution onto a surface is quite simple. The surface of interest may be the IRE material (adsorbant) itself, which is put into an appropriate liquid cell with a solution of the adsorbate. The increase in concentration of the adsorbate molecule is monitored within the ATR detection volume as a function of time. The adsorption rate can be calculated from the appropriate theory with the additional consideration of the decay of the evanescent wave [21]. Couzis and Gulari studied the adsorption of polystyrene in carbon tetrachloride solution onto a germanium IRE [21]. These workers found that the rate-limiting step of the adsorption process was not the diffusion of the polymer chains from solution (which was relatively fast) but rather the diffusion of the chains through a constrained layer of polymer at the IRE surface that formed immediately on contact of the solution with the IRE [21].

Surface Orientation

It was shown in Fig. 4.19 that the incident beam in transmission is a transverse wave: that is to say, it has components only in the plane perpendicular to the direction of propagation. On the other hand, Fig. 4.19 shows that the incident beam in ATR has components in all spatial directions.

All of the considerations discussed previously for designing ATR experiments apply, in general, to designing orientation measurements by ATR. However, the additional requirement is that the polarization of the incident radiation be known and reproducible.

It was shown in Fig. 4.1 that the electric vectors in three dimensions of the evanescent wave at the surface of the rarer medium are defined as E_{x0}, E_{y0}, and E_{z0}. The plane that the reflected incident beam forms as it propagates through the IRE is called the plane of incidence. When the polarizer is set to select the polarization perpendicular to the plane of incidence, it may be noted in Fig. 4.1 that the electric vector E_{y0} will be

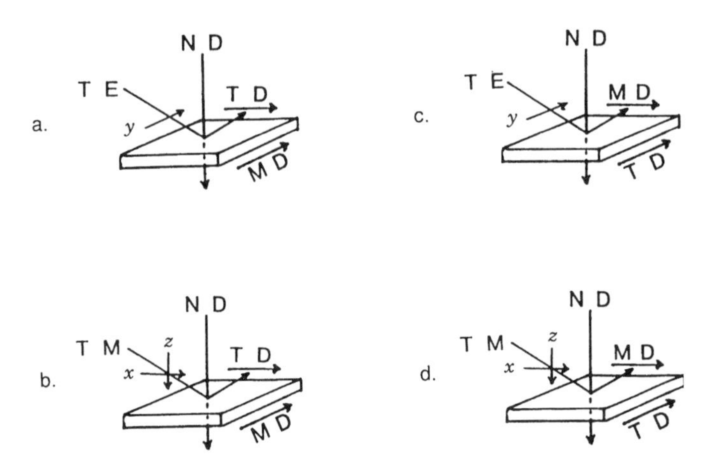

Fig. 4.31 Four configurations of the electric field vectors of the polarized radiation and the sample on the IRE in the ATR technique. (*a*) TE electric vector parallel to the MD (machine direction) in the sample, (*b*) TM electric vector parallel to the ND (normal direction) in the sample, (*c*) TE electric vector parallel to the TD (transverse direction) in the sample, and (*d*) TM electric vector parallel to the ND direction in the sample.

selected, and this is called the transverse electric (TE) wave. When the polarizer is turned 90°, it is set to select the polarization parallel to the plane of incidence. It may be noted in the figure that both the E_{x0} and E_{z0} vectors will be selected, and this is called the transverse magnetic (TM) wave. The four possible primary orientations of the polarizer and the sample on the IRE surface are shown in Fig. 4.31. The electric vectors active in each configuration are designated by x, y, and z, corresponding to E_{x0}, E_{y0}, and E_{z0}, respectively, in Fig. 4.31. The polarizer can be put between the sample and source or between the sample and detector. The easiest way to determine the experimental setup for obtaining the four primary polarizations shown in Fig. 4.31 is to acquire a highly oriented specimen of interest. This is easily done, for example, by heating a thermoplastic and stretching it in one direction only. This will yield a uniaxially oriented specimen. A highly, uniaxially oriented polypropylene specimen was first placed on the IRE with its primary orientation direction (machine direction, MD) perpendicular to the plane of incidence as in Fig. 4.31*a* and *b*. The spectra obtained when the polarizer was set to give the TE and TM wave, as in Fig. 4.31*a* and *b*, are shown in Fig. 4.32*a* and *b*, respectively. Likewise the configurations in Fig. 4.31*c* and *d* with the sample turned 90° so that the MD is parallel to the plane of incidence yielded the spectra in Fig. 4.32*c* and *d*. This type of experiment will permit the determination of the correct configuration of the apparatus.

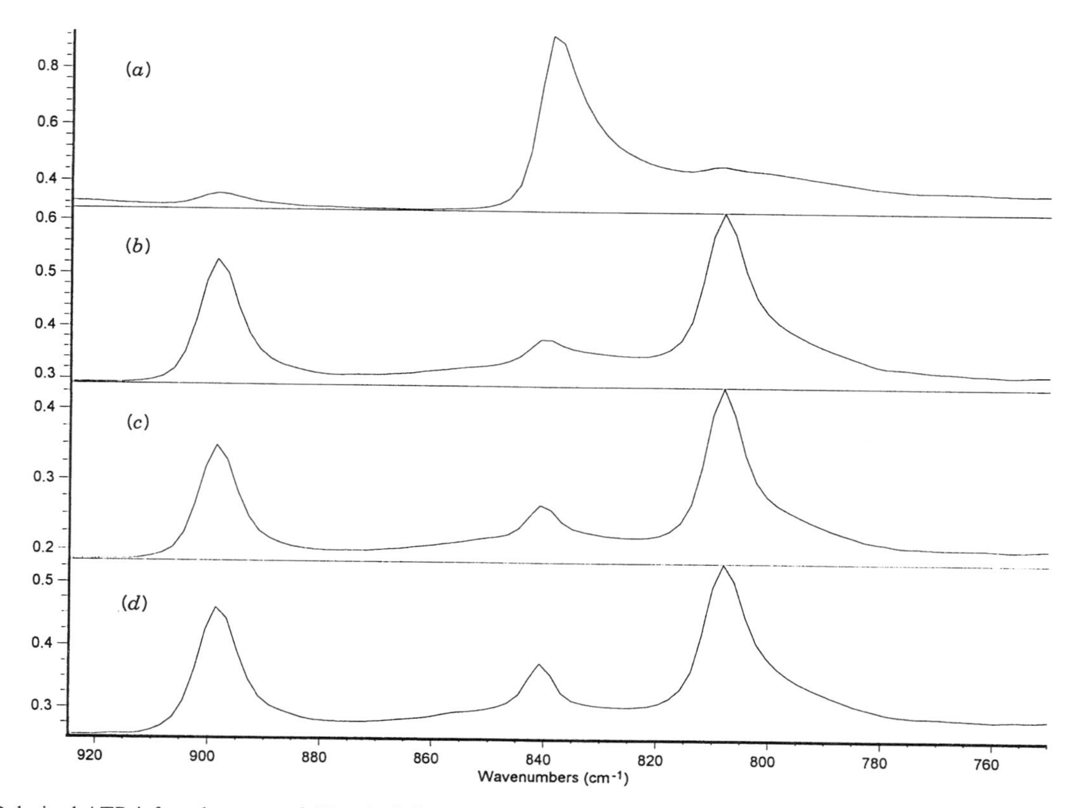

Fig. 4.32 Polarized ATR infrared spectra of 6X uniaxially oriented polypropylene film on 40° KRS-5 IRE: (1) TE vector parallel to the MD, (2) TM vector parallel to the ND, (3) TE electric vector parallel to the TD, and (4) TM electric vector parallel to the ND.

Specific Experimental Details

There may be some initial confusion in setting up the ATR orientation experiment. This is because the direction of the polarized radiation exiting the polarizer may be different from what is intuitively expected. Polarizers may be made from germanium, silicon, KRS-5, and other materials. Typically, when polarizer is set at 0°, the material of the polarizer appears to be vertical; that is, the leaves or wires run vertically. However, the polarized radiation does not run vertically out of the polarizer set in this way but rather horizontally. Similarly, when the polarizer is set at 90°, the radiation runs vertically. The reason for the orientation of the polarized radiation being perpendicular to the direction that is intuitively expected can be demonstrated as follows: A wire-grid polarizer consists of an array of very thin wires arranged parallel to, and spaced close to, one another. The radiation impinging on this array produces electric currents in the highly conductive wires. The currents are converted to heat by the small, but significant, resistance of the wires, and therefore the energy of the electric fields of the impinging radiation is dissipated in the direction parallel to the wires. The spaces between the wires are nonconducting and therefore allow the electric fields perpendicular to the wires to pass through without losing energy. The wire-grid dissipates the energy of the electric fields parallel to the grid but allows the electric fields perpendicular to the grid to pass through without energy loss. This polarizes the impinging radiation perpendicular to the direction of the grid. The initial confusion is cleared up readily if the previously mentioned ATR experiment is done with a highly oriented specimen. Alternatively, a similar experiment can be done in transmission with a uniaxially oriented specimen, and the polarized radiation orientation emitting from the polarizer can be readily determined.

The spectra in Fig. 4.32 were obtained on a 6X (500%) uniaxially oriented polypropylene film with a 40° KRS-5 IRE, and using polarized infrared radiation. The uniaxially oriented specimen has cylindrical symmetry; that is, the level of macromolecular chain orientation is essentially equal in the direction perpendicular to the MD, which is the transverse direction (TD), and the direction normal to the film surface (ND). Of course the MD has high orientation. The use of the 40° angle of incidence, which is slightly higher than the critical angle for polypropylene on KRS-5, yields electric vectors of the infrared radiation that are purely parallel to the film surface in the case of 90° polarization and perpendicular to the film surface in the case of 0° polarization. In other words, the x vector of the TM wave has effectively zero amplitude. In the spectra in Fig. 4.32 the 841-cm^{-1} band vibrates parallel to the polypropylene chain's long axis and the 809-cm^{-1} band vibrates perpendicular to the chain's long axis. The 6X uniaxially oriented specimen has the long axis of the polypropylene chain oriented largely in the MD direction. Due to cylindrical symmetry the TD and ND directions have much lower and equal orientation in those directions. Therefore the spectrum in Fig. 4.32*a* (corresponding to the configuration in Fig. 4.31*a*) exhibits a large 841-cm^{-1} band and a very small 809-cm^{-1} band. The three spectra in Figures 4.32*b*, *c*, and *d* (corresponding to the configurations in Fig. 4.31*b*, *c*, and *d*, respectively) are virtually identical in the absorbances of the 841 and 809 cm^{-1} bands due to the cylindrical symmetry of the uniaxially oriented polypropylene specimen. This is explained in further detail in [22].

A method for the quantitative determination of the orientation of polypropylene based on measurements as described above (40°KRS-5 IRE) was developed, and also described in detail in the same publication [22]. Figure 4.33 shows the orientation fractions, P_{MD}, P_{TD} and P_{ND} (where $P_{MD} + P_{TD} + P_{ND} = 1$), for a series of 1X, 2X, 3X, 4X, 5X, and 6X uniaxially oriented polypropylene specimens [22]. As can be seen in Fig. 4.33, P_{MD} increases dramatically in the series while P_{TD} and P_{ND} decrease, and $P_{TD} \cong P_{ND}$ in all cases. Further $P_{MD} \cong P_{TD} \cong P_{ND} \cong 0.33$ as expected for the unoriented, 1X, specimen.

It is worth noting that at angles of incidence much higher than the critical angle the 0° polarization (TM wave) has two significant electric vector components: one perpendicular to the IRE surface (the Z vector) and the other parallel to the IRE surface and parallel to the long axis of the IRE (the X vector) as shown in Fig. 4.31*b*. The consequences of this may be seen for spectra taken on 45° KRS-5 IRE in Fig. 4.34. Figure 4.34*a* and *c* are essentially identical to Fig. 4.32*a* and *c*, because the TE wave has only one component (the y vector) at 45° or 40°; namely parallel to the IRE surface and perpendicular to the long axis of the IRE. Fig. 4.34*b* is also essentially identical to Fig. 4.32*b*, since the components of the TM wave (the x and z vectors) are along the TD and ND directions of the film, respectively, which have the same orientation due to cylindrical symmetry. However, Fig. 4.34*d* is different from Fig. 4.32*d*, since the TM wave in this configuration also has a component (the x vector) in the direction of the MD. The high orientation in the MD direction of the 6*X* polypropylene film causes an increase in the absorbance of the 841-cm^{-1} band and a corresponding decrease in the absorbance of the 809-cm^{-1} band.

Figure 4.33 shows that the orientation fractions obtained at 45° on KRS-5 IRE are in agreement with those obtained at 40° on KRS-5 IRE [22]. For the case of relatively thin film specimens, a correspondence is expected between the macromolecular orientation determined by ATR and that determined by transmission spectroscopy. The orientation determined by transmission is typically expressed in terms of the Herman's orientation

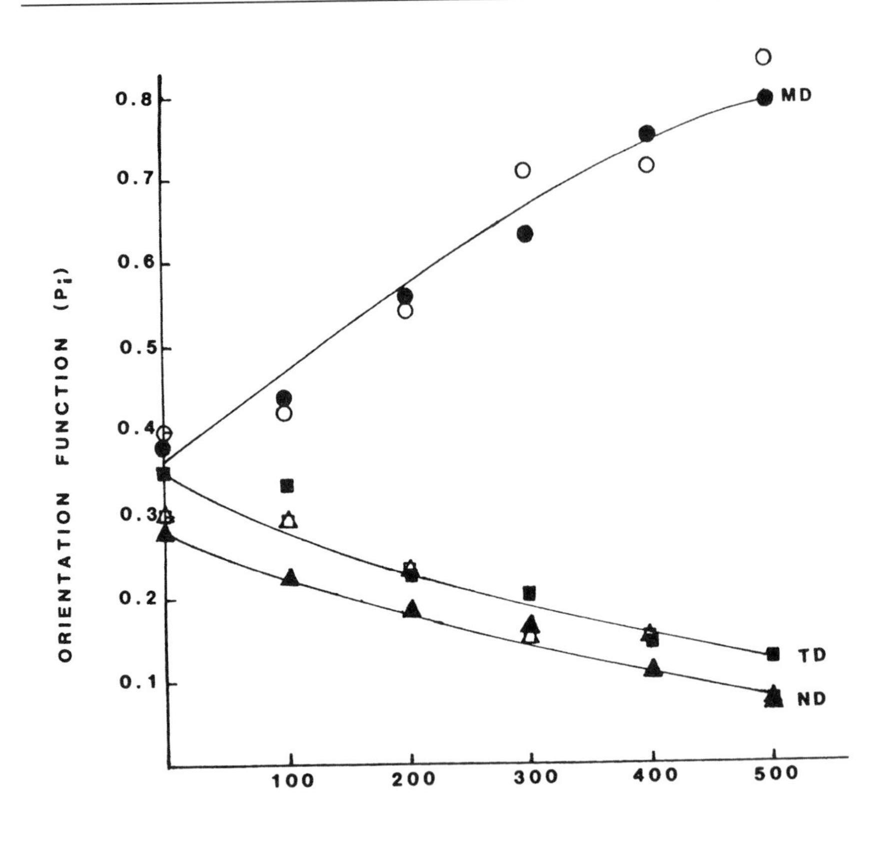

Fig. 4.33 Comparison of fractional orthogonal orientation fractions for a series of uniaxially drawn polypropylene sheets determined by IRS using 40° single-cut IRE and 45° double-cut IRE. (The lines drawn are smooth curves arbitrarily fit to the original IRS data.)

	40° Single-cut IRE	45° Double-cut IRE
P_{MD}	●	○
P_{TD}	■	□
P_{ND}	▲	△

function (f). This is determined from the equation

$$f = \frac{(D - 1)(D_0 + 2)}{(D + 2)(D_0 - 1)}. \tag{4.10}$$

Where $D = A_p/A_s$, A_p is the absorbance parallel to the major deformation direction (normally MD) and A_s is perpendicular to this direction,

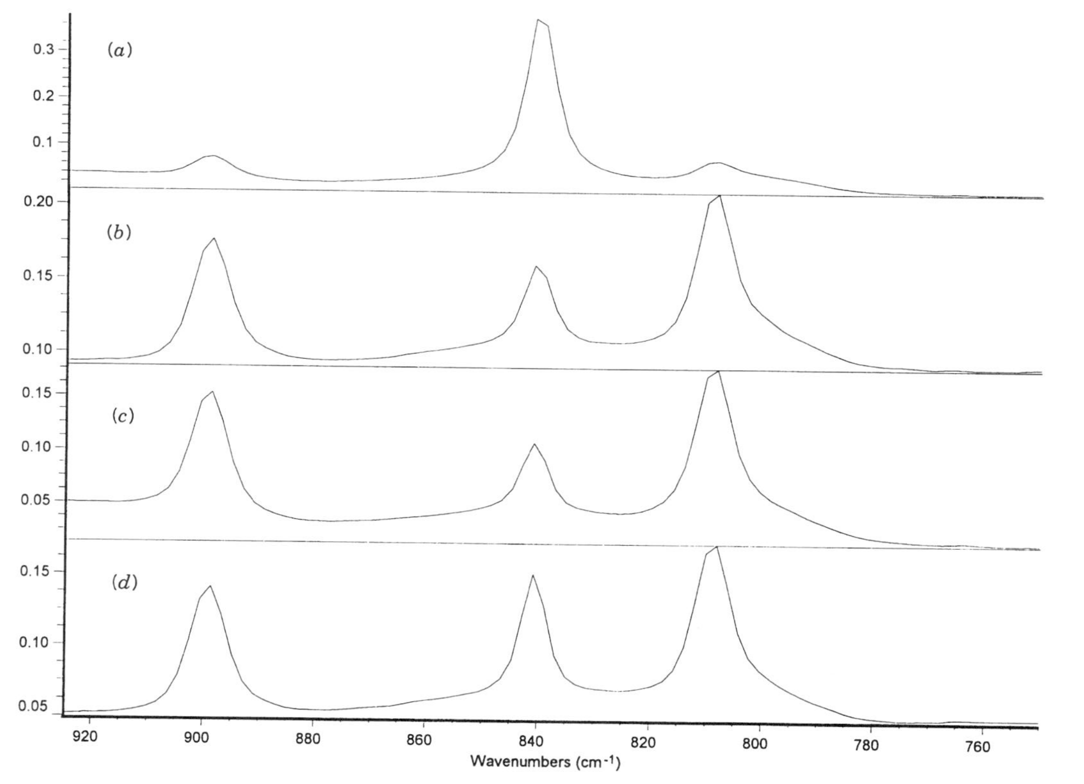

Fig. 4.34 Polarized ATR infrared spectra of 6X uniaxially oriented polypropylene film on 45° KRS-5 IRE: (1) TE vector parallel to the MD, (2) TM vector parallel to the ND, (3) TE electric vector parallel to the TD, and (4) TM electric vector parallel to the ND.

$D_0 = 2\cot^2\phi$, and ϕ is the transition moment angle of the spectral band being measured. In the case of the 841-cm^{-1} band, it was mentioned that its transition moment angle is about 0°. This reduces equation (4.10) to

$$f = \frac{D-1}{D+2}. \tag{4.11}$$

Two polypropylene films with 5X and 6X uniaxial deformation were measured by polarized transmission spectroscopy, and Herman's orientation functions were calculated from equation (4.11). Measurements were also done by polarized ATR spectroscopy and the orientation fractions (P_{MD}, P_{TD} and P_{ND}) determined. It has been shown that the P_{MD} orientation fraction is related to the Herman's orientation function by [22]

$$f = \frac{3P_{MD}-1}{2}. \tag{4.12}$$

The Herman's orientation functions for the 5X and 6X uniaxially oriented polypropylene films determined from polarized transmission and ATR spectroscopy are presented in Table 4.10. It can be seen that the agreement is quite good between the two methods. It should be borne in mind, however, that the transmission measurement represents the bulk, while the ATR measurement represents the near-surface layers of the films, and this may be responsible for some of the difference in the two orientation measurements.

TABLE 4.10 Comparison of the Herman's Orientation Functions for 5X and 6X Uniaxially Oriented Films as Determined by Transmission and ATR Spectroscopy

Transmission	5X	6X
D	4.14	5.71
f	0.51	0.61
ATR		
P_{MD}	0.69	0.73
P_{TD}	0.15	0.13
P_{ND}	0.16	0.14
f	0.54	0.59

4.8. CONCLUSION

The few applications of ATR spectroscopy that have been given here demonstrate the power of this technique for the solution of a wide variety of analytical problems. The reader will no doubt recognize many other applications of the general techniques described. This observation is confirmed by simply consulting the published literature, where the number of articles describing the application of ATR spectroscopy in a wide variety of fields is continually increasing.

Some further examples that demonstrate the wide applicability of ATR spectroscopy are liquid flow streams involving soft drink components [23], paper mill pulping liquors [24], and highly corrosive liquids [25, 26] analyzed by flow cells with ATR optics and ATR diamond probes. Diffusion, sorption, and exudation of water [27, 28] and organics [29, 30] can be determined as can polymerization reactions be monitored to determine the mechanism and kinetics of the reactions [31–34]. Activity in studies involving biological systems [35, 36] and biomaterials [37–42] utilizing ATR spectroscopy have increased dramatically. ATR has found application for the study of adhesion [43,44], polymer film drawing [45], sugar powders [46], polymer photodegradation [47], laundry detergent [48], papermaking [49], and textiles [50].

REFERENCES

1. A. M. Taylor and A. M. Glover, *J. Opt. Soc. Am.*, **23**, 206 (1933).
2. A. M. Taylor and D. A. Durfee, *J. Opt. Soc. Am.*, **23**, 263 (1933).
3. A. M. Taylor and A. King, *J. Opt. Soc. Am.*, **23**, 308 (1933).
4. N. J. Harrick, *Phys. Rev.*, **103**, 1173 (1956).
5. N. J. Harrick, *Phys. Rev. Letters*, **4**, 224 (1960).
6. N. J. Harrick, *J. Phys. Chem.*, **64**, 1110 (1960).
7. N. J. Harrick, *Ann. N. Y. Acad. Sci.*, **101**, 928 (1963).
8. J. Fahrenfort, *Spectrochim. Acta*, **17**, 698 (1961).
9. G. Zhang, W. G. Pitt, S. R. Goates, and N. L. Owen, *J. Appl. Polym. Sci.*, **24**, 519 (1994).
10. N. Inagaki, S. Tanaka, and Y. Suzuki, *J. Appl. Polym. Sci.*, **51**, 2131 (1994).
11. A. Garton, K. Ha, and M. Adams, *Proc. ACS Division of Polymeric Materials Science and Engineering*, **64**, 36 (1991).
12. M. R. Pereira and J. Yarwood, *J. Polym. Sci.: Part B: Polym. Phys.*, **32**, 1881 (1994).
13. G. Chen and L. J. Fina, *J. Appl. Polym. Sci.*, **48**, 1229 (1993).
14. J. Huang and M. W. Urban, *Appl. Spectry.*, **47**, 973 (1993).
15. R. A. Schick, J. L. Koenig, and H. Isido, *Appl. Spectry.*, **47**, 1237 (1993).
16. R. Iwamoto, M. Miya, K. Ohta, and S. Mima, *J. Chem. Phys.*, **74**(9), 4780 (1981).

17. G. L. Grobe, A. S. Nagel, J. A. Gardella, R. L. Chin, and L. Salvati, *Appl. Spectrosc.*, **42**, 980 (1988).
18. F. M. Mirabella, *J. Polym. Sci., Polym. Phys. Ed.*, **20**, 2309 (1982).
19. F. M. Mirabella, *J. Polym. Sci., Polym. Phys. Ed.*, **23**, 861 (1985).
20. J. G. Van Alsten, *Trends Polym. Sci.*, **3**, 272 (1995).
21. U. A. Couzis and E. Gulari, *Macromolecules*, **27**, 3580 (1994).
22. F. M. Mirabella, *Appl. Spectry.*, **42**, 1258 (1988).
23. E. K. Kemsley, R. H. Wilson, G. Pulter, and L. L. Day, *Appl. Spectry.*, **47**, 1651 (1993).
24. R. L. Yeager, *Pulp Pap.*, p. 111, June 1996.
25. M. Milosevic, D. Sting, and A. Rein, *Spectrosc.*, **10**, 44 (1995).
26. W. M. Doyle and B. Nadel, *Spectrosc.*, **11**, 35 (1996).
27. T. Nguyen, D. Bentz, and E. Byrd, *J. Coatings Tech.*, **67**, 37 (1995).
28. T. Nguyen, E. Byrd, and D. Bentz, *J. Adhesion*, **48**, 169 (1995).
29. M. He, M. W. Urban, and R. S. Bauer, *J. Appl. Polym. Sci.*, **49**, 345 (1993).
30. R. P. Semwal, S. Banerjee, L. R. Chauhan, and A. Bhattacharya, *J. Appl. Polym. Sci.*, **60**, 29 (1996).
31. L. Xu and J. R. Schalup, *Appl. Spectry.*, **50**, 109 (1996).
32. J. E. Dietz, B. J. Elliott, and N. A. Peppas, *Macromolecules*, **28**, 5163 (1995).
33. Q. Fan and L. M. Ng, *J. Vac. Sci. Technol.*, **A14**, 1326 (1996).
34. B. Kolthammer, D. J. Mangold, and D. R. Gifford, *J. Polym. Sci.: Part A: Polym. Chem.*, **30**, 1017 (1992).
35. D. Necsoiu and D. Ciotaru, *Spectrosc.*, **9**, 45 (1994).
36. M. G. Sowa, J. Wang, C. P. Schultz, M. K. Ahmed, and H. H. Mantsch, *Bibr. Spectry.*, **10**, 49 (1995).
37. A. Sawyer, J. Bandekar, and H. Li, *J. Vac. Sci. Technol.*, **A12**, 2966 (1994).
38. M. Kiremitci-Gumusderelioglu and A. Pesmen, *Biomaterials*, **17**, 443 (1996).
39. D. Klee, B. Severich, and H. Hocker, *Macromol. Symp.*, **103**, 19 (1996).
40. M. Muller, L. Werner, K. Grundke, K. J. Eichhorn, and H. J. Jacobasch, *Macromol. Symp.*, **103**, 55 (1996).
41. A. Magnani, E. Busi, and R. Barbucci, *J. Mater. Sci.: Mater. in Medicine*, **5**, 839 (1994).
42. D. E. Niveus, J. Schmit, J. Sniatecki, T. Anderson, J. Q. Chambers, and D. C. White, *Appl. Spectry.*, **47**, 668 (1993).
43. M. X. Xu, W. H. Zhang, P. X. Xue, W. Gao, and K. D. Yao, *J. Appl. Polym. Sci.*, **58**, 1047 (1995).
44. G. Hong and F. J. Boerio, *J. Appl. Polym. Sci.*, **55**, 437 (1995).
45. D. J. Walls and J. C. Coburn, *J. Polym. Sci.: Part B: Polym. Phys.*, **30**, 887 (1992).
46. N. Dupuy, M. Meureus, B. Sombret, P. Legrand, and J. P. Huvenne, *Appl. Spectry.*, **47**, 452 (1993).
47. G. Zhang, W. G. Pitt, S. R. Goates, and N. L. Owens, *J. Appl. Polym. Sci.*, **54**, 419 (1994).
48. D. P. DeSalvo, *International Lab.*, p. 24, May 1994.
49. I. Forsskahl, E. Kemtta, P. Kyyronen, and O. Sundstrum, *Appl. Spectry.*, **49**, 163 (1995).
50. J. S. Church and D. J. Evans, *J. Appl. Polym. Sci.*, **57**, 1585 (1995).

5 Diffuse Reflectance Spectroscopy

Jonathan P. Blitz

5.1 INTRODUCTION

The optical phenomenon known as diffuse reflectance is commonly used in the UV-visible, near-infrared (NIR), and mid-infrared (sometimes called DRIFT or DRIFTS) regions to obtain molecular spectroscopic information. It is usually used to obtain spectra of powders with minimum sample preparation. A reflectance spectrum is obtained by the collection and analysis of surface-reflected electromagnetic radiation as a function of frequency (ν, usually in wavenumbers, cm^{-1}) or wavelength (λ, usually in nanometers, nm). Two different types of reflection can occur: regular or specular reflection usually associated with reflection from smooth, polished surfaces like mirrors, and diffuse reflection associated with reflection from so-called mat or dull surfaces textured like powders. Techniques such as external reflectance and total internal reflectance spectroscopy use the phenomenon of specular reflection to obtain spectroscopic information. In

Modern Techniques in Applied Molecular Spectroscopy, Edited by Francis M. Mirabella.
Techniques in Analytical Chemistry Series.
ISBN 0-471-12359-5

diffuse reflectance spectroscopy, electromagnetic radiation reflected from dull surfaces is collected and analyzed. If a sample to be analyzed is not shiny, and for whatever reason is not amenable to conventional transmission spectroscopy, diffuse reflectance spectroscopy is a logical alternative.

To explain what diffuse reflectance is, consider a white wall illuminated by sunlight. The wall appears equally bright regardless of whether you stand directly in front of it, or at any other angle. The surface of the wall reflects the incident sunlight at angles independent of the angle of incidence. In other words, the surface of the wall scatters the incoming radiation. Ideal diffuse reflection requires that the angular distribution of the reflected radiation be independent of the angle of incidence. In contrast to this example, consider a mirror illuminated by sunlight. At most angles of observation, the sunlight will not be visible to the observer. However, at a very specific angle, the reflected sunlight will provide a direct image of the source (the sun). This specularly reflected radiation is characterized by the angle of reflection being equal to the angle of incidence. Specular and diffuse reflections are illustrated in Fig. 5.1.

Whereas specular reflectance can be rigorously treated theoretically using the Fresnel equations, many complex processes produce the phenomenon called diffuse reflection. When light is shined on a dull surface, such as a densely packed powdered sample for diffuse reflectance spectroscopy, the sample will result in a combination of reflection, refraction, and diffraction (i.e., scattering) of the impinging light. Of course, if the sample is of spectroscopic interest, light will also be absorbed by the sample at selected wavelengths or frequencies. Samples of interest for diffuse reflectance spectroscopy are therefore simultaneous scatterers and absorbers of electromagnetic radiation. These complicated and related phenomenon are generally treated with so-called two-constant theories, where the two constants

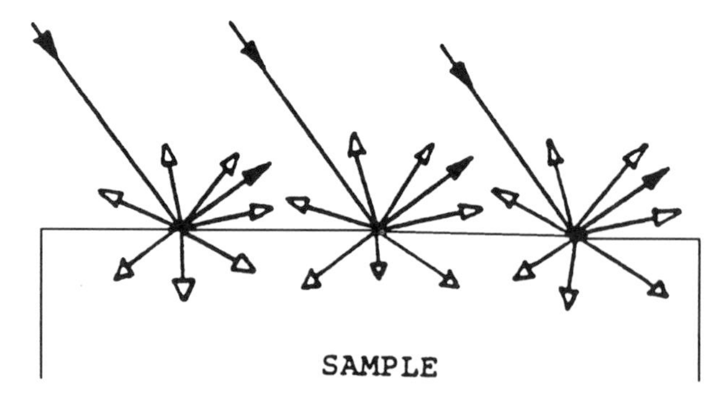

Fig. 5.1 Specular reflection, denoted by the shaded arrows, is radiation whose angle of reflection equals the angle of incidence. Diffuse reflection, denoted by hollow arrows, is radiation whose angle of reflection is independent of the angle of incidence.

characterize the scattering and absorbance characteristics of the sample. The most often used theory to describe and analyze diffuse reflectance spectra is the Kubelka-Munk theory. The two-constant Kubelka-Munk theory leads to conclusions that are qualitatively confirmed by experiment and can be used for quantitative work in many cases. Most molecular spectroscopic software can convert spectra to Kubelka-Munk units automatically. The best description of the fundamentals of diffuse reflectance spectroscopy can be found in a book by Kortüm [1].

The first serious diffuse reflectance studies were carried out around 1940 in the visible region at about the same time that precision spectrophotometers became available. Since then diffuse reflectance spectroscopy has been expanded to the near-infrared (NIR) and the mid-infrared regions. Applications of diffuse reflectance spectroscopy, especially in the latter two regions, have truly exploded in the past 10 years. Uses of this technique range from quality control of products on the factory floor to a sophisticated research tool for characterizing catalytic surfaces.

Using diffuse reflectance spectroscopy in the various regions of the electromagnetic spectrum involves different possibilities, trade-offs, and experimental considerations. On the more theoretical side, in the UV-visible and mid-infrared regions, fundamental electronic absorptions and vibrational modes are excited, respectively. The potentially large absorptivities (i.e., the sample can absorb much of the incoming radiation) of samples in these regions can require special sample preparation procedures to avoid spectral artifacts. On the practical side, different optical accessories are used to irradiate the sample and collect the diffusely reflected radiation in the mid-infrared region as compared to the UV-visible and NIR regions.

More detailed discussions of how diffuse reflectance fits into the myriad techniques in modern molecular spectroscopy, the interplay between theoretical and experimental considerations, and selected applications follow. Competitive techniques to diffuse reflectance spectroscopy will be compared and contrasted. A discussion of diffuse reflectance, the assumptions inherent in the widely applied Kubelka-Munk theory, and practical experimental considerations necessary to acquire reliable spectra will then be discussed. Finally, a review of selected applications using diffuse reflectance spectroscopy will be presented.

5.2 ROLE OF DIFFUSE REFLECTANCE IN MOLECULAR SPECTROSCOPY

There are many different sample-handling techniques available to the molecular spectroscopist including diffuse reflectance, external reflectance, internal reflectance, and photoacoustic spectroscopy. Given all these choices, it is easy to lose sight of the fact that these different techniques should all be compared to the benchmark transmission method. There is one overriding factor that should always be kept in mind: *If it can be done using transmission spectroscopy then do it*! There are a variety of reasons for saying this. First, transmission spectroscopy is experimentally simpler and less expensive than other methods. There are no optical accessories to align, and the theory governing transmission spectroscopy is firmly grounded in the fundamentals. Also in a transmission experiment most of the energy emanating from the source reaches the detector. Thus the signal-to-noise (S/N) ratio of a transmission spectrum will usually be greater than when using competitive techniques if everything else is equal. Since much of the energy emanating from the source does not reach the detector using alternative sampling techniques, these spectra are often detector noise limited. In other words, the sensitivity of the detector limits the S/N ratio of the resulting spectrum. For example, a perfectly aligned diffuse reflectance cell in the mid-infrared containing a nonabsorbing, diffusely reflecting sample will pass only 15–20% of the energy from the source to the detector. That is why high-sensitivity detectors, such as liquid nitrogen cooled MCT (mercury-cadmium-telluride) detectors are used in many applications of diffuse reflectance mid-infrared spectroscopy. If one can get around these inherent limitations by using transmission spectroscopy why not do it?

In many instances, some obvious and some not so obvious, transmission spectroscopy is not viable. Restricting our discussion to solid samples, if a beam of light cannot be passed through the sample using a transmission method like the KBr pellet technique, then one has no choice but to consider alternate means of spectral acquisition. Just because a transmission spectrum can be obtained does not necessarily mean it is advisable. The high pressures required to press a KBr pellet for transmission spectroscopy can alter the sample [2] or even cause unwanted reactions to occur [3]. For these and other reasons, alternate methods of spectral acquisition have been developed.

The pertinent sampling techniques for our discussion (other than transmission and diffuse reflectance spectroscopy) have been previously listed as external reflectance, internal reflectance, and photoacoustic spectroscopy. If the sample has a smooth, polished surface then external reflectance should be considered (Chapter 3). Internal reflectance (or attenuated total reflectance, ATR) is also a possibility if the sample can be placed in intimate contact with the internal reflection element (Chapter 4). Samples that exist as powders, or can be conveniently converted to powder form, are ideal candidates for either diffuse reflectance or photoacoustic spectroscopy. In other words, diffuse reflectance spectroscopy is not competitive with either internal reflectance or external reflectance spectroscopy. Diffuse reflectance and photoacoustic spectroscopy are often applied to the same types of samples and are therefore competitive.

Since photoacoustic and diffuse reflectance spectroscopy are competitive techniques, Childers et al. have compared these two methods in the

UV-visible and NIR [4] and in the mid-infrared [5] under precisely controlled conditions. These authors' conclusions were similar whatever the spectral region considered. Diffuse reflectance spectroscopy exhibits a superior S/N ratio and is therefore more sensitive. It is also superior for samples that are not intensely absorbing. If sample particle size is $> 100\,\mu m$ and is not amenable to grinding and/or dispersion in a nonabsorbing matrix, photoacoustic spectroscopy exhibits certain advantages. Often it is also possible to perform depth-profiling studies using photoacoustic spectroscopy. The diffuse reflectance method has thus far exhibited a limited capacity to perform such studies.

Despite these comparisons, photoacoustic spectroscopy has not been nearly as widely applied as has diffuse reflectance. Although similar information can often be obtained using either technique, as a practical matter diffuse reflectance has proved more generally useful. At one level, due to the way in which a photoacoustic signal is generated, spectral acquisition times are generally greater than when using the diffuse reflectance technique. Perhaps even more important, mechanical vibrations play havoc on a photoacoustic signal. The best photoacoustic spectra are often obtained in the middle of the night when the laboratory is deserted and external vibrations are at a minimum. Given the trend of molecular spectroscopy being applied more frequently in nonlaboratory environments, such as the factory floor, it is not surprising that diffuse reflectance has received more attention. The vast majority of diffuse reflectance near infrared spectroscopic work has been and will continue to be aimed at these types of process environments.

In summary, if one has an application in which a sample that scatters radiation must be analyzed, and the sample cannot be conveniently made to contact with an internal reflection element, diffuse reflectance and photoacoustic spectroscopy are the two choices. This of course assumes that transmission spectroscopy, for whatever reason, has been ruled out. Unless it is a specialized case, diffuse reflectance spectroscopy is a more practical alternative as compared to photoacoustic spectroscopy for reasons that have been previously described.

Before discussing the specific nuts and bolts of obtaining a diffuse reflectance spectrum, the first part of the next section will describe in more detail the theory of diffuse reflectance spectroscopy.

5.3 DIFFUSE REFLECTANCE THEORY AND PRACTICE

Fresnel and Kubelka-Munk Reflectance

The concepts of specular and diffuse reflectance have been previously discussed. The optical phenomena resulting in diffuse reflectance are many and complex; far from a comprehensive treatment will be given here.

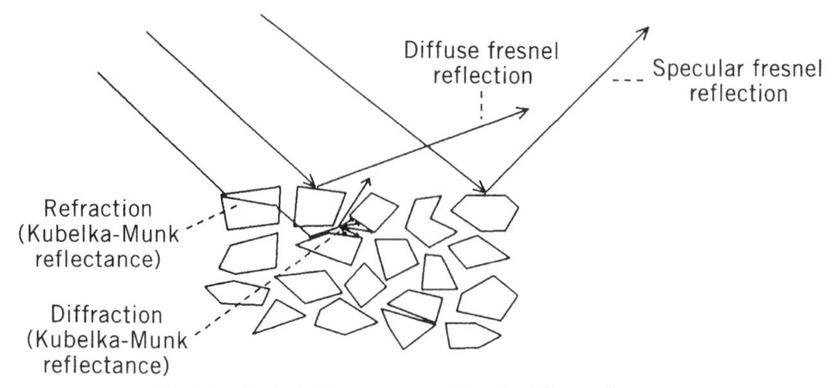

Fig. 5.2 Optical phenomena resulting in diffuse reflectance.

However, it is important that a requisite level of understanding be attained. The way in which samples are prepared and analyzed are inextricably related to diffuse reflectance theory. When diffuse reflectance spectra are obtained without considering the assumed optical phenomena, misleading and spurious results can be obtained. In this section only the theory that has direct, practical implications to the collection and interpretation of diffuse reflectance spectra will be discussed. The relationship between theory and experiment will be made as explicit as possible.

Figure 5.2 schematically shows some optical phenomena involved in a diffuse reflectance experiment. Consider the sample as a randomly distributed powder. The incoming radiation can undergo specular reflectance from the individual particles. The first theoretical treatment to explain the phenomenon of diffuse reflectance was given by Bouguer. It was assumed that diffuse reflection occurs from specular reflection from elementary mirrors of the sample surfaces with planes statistically distributed at all angles. Specular reflection can thus be further differentiated into diffuse Fresnel reflection and specular Fresnel reflection. Since the individual sample particle surfaces are randomly distributed, the surface of the particle is not necessarily parallel with the macroscopic sample surface. So although the radiation reflected from such a surface undergoes specular reflection (angle of incidence = angle of reflection), the reflected radiation appears diffuse with respect to the macroscopic sample surface. Therefore, we have the term diffuse Fresnel reflection. If the surface of the particle is parallel with the macroscopic sample surface, and specular reflection occurs, this is specular Fresnel reflection. Since both diffuse and specular Fresnel reflectance result from the same phenomenon, spectroscopically the same information content is contained in each. The beam of light in Fresnel reflectance interacts only once with the sample surface. If the source beam interacts more intimately with the sample, say, by undergoing multiple reflections, it stands to reason that more absorption by the sample will occur. This is the case when the phenomenon known as diffuse reflectance occurs.

The Bouguer elementary mirror hypothesis discussed above did not allow for the possibility of refraction. Due to refraction the incoming beam will also penetrate inside the particle and, if not absorbed, will exit from the surface after many reflections and diffractions. This radiation can be termed Kubelka-Munk reflectance. A competitive theory to that proposed by Bouguer assumed that Kubekla-Munk reflectance is solely responsible for diffuse reflectance. (At the time this theory was proposed, Kubelka and Munk were not yet born. The term Kubelka-Munk reflectance was adopted later.) After much experimental work early in the visible region [1], and more recently in the mid-infrared region [6–8], it has been found that both Fresnel and Kubelka-Munk reflectance nearly always occur simultaneously to greater or lesser extents, based on the nature of the sample and sample preparation methods.

To understand why there is usually a combination of Fresnel and Kubelka-Munk reflectance, one needs to understand what governs Fresnel reflectance. The Fresnel equations for specular reflectance show that as absorptivity (the likelihood that light will be absorbed by the sample) increases, so does specular reflectance. Under the simplest conditions, the Fresnel reflectance can be described as follows:

$$R_{\mathrm{F}} = (\eta - 1)^2 + \frac{\eta^2\kappa^2}{(\eta + 1)^2} + \eta^2\kappa^2, \qquad (5.1)$$

where R_{F} is the Fresnel reflectance, η is the sample's refractive index, and κ is proportional to the absorptivity. As κ increases, which will happen near an absorption band, there is an increase in Fresnel reflectance. Only nonabsorbing materials, which are uninteresting spectroscopically, can even in principle be ideal Kubelka-Munk reflectors. Thus, when radiation is reflected from a dull or mat surface, diffuse and Fresnel fractions are superimposed. This is a real problem in diffuse reflectance spectroscopy for a variety of reasons. First, Fresnel reflection contains less spectroscopic information than the diffusely reflected component because the latter interacts more with the sample increasing the likelihood of absorption. Second, a large Fresnel component means spectral distortion in diffuse reflectance spectroscopy.

Many seeming anomalies of diffuse reflectance spectra arise from the almost inevitable phenomenon of specular reflectance. However, the Kubelka-Munk theory to quantitatively describe diffuse reflectance spectroscopy assumes no specular reflection component. Since very complex processes are involved in diffuse reflectance spectroscopy, other simplifying assumptions are also inherent in the Kubelka-Munk theory.

Before plunging into the Kubelka-Munk theory, a cautionary note is warranted. Although the Kubelka-Munk theory is the most widely applied theory to describe diffuse reflectance, and can be thought of as analogous to Beer's law for transmission spectroscopy, there are important differences. Beer's law is a rigorously proven theory soundly based in the fundamentals of spectroscopy. The Kubelka-Munk theory is little more than an empirical model of diffuse reflectance that often works if care is taken to fulfill the assumptions of the model. Other empirical models exist to describe diffuse reflectance, but a discussion of these models is beyond the scope of this chapter.

Kubelka-Munk Theory

In this section Kubelka-Munk theory is described in terms of how it is used to analyze diffuse reflectance spectra. Many assumptions, both explicit and implicit, will be discussed as they relate to practical experimental considerations. First, definitions and analogies to transmission spectroscopy are provided. This is followed by the form of the Kubelka-Munk function. Important experimental procedures to enhance the validity of the Kubelka-Munk function are then considered. Finally, a few examples are given to illustrate what can happen if caution is not exercised when using diffuse reflectance spectroscopy.

In transmission spectroscopy, where a beam of light is passed through a sample, the transmittance is the ratio of intensities of transmitted to incident light:

$$T = \frac{I}{I_0}, \tag{5.2}$$

where T is the transmittance (1 for a completely transparent sample), I is the intensity of transmitted light, I_0 the intensity of the incident beam. In an analogous fashion the remittance of a diffusely reflecting sample is the ratio of intensities of reflected to incident light:

$$R_\infty = \frac{J}{I_0}, \tag{5.3}$$

where R_∞ is the absolute remittance, J is the intensity of the reflected radiation, and I_0 is again the intensity of the incident beam. (The ∞ subscript denotes that the sample is "infinitely thick"; in other words, none of the light irradiating the sample penetrates to the bottom of the sample holder. This is usually the case when the sample thickness is approximately 5 mm or more). A perfect diffusely reflecting substance, practically never attained, would have $R_\infty = 1$.

Since it is not practical to measure R_∞, the absolute remittance, the measured quantity is usually the relative remittance, R'_∞:

$$R'_\infty = \frac{R_{\infty\,\text{sample}}}{R_{\infty\,\text{standard}}} \tag{5.4}$$

Notice that if $R_{\infty\,\text{standard}} = 1$, then the absolute and relative remittances must be equal.

The relative remittance is analogous to transmittance in transmission spectroscopy. In transmission spectroscopy it is often convenient to present data in absorbance units, A, where $A = \log(1/T)$. This is because Beer's law says that absorbance is linearly related to the concentration of absorbing species. In an analogous fashion, it is possible to plot $\log(1/R'_\infty)$ against wavelength or frequency. This is called apparent absorbance units. This does not imply, however, that Beer's law is valid for diffuse reflectance spectroscopy.

Ideally one would like a function, like Beer's law in transmission spectroscopy, to linearly relate analyte concentration with the reflectance characteristics of a diffusely reflecting sample. The function must used is that derived by Kubelka and Munk:

$$F(R_\infty) = \frac{(1 - R_\infty)^2}{2R_\infty} = \frac{K}{S} = \frac{2.303\varepsilon C}{S} \tag{5.5}$$

where K is the absorption coefficient (twice the Beer's law absorption coefficient), S is twice the scattering coefficient of the sample, ε is the absorptivity, and C is the analyte concentration. One can see why the Kubelka-Munk theory is a two-constant theory: the reflectance characteristics are related to the ratio of the absorption to scattering coefficients which are the two constants.

Table 5.1 gives a better feel for the behavior of the Kubelka-Munk function. The first column, % reflectance, is analogous to % transmittance in transmission spectroscopy. The second column, apparent absorbance, is calculated from the % reflectance as one would using Beer's law from transmission spectroscopy. The third column is the corresponding value in Kubelka-Munk units corresponding to the % reflectance data in the first column. It can be seen from these data that the Kubelka-Munk function slights the weak bands (high reflectance) and enhances the strong bands (low reflectance) as compared to a classical absorption spectrum.

Let us take a small detour. If the accuracy of measurements depends only on the intensity difference of the diffusely reflected radiation (no instrumental error), the relative error in the Kubelka-Munk function can be

TABLE 5.1 Comparison of Absorbance and Kubelka-Munk Units at Various % Reflectance Values

% Reflectance	Apparent Aborbance ($\log 1/R_\infty$)	Kubelka-Munk Units $[(1 - R_\infty)^2/2R_\infty]$
100	0	0
90	0.046	0.0056
80	0.097	0.025
70	0.15	0.064
60	0.22	0.13
50	0.30	0.25
40	0.40	0.45
30	0.52	0.82
20	0.70	1.6
10	1.0	4.0
1	2.0	49

obtained. Results of this mathematical analysis are shown in Fig. 5.3. From these results the most favorable range for measurement lies between $0.2 < R_\infty < 0.7$. In other words, best results are obtained between 20 and 70% reflectance (remittance). As seen from Fig. 5.3, the error at larger and smaller reflectances increases very rapidly.

It is apparent from equation (5.5) that the Kubelka-Munk function linearly relates analyte concentration with band intensity provided that S

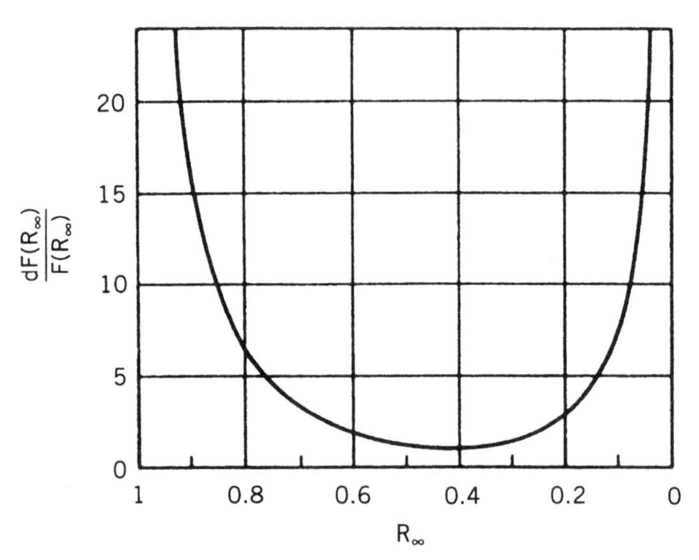

Fig. 5.3 Inherent error in the Kubelka-Munk function versus R_∞.

remains constant and the assumptions made in deriving the Kubelka-Munk equation are valid. To obtain diffuse reflectance spectra that have the best chance of obeying the Kubelka-Munk equation with quantitative validity, the question becomes what are the relevant assumptions made in deriving *and using* the Kubelka-Munk equation?

The assumptions in deriving and using the Kubelka-Munk equation are many. Those assumptions with practical implications will now be discussed. (1) All Fresnel reflection is ignored. This is the single most important consideration when obtaining any diffuse reflectance spectrum [9]. As has been previously discussed, Fresnel reflection increases with absorptivity. When collecting diffuse reflectance spectra of samples with high absorptivity, take care to reduce Fresnel reflectance. (2) The particles in the sample are much smaller than the thickness of the entire sample. (3) The sample thickness should be greater than the beam penetration depth to meet the "infinitely thick" criterion. (4) The sample diameter should be much greater than the focus of the incident beam to avoid optical effects that are not due to the sample. Also the sample area detected should be smaller than the area illuminated.

For qualitative diffuse reflectance work, the only real concern is to reduce Fresnel reflection to avoid spectral artifacts. An artifact known as anomalous dispersion is caused by an increase in Fresnel reflectance near strong absorption bands (5.1). This can be detected by either derivative shaped features or a shift in the peak maximum to higher frequency or lower wavelength. For bands of very high absorptivity, the increase in Fresnel reflectance around the absorption actually increases the reflectance in this region. The resulting large reflectance maximum results in a reststrahlen band. These types of spectral artifacts only occur in regions of high absorptivity. In the mid-IR region, where fundamental vibrational modes predominate, very high absorptivities are not uncommon. These artifacts can be eliminated by diluting the sample with a nonabsorbing matrix. In mid-IR diffuse reflectance spectroscopy, this is commonly done by diluting the sample in a solid mixture of KCl. The mid-IR spectrum of pure silica powder is compared to that of a 5% mixture of silica in KCl in Fig. 5.4. Since 95% of the sample is nonabsorbing in the 5% mixture (Fig. 5.4*b*), the decrease in the sample's absorptivity results in a decrease in Fresnel reflection. The reststrahlen band is thus eliminated.

Due to the very high absorptivities, if care is not taken to reduce Fresnel reflectance, these types of artifacts will be common in the mid-IR. In the NIR vibrational overtone and combination bands that have orders of magnitude lower absorptivities. These spectral artifacts are therefore not a problem in the NIR. In the UV-visible region, fundamental electronic absorptions predominate. High absorptivities are again common, though these highly distorted spectral features are not nearly as common in the UV-visible region as compared to the mid-IR.

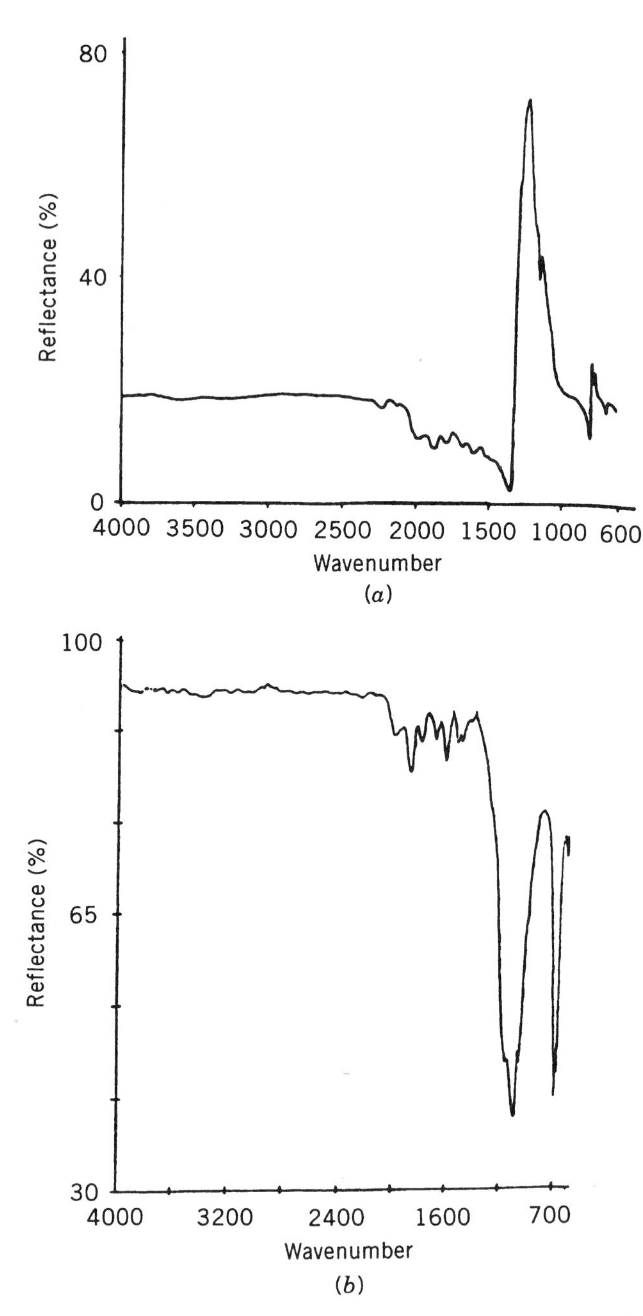

Fig. 5.4 (*a*) Diffuse reflectance spectrum of pure silica illustrates the *reststrahlen* band at $1100\,\text{cm}^{-1}$ from an increase in Fresnel reflectance. (*b*) Dilution of the silica sample in KCl eliminates the spectral artifact.

When using diffuse reflectance spectroscopy for quantitative purposes, beyond the above considerations, precise methods of sample preparation are necessary. The scattering coefficient must be kept constant from sample to sample to perform quantitative diffuse reflectance work. To achieve this, two important parameters must remain constant from sample to sample: particle size and sample packing. Particle size can be controlled by sieving, or more quickly and practically by a ball and mill grinder to powder the sample (provided that grinding time is kept constant). Depending on the nature of the sample, adequate sample packing reproducibility can be attained by gentle tapping of the sample in the sample cup, or more precisely by applying a known amount of pressure to the sample for a specified time. High pressures should be avoided, however, since resulting surface gloss is an indication of Fresnel reflection.

Changes in the scattering coefficient impact quantitative work in unforeseen ways. Take, for instance, crystals of $CuSO_4 \cdot 5H_2O$, which have a fairly strong blue color. Grind these crystals to smaller particle size, and the material becomes a pale blue. At very small particle sizes, $CuSO_4 \cdot 5H_2O$ appears practically white. In other words, the sample absorbs less light, in the visible region, with decreasing particle size. Since the absorptivity in fact increases with decreasing particle size [10], which should result in a *more* colored material, the scattering properties of the sample must be changing. As particle size increases scattering decreases. The radiation can thus penetrate deeper into the sample, resulting in more Kubelka-Munk reflectance with corresponding increased absorbance and more intense color. This is a strikingly visual example of how the scattering properties of the sample affect reflectance spectra.

In contrast to the previous example, strongly absorbing materials such as $KMnO_4$ in the visible region, exhibit an increase in apparent absorbance ($KMnO_4$ is more colored) as particle size is decreased. This can be explained by the fact that the reflection characteristics of strongly absorbing materials contain a large contribution from specular reflectance. As particle size increases, so does specular reflectance. Thus, by decreasing particle size, the specular reflectance component decreases, resulting in an increase in apparent absorbance. Again the sample's scattering constant changes as a function of particle size, but now the opposite effect is obtained. Another anomaly from specular reflectance.

An additional characteristic of diffuse reflectance spectra of strongly absorbing species is that when such samples are diluted with a nonabsorbing matrix, reflectance characteristics are largely unchanged over a broad concentration range. A plot of band intensity in Kubelka-Munk units versus analyte concentration exhibits negative deviations from linearity at high analyte concentration. (This is often also the case in Beer's law calibration curves, but for different reasons). Since the Kubelka-Munk theory attempts to describe the reflectance behavior of weakly absorbing samples, the implicit assumption is that the beam penetration depth does not change as analyte concentration increases. In effect it is assumed that the pathlength (as it is called in transmission spectroscopy) is constant. With strongly absorbing samples at high analyte concentrations, the beam penetration depth decreases with increasing analyte concentration. At high analyte concentration therefore, the intensity of strong absorption bands does not linearly vary with concentration, simply because of the reduction in beam penetration depth or pathlength. Dilution of the sample in a nonabsorbing matrix cures this problem.

The scattering properties and reflectance characteristics of the sample must remain constant for quantitative diffuse reflectance work in any spectral region. For qualitative work the requirements are less stringent, but for strongly absorbing samples spectral artifacts are still a possibility.

The Practice of Diffuse Reflectance Spectroscopy

The instrumental requirements for UV-visible, NIR, and mid-IR spectroscopy vary. For our discussion, probably the most important difference between doing spectroscopy in the various regions is detector technology. UV-visible spectrometers use photomultiplier tubes that are highly sensitive devices. Lead sulfide (and other) detectors are also very sensitive in the NIR. Mid-infrared detectors are, however, much less sensitive. Consequently the collection of diffuse reflectance spectra in the UV-visible–NIR spectral regions is done differently than in the mid-infrared region. This section will thus be divided into two parts: UV-visible–NIR diffuse reflectance spectroscopy and mid-infrared diffuse reflectance spectroscopy.

UV-Visible–NIR Diffuse Reflectance Spectroscopy

Most diffuse reflectance spectra collected in these regions are done with an optical accessory called an integrating sphere. The integrating sphere is a hollow enclosure with walls constructed of a diffusely reflecting material that reflects all wavelengths of interest with a high reflecting power. Traditionally freshly prepared MgO was used to coat the inner sphere surface. There are many practical problems with this material, however. The surface is difficult to prepare, fragile, and relatively unstable. Fortunately, modern integrating spheres come with their own durable coatings already applied. A thermoplastic resin with a trade-name of Spectralon^{R} (Labsphere, Sutton, NH) exhibits an absolute reflectance of $>95\%$ from 250–2500 nm (UV $\rightarrow$ NIR range). This material is used to coat many integrating spheres for diffuse reflectance spectroscopy in these spectral regions.

There are many experimental arrangements used for diffuse reflectance spectroscopy with an integrating sphere. Most of these arrangements can be

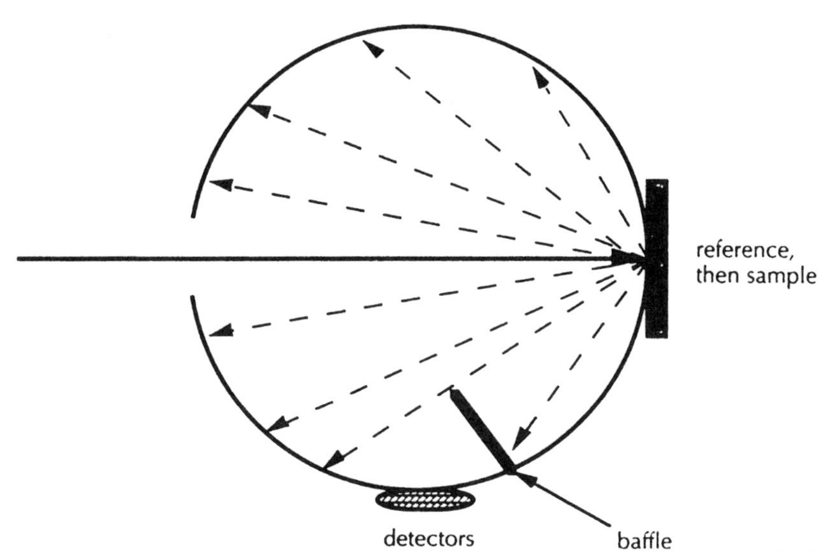

Fig. 5.5 Integrating sphere used to obtain a diffuse reflectance spectrum by the substitution method. Reprinted with permission from [16].

classified as either a substitution or a comparison method. The way to obtain a diffuse reflectance spectrum using the substitution method with a dispersive single beam spectrophotometer is illustrated in Fig. 5.5. Monochromatic radiation enters the sphere from an external source through the aperture. First a highly reflective reference material, such as Spectralon^{R} which should be provided with the sphere, is illuminated. The radiation intensity is measured by a detector mounted on the sphere wall. The standard is then replaced by the sample and the measurement repeated. The relative reflectance values of sample/standard thus provide the raw data for the sample's spectrum. A single port is provided in the integrating sphere for measurements made using the substitution method. So-called sphere error occurs in this method due to the sample becoming part of the sphere wall. When the sample and reference exhibit different reflectance values, which is inevitable for a spectroscopically interesting sample, the ratio of sphere wall radiances is not directly equal to the ratio of reflectances. In effect the blank and sample scans are measured as if they were done in two separate integrating spheres. The sphere or substitution error is small (roughly 5%) in cases where the sample and standard are approximately equal in reflectivity. If the sample and standard have very different reflectivities, substitution error can be significant.

Substitution error can be dealt with in many ways. First, it can be reduced by making the sample and reference areas small with respect to the

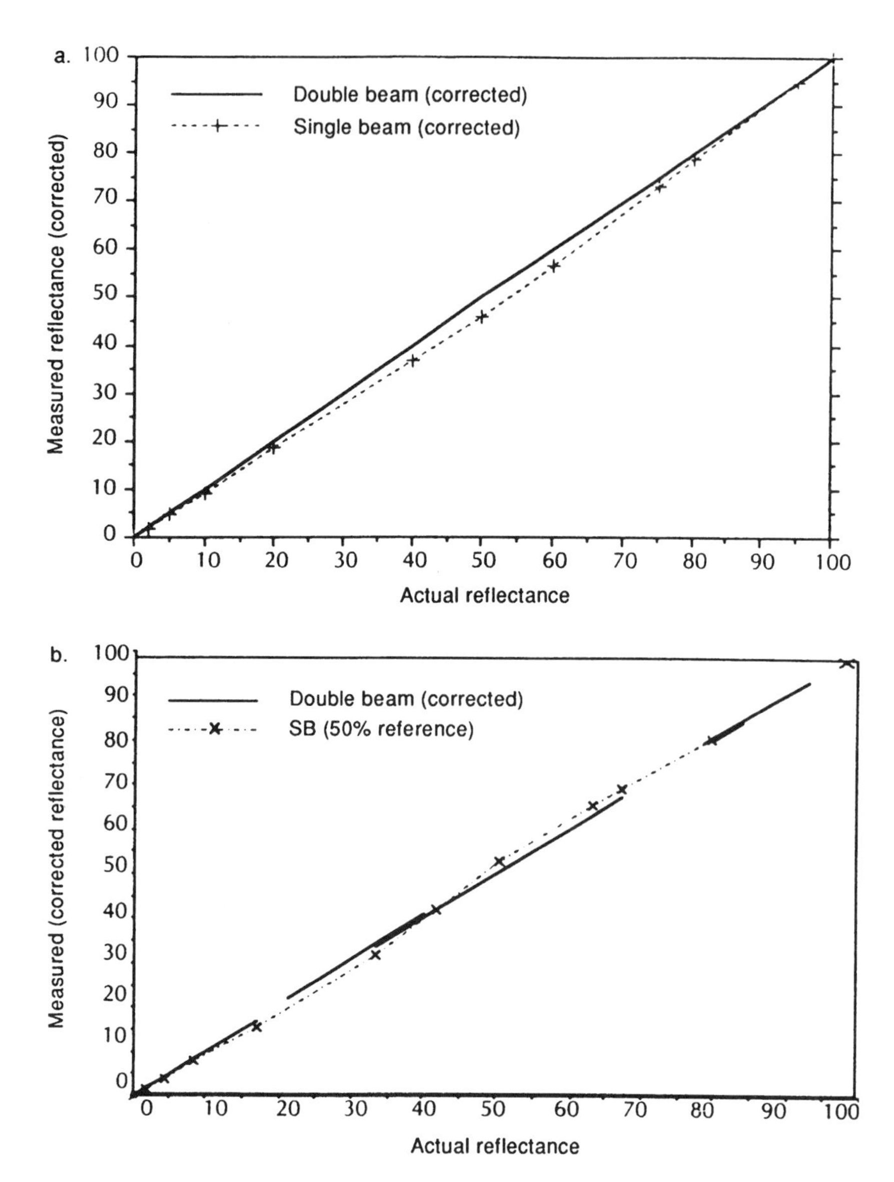

Fig. 5.6 (*a*) Single-beam substitution error with a 100% reflector as a reference. Deviation from the solid line is maximum at 50% reflectance. (*b*) Single-beam substitution error with a 50% reflector as a reference. Reprinted with permission from [16].

total surface. In other words, use a larger sphere. This can only be taken so far, however. There is a loss in sensitivity as sphere size is increased. Thus a compromise must be reached in which one trades off sphere error for sensitivity (spectral S/N). Another way to reduce substitution error is to use a reference that is very close in reflectance to the sample. Figure 5.6*a* illustrates a single-beam substitution error with a 100% reflector as a reference. Deviation from the actual reflectance is maximum as the sample approaches 50%. Figure 5.6*b* illustrates single beam substitution error with a 50% reflector as a reference. Comparison of Figs. 5.6*a* and 5.6*b* shows less substitution error using the 50% reflectance material. Thus, by matching the reflectance of the reference and sample, an accurate measurement of reflectance using the substitution method can be developed. Calibrated standards also exist that can be used to calibrate the sphere for substitution error. Sets of reflectance standards can be obtained [11] that have been measured on a sphere without substitution error. A table of measured versus actual readings can then be generated and used to correct for substitution error.

When should you be concerned with substitution error? If you are doing strictly qualitative work, identifying the location of peaks, substitution error should be of no concern. If you are trying to extract quantitative data from your spectra, however, substitution error is something well worth considering.

The second method for obtaining diffuse reflectance UV-vis–NIR spectra is the comparison method. Now both the sample and the standard make up a part of the sphere wall throughout the measurements. Two sphere ports are provided for simultaneous mounting of sample and reference. First the sample to be measured is placed in a "dummy" port, and the calibrated standard in the sample port. After obtaining the blank scan, the sample is substituted for the standard, and the reflectance spectrum is obtained. The comparison method eliminates substitution or sphere error because the geometry of the sphere is identical in measuring both the blank and the sample. The comparison method is more time-consuming than the substitution method, however, since the blank must be rescanned with each new sample.

Unlike FT-IR instruments which all collect spectra in fundamentally the same way, UV-vis–NIR instruments can work in very different ways. The discussion so far has been concerned with single-beam instruments. Most UV-vis–NIR spectrometers are now double-beam, dual-beam, or diode array instruments. An integrating sphere accessory for a double-beam spectrophotometer is shown in Fig. 5.7. The chopped signal from the source is separated into sample and reference beams. The sample and reference are measured concurrently, thus allowing for a comparison measurement without sacrificing analysis time. An integrating sphere accessory for a dual beam spectrophotometer requires a substitution measurement. Instead of a

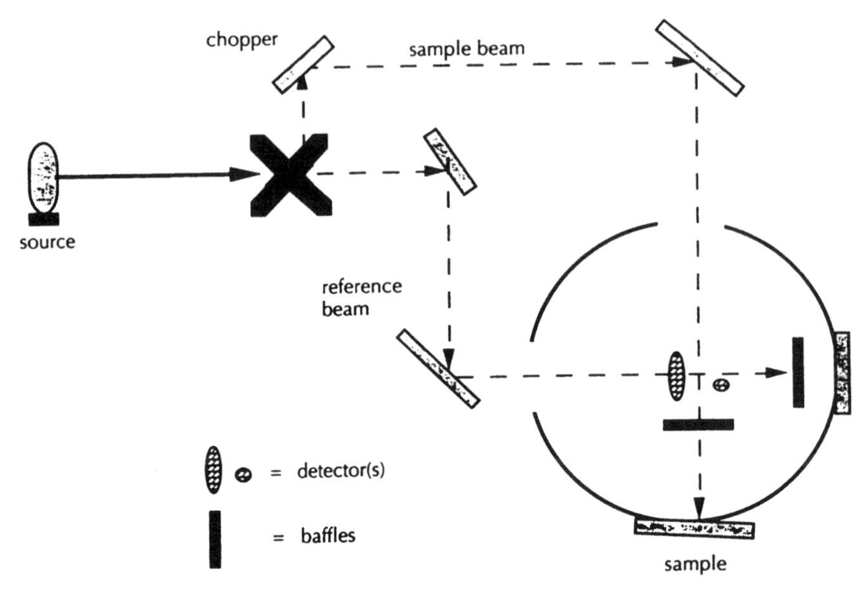

Fig. 5.7 Interesting sphere for a double-beam spectrophotometer. Reprinted with permission from [16].

mechanical chopper the beam is split using a beam splitter. Half the beam is sent toward the original instrument detector, while the other half is sent to the integrating sphere set up for a substitution measurement. Diode array spectrometers have become increasingly popular because of the speed in which a spectrum can be taken. A reflectance spectrum using a diode array spectrometer contains the source inside the sphere to produce polychromatic illumination (Fig. 5.8). Collection of a diffuse reflectance spectrum using a diode array instrument, like a dual-beam instrument, requires a substitution measurement.

The analysis of powdered samples using the diffuse reflectance technique is easiest. Sample holders are usually either a Teflon or aluminum block at least 5 mm deep to meet the "infinitely thick" criterion discussed in the theory section. A quartz window can be used for UV → NIR measurements, although a less-expensive glass window is fine if UV measurements need not be done.

The collection of diffuse reflectance spectra for qualitative information, such as color matching by visible spectrometry, is straightforward. Place the powder in the sample holder and collect the spectrum using either the substitution or comparison method, whichever is appropriate for the instru-

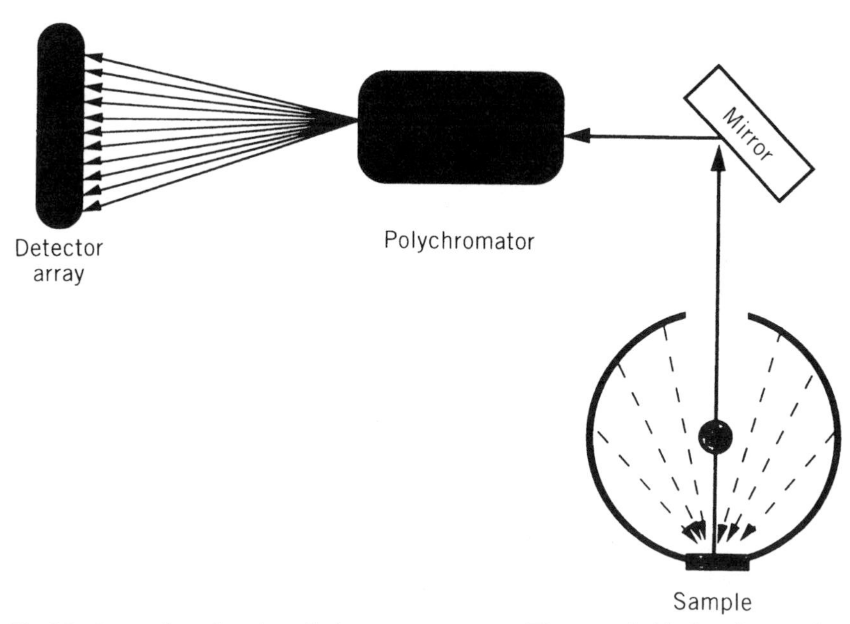

Fig. 5.8 Integrating sphere for a diode array spectrometer. The source inside the sphere results in diffuse illumination of the sample. Reprinted with permission from [16].

ment type. The only concern should be the possibility of spectral artifacts for intensely absorbing samples, as was discussed in the previous section. Dilution of the powdered sample in a nonabsorbing matrix such as MgO or $BaSO_4$ may be necessary in extreme cases. Loosely packed powders can also be measured in a cuvette, but the reference should be measured in a matched cuvette. Another alternative is to press the powder into a pellet, but take care to avoid a glossy surface appearance which suggests specular reflectance.

Using integrating spheres to obtain quantitatively valid diffuse reflectance spectra requires more care. Substitution error is only one source of error associated with the use of integrating spheres. There are further factors to discuss, and for a full consideration, see Clarke and Compton [12].

From the theory section, it can be surmised that since the packing density affects the scattering coefficient, a constant packing density must be attained. The nature of the particle surface also has an effect on the reflectance spectrum. If possible, the surface should be smooth and even but free of gloss. Particle size is also very important since this parameter also affects the scattering coefficient. A grinding device such as a Wig-L-Bug should be used to grind the sample. Very fine particles agglomerate, so a stationary state is

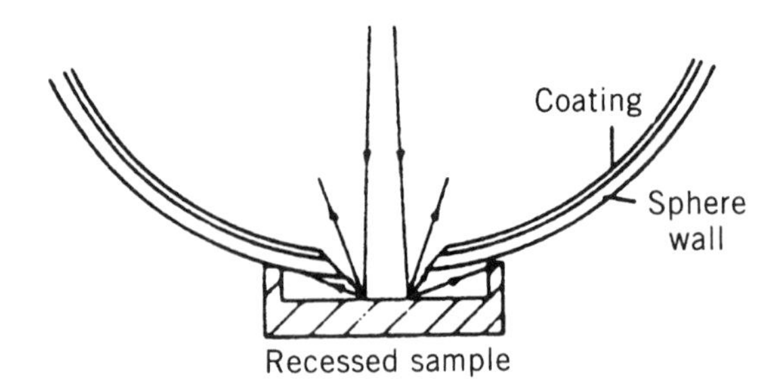

Fig. 5.9 Sample recess error. Some diffusely reflected radiation from the sample does not reenter the sphere if the sample is not flush with the sphere.

reached in the $0.2 \rightarrow 2\,\mu m$ range. Sample recess error is less obvious. Different distances of the sample surface from the sphere wall causes difference in reflectance. Some diffusely reflected radiation will not enter the sphere if the sample is set back from the sphere wall (Fig 5.9). If the sample and reference are recessed by the same amount, the resulting error will be much smaller. Sample preparation, sphere geometry, sphere alignment, and sample alignment all have an effect on diffuse reflectance spectra. Quantitative diffuse reflectance work using integrating spheres in the UV-vis–NIR, or biconical devices in the mid-infrared (to be discussed), should not be done in a cavalier fashion. Despite these concerns, quantitative work has and will continue to be done. Quantitative diffuse reflectance work in the NIR abounds, as will be discussed.

Diffuse reflectance spectra of samples other than powders can also be obtained. Spectra of solid, nonglossy materials can be easily obtained by placing the sample flush with the sample port. Spectra of samples that are smaller than the sample port should be collected using a masking technique. The masked sample, and an identically masked reference, should fill the sample port. The usually black mask is fabricated such that stray light does not enter the sphere from the outside, and reflectance from the mask within the sphere does not add to the reflectance of the sample. An effective mask can be easily prepared from a manila file folder painted with flat black paint cut such that it is large enough to fill the sample port. A hole is cut out of the mask so that the sample and reference can be mounted behind the mask for spectral acquisition.

Spectra of nontraditional samples such as fabrics can also be obtained. The fabric sample is stretched to introduce a flat, nonwrinkled surface to the sphere port. The sample must be backed by a light trap (similar in function to a mask) to ensure that external light does not enter the sphere to affect the spectrum. Since spectral acquisition of so many different sample types can be obtained, procedures are commonly devised as they are needed.

Mid-Infrared Diffuse Reflectance Spectroscopy

Although an integrating sphere is the most commonly used device to collect diffuse reflectance spectra in the UV-visible–NIR regions, the integrating sphere presents several problems for mid-infrared work. First, coatings for the integrating sphere suitable to mid-infrared spectroscopy cannot be made with the same high reflectance characteristics as in the UV-visible region. Second, infrared detectors are orders of magnitude less sensitive than photomultiplier tubes used in UV-visible spectroscopy. Given these two reasons, and the fact that integrating spheres are intrinsically inefficient, the S/N ratio of mid-infrared diffuse reflectance spectra will be low. This is not to say that integrating spheres cannot be used in modern FT-IR spectrometers, but thus far their application has been limited. The fact remains that if an absolute reflectance measurement must be made, an integrating sphere is the accessory to use.

With the advent and widespread availability of Fourier transform infrared (FT-IR) spectrometers and high-sensitivity infrared (liquid nitrogen cooled mercury-cadmium-telluride or MCT) detectors, Fuller and Griffiths [13] show the feasibility of collecting practical mid-infrared diffuse reflectance spectra. The optical configuration that they have devised is no longer used in commercially available accessories, but their work has paved the way for this technique which is now almost commonplace. Diffuse reflectance accessories now available for the mid-infrared use ellipsoidal mirrors to focus and collect the infrared energy. Optical configurations for the two main types of commercially available accessories, typified by Spectra-Tech's Collector and Harrick's Praying Mantis are shown in Fig. 5.10. The main difference between these two types of accessories involves the reduction of specular reflection. The Spectra-Tech accessory has no reasonable way to reduce the specular component, which can be a problem with highly absorbing samples, as has been discussed. The Harrick-type design is superior in this respect. With these accessories, sample position or height with respect to the ellipsoidal mirrors is crucial for obtaining reliable spectra with high S/N ratio. In this regard the Spectra-Tech accessory is superior. If one is to obtain routine diffuse reflectance spectra, either type of accessory will afford the opportunity to obtain quality spectra.

Whatever the accessory you choose, proper alignment of the accessory in the FT-IR spectrometer is the single most critical factor. Follow the instructions that come with the accessory, but do not think it will be easy. Alignment is a painstaking process, nevertheless, it is absolutely necessary that it should be done properly. Many accessories collect dust inside cabinets because they are deemed useless as a result of improper alignment. The idea is to obtain the maximum amount of signal by maximizing the

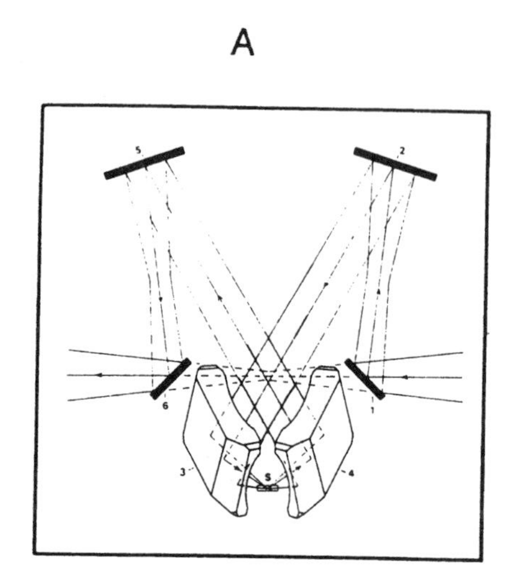

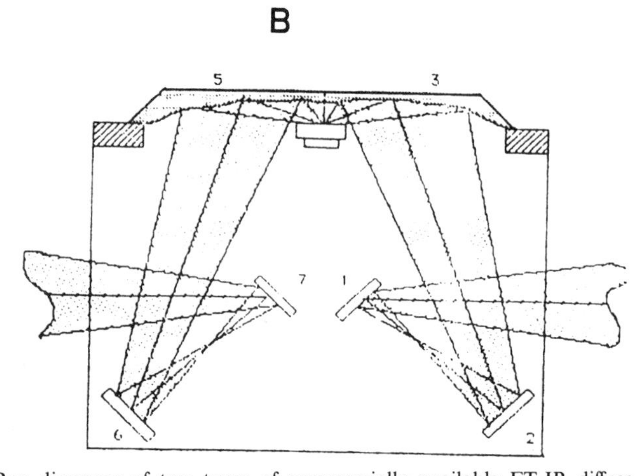

Fig. 5.10 Ray diagrams of two types of commercially available FT-IR diffuse reflectance accessories. (*a*) The accessory from Harrick uses an off-line optical geometry to eliminate specular Fresnel reflectance. (*b*) The accessory from Spectra-Tech uses an on-line optical geometry that collects the specular Fresnel component.

interferogram centerburst. This is usually done by monitoring the signal reaching the detector in the spectrometer's alignment mode while aligning the accessory. Be sure that any apertures present in the FT-IR spectrometer are fully open and that the detector gain is increased when beginning alignment procedures to enhance any potential signal.

Since most of the signal will, in any event, not reach the detector in diffuse reflectance spectroscopy, for best results it is important to have an FT-IR spectrometer that is properly configured. There are two primary considerations. First, the spectrometer should be equipped with a high-energy infrared source. Second, and probably more important, the spectrometer should be equipped with a high-sensitivity detector. Although a room temperature DTGS (deuterated triglycine sulfate) detector can be used, a liquid nitrogen cooled MCT detector is highly recommended.

It is not always necessary, but most diffuse reflectance mid-infrared work is performed by diluting the sample in a nonabsorbing matrix. It is a prudent practice to avoid spectral artifacts. This matrix is usually either powdered KBr or KCl. The particle size should be less than about 40 μm. Such small particles cannot be easily obtained by grinding in a mortar and pestle. The KCl should be first ground by hand in a mortar and pestle, followed by two minutes in a Wig-L-Bug grinder. This gives a particle size of about 5–10 μm. The KCl should then be dried in an oven for a few days and stored in a desiccator for use. Adsorbed water can have significant adverse effects on diffuse reflectance spectra.

Once the FT-IR is properly configured, the diffuse reflectance optical accessory is properly aligned, and the nonabsorbing matrix is prepared and dry, a background spectrum of the pure nonabsorbing matrix can be obtained. The sample cup should be filled with KCl and leveled off. For quantitative results if the KCl is to be packed, the sample spectrum must also be packed with the same force and duration. The sample height is then adjusted to maximize the signal at the detector, and the spectrum is ready to be obtained. Sample height is very important; changes in sample height of 0.1 mm or less from the optimum can cause significant distortions in the spectra [14].

Now that the background spectrum has been collected, it is time to collect the sample spectrum. Usually anywhere from a 2–15% by weight dispersion of the sample in KCl is adequate. The sample is mixed with KCl in a Wig-L-Bug for about one minute to provide a good dispersion. If grinding the sample is to be avoided, use the Wig-L-Bug without the grinding ball. The sample cup is then filled and leveled, the sample height is optimized as with the KCl background, and the spectrum is obtained. If qualitative or functional group information is all that is required, there is no need to convert the spectrum to Kubelka-Munk units. Simply analyze the spectrum as percent reflectance (analogous to percent transmittance) as a function of wavenumber.

To maintain quality spectra over time, one should realign the optical accessory weekly. This assumes the accessory has not been moved and no significant adjustments to the FT-IR have been made. New background spectra should be obtained daily in order to obtain the best possible spectra.

The aforementioned procedures represent what is deemed to be the best way to obtain diffuse reflectance mid-infrared spectra. This is not to preclude a hurried analyst from simply taking a powdered sample, placing it in the sample cup, optimizing sample height, and obtaining a spectrum (ratioed to pure KCl). If, however, the spectrum exhibits broad peaks, appears undifferentiated and/or distorted, diluting the sample in KCl and decreasing the sample's particle size will almost certainly improve the quality of the resulting spectrum.

5.4 APPLICATIONS OF DIFFUSE REFLECTANCE SPECTROSCOPY

Diffuse Reflectance UV-Visible Spectroscopy

Diffuse reflectance UV-visible spectroscopy is mostly used in the applied sciences for the analysis of dyestuffs, printing inks, paints and pigments, paper, ceramics, and so on. For color matching/analysis, visible diffuse reflectance spectroscopy is used for denture materials, tiles, cosmetics, and medicinal tablets, to name a few applications. Most of these applications do not reach the traditional chemical or spectroscopic literature for a variety of reasons. Much of this work is proprietary, and even if it is not proprietary it is industrially driven, so the desire or need to publish is not present. Also much of this work is not considered cutting edge research, since these applications have been around for many years. Even when the work is of excellent quality, it may be hard to publish. An excellent but older review of UV-visible diffuse reflectance applations is given by Wendlandt and Hecht [15]. A more recent general review of applications does not exist. However companies such as Labsphere, which are in the business, put out, useful literature [16] and are happy to offer assistance to someone with an interest in the technique. Although some capabilities and methods have changed, many applications have not. A few applications worth mentioning specifically follow.

Schmidt and Heitner [17] review the use of UV-visible diffuse reflectance spectroscopy for chromophore research on wood fibers. These workers investigated the chemistry of light-induced yellowing and reduction. UV-visible diffuse reflectance spectroscopy is also useful in applications where transition metal oxidation state is important, such as studies of metal oxidation and heterogeneous catalysis. One recent catalysis study [18] involved an *in situ* UV diffuse reflectance study of silica supported noble metal catalysts. The stainless steel cell, with gas inlet and outlet ports, is capable of temperatures as high as 500°C. The integrating sphere is externally interfaced with the spectrometer through fiber optic cables, so the measurement is made remote to the spectrometer. The integrating sphere is directly attached to the quartz window of the reactor. The use of fiber optics for remote measurements in the UV-visible regions is now a practical option. For applications in which it is not feasible or wise to place the spectrometer where the sample is, fiber optics allows these measurements to be performed remotely.

Diffuse Reflectance Near-Infrared Spectroscopy

To say that NIR spectroscopic applications have in the past 10 years exploded is almost an understatement. A review article on NIR reflectance analysis (really diffuse reflectance, although not usually named as such in NIR parlance) in 1983 [19] called it a "Sleeper among Spectroscopic Techniques." Approximately 10 years later, an updated review article on NIR spectroscopy [20] proclaimed that "The Giant Is Running Strong." NIR reflectance spectroscopy has become an accepted tool for monitoring many industrial processes because it is fast, involves little sample preparation, is nondestructive, and can perform multicomponent analyses.

Absorption of radiation in the near-infrared region results predominantly from vibrations of light atoms with strong bonds. Primarily N–H, O–H, and C–H bonds are probed by NIR spectroscopy. This limits the chemical structures observed to mostly organic compounds. Since near-infrared absorption consists of overtones and combinations of fundamental (mid-infrared) absorptions, unique functional group type structural information is obscured. This fact makes it impossible to visually observe a spectrum to distinguish structural features. This has not deterred NIR spectroscopists because statistical methods allow one to perform quantitative analyses of extraordinarily complex samples with astounding success.

The NIR literature abounds with various statistical analyses of data sets such as partial least squares, principal components analysis, multiple linear regression, and neural networks. A sampling of a very small portion of this literature is provided [21–23]. The presumed problem with NIR reflectance analysis is that it relies so heavily on chemometrics. A spectroscopist cannot directly interpret an NIR reflectance spectrum. The quantitative data generated (which is pretty much what NIR spectroscopy is used for) comes from an algorithm that to most is a black box. This can lead to a feeling of a lack of control over the analysis, but it works! It should be noted that the ultimate success of a quantitative NIR reflectance analysis is dependent on intricate calibration procedures involving a training sample set that must be chosen and analyzed very carefully. The data generated are limited by the quality of the calibration. The interpretation of NIR spectra has advanced recently because of yet another sophisticated statistical correlation method described by Barton et al. [24].

A primary reason for the applicability of NIR reflectance spectroscopy is that since overtone and combination bands have extremely low absorptivities, this is the ideal region for diffuse reflectance applications. Recall from (5.1) that as absorptivity increases so does Fresnel or specular reflectance, the bane of diffuse reflectance spectroscopy. In the mid-IR, and to a lesser extent in the UV-visible, specular reflectance is reduced by dilution of the sample in a nonabsorbing matrix. Due to the low absorptivities in the NIR, specular reflectance is not a problem. This simplifies sample preparation requirements considerably. On the negative side, however, low absorptivities make NIR diffuse reflectance analysis a poor tool for trace analysis. If components are present in a sample at levels below 1%, NIR becomes a less than ideal tool.

NIR spectroscopy is dominated by necessarily pragmatic industrial scientists. Therefore the practice of NIR diffuse reflectance spectroscopy is for the most part dissimilar to work in the other spectral regions. For instance, since the seminal work of Norris [25], quantitative diffuse reflectance measurements in the NIR has become routine for major components in agricultural commodities. Almost all NIR diffuse reflectance spectra are converted to log $1/R$ (apparent absorbance) before statistical analysis for quantitative data extraction. This implies that Beer's law is valid for NIR diffuse reflectance spectroscopy. Since Beer's law assumes that reflectance and scattering of radiation by the sample is insignificant, this is surprising. Surely many practitioners of quantitative NIR diffuse reflectance spectroscopy recognize this contradiction, but the use of log $1/R$ values for quantitative data interpretation works. Olinger and Griffiths [26], a full 20 years after the widespread use of this technique, attempt to explain this apparent contradiction.

Although the fact that apparent absorbance works in this application is of academic interest, it does not address the real reason for using log $1/R$ values. The application of NIR for quantitative work, as has been discussed, requires the use of a complicated chemometric model to extract spectroscopic information. The generation of a statistical model is difficult, costly, and time-consuming. For NIR reflectance spectroscopy to be of practical use, recalibration of the instrument should be rarely, if ever, required. The mathematical model should be as robust as possible. The use of the log $1/R$ function minimizes the effect of most of the inevitable changes in the instrument over time on the chemometric model.

Another significant difference between NIR diffuse reflectance spectroscopy and the technique in other spectral regions is how the scattered radiation is collected. Traditional laboratory researchers in NIR reflectance spectroscopy, as has been previously discussed, often use integrating spheres to collect spectra. Since this is a highly applied field, however, ease of sample analysis is very important. In the purchase of an NIR reflectance spectrometer for a typical quantitative application, there is no integrating sphere to be found. A holder is provided that you simply fill with sample. The sample scatters the radiation, and the detector collects the scattered radiation at some set of angles determined by the geometry of the spectrometer. Many NIR diffuse reflectance fiber optics applications work on the same principle.

The sheer number and variety of NIR applications published each year is astounding. More than 500 articles a year are now published ranging from agricultural to pharmaceutical to imaging applications. This is not even counting all of the proprietary work that never gets published. Although not all of these involve diffuse reflectance, many NIR applications use this phenomenon. Just a few such applications will be discussed.

One source containing a wealth of recent work in NIR spectroscopy is a book edited by Murray and Cowe [27]. One can find instrumental advances in fiber optics, studies on recent statistical approaches to analyze data sets and NIR reflectance studies of plants, agricultural commodities, polymers, textiles, and so on.

An interesting example of NIR reflectance spectroscopy, described by Lodder et al. [28], involves the direct analysis of single pharmaceutical tablets. A specially designed sample holder was described to enable rapid and convenient measurements of capsules. The sample holder fits into the sample compartment of a commercial NIR spectrometer.

A couple of interesting biomedical applications have been recently published. Tamura [29] used fiber optic bundles to measure absorption changes of rat brains *in vitro*. Marbach and Heise [30] describe a diffuse reflectance accessory for *in vivo* measurements of human skin tissue. These workers demonstrated the ability to perform noninvasive measurements of blood glucose level. Another accessory, not necessarily for biomedical work, is described by Hogue [31]. This accessory allows samples to be mounted horizontally rather than vertically as in an integrating sphere, which is useful for certain applications.

Sample preparation for NIR diffuse reflectance spectroscopy involves just a couple of considerations. Sample particle size and packing density should remain constant. This is because both parameters affect the scattering properties of the sample. One way to keep particle size constant is to grind the sample in a Wig-L-Bug grinder. Many samples, such as agricultural commodities like wheats, should be ground for this reason. Another good reason to grind complex samples such as these is for homogenization. Many complex samples are inherently heterogeneous, so depending on where the sample beam hits the specimen, one may obtain different spectra. For quantitative analysis this state of affairs is not recommended. A relatively new field that has begun to find its way to NIR reflectance spectroscopy is spectral imaging. Imaging spectroscopy determines the distribution of constituents in a heterogeneous sample. Robert et al. [32] describe an NIR reflectance spectral imaging system to determine the components of wheat. NIR reflectance spectroscopy is a broad and currently very active field.

Diffuse Reflectance Mid-Infrared Spectroscopy

General Considerations

It took a few years, but the work in 1978 by Fuller and Griffiths [13] has resulted in an explosion of applications in diffuse reflectance mid-infrared spectroscopy, many of which were predicted in their original publication. The acronym for this technique, coined by Griffiths, is *D*iffuse *R*eflectance *I*nfrared *F*ourier *T*ransform *S*pectroscopy, or DRIFTS. While this is arguably an unfortunate acronym, the technique has proved to be widely applicable. The wealth of direct spectroscopic information inherent in a mid-infrared spectrum, which results mainly from fundamental vibrational modes, made the promise of extending this already proven technique to mid-infrared spectroscopy highly promising. The technique has come to fulfill many of these expectations, and in this section a smattering of applications of DRIFTS will be discussed.

Undoubtedly the most widely used application of DRIFTS, not to be found in the literature, is the rapid identification of powders with little or no sample preparation. If a lot of sample is available, the option exists to simply fill the sample cup and obtain the spectrum. This is not recommended, however; it is worth taking the extra two minutes to make a dispersion in KCl to collect the spectrum properly. If only a small amount of sample is available, it is necessary to prepare the sample as a dispersion in KCl. An important but little recognized fact is that very little sample is required to obtain a spectrum of high S/N ratio. Most samples will yield a quality spectrum with just 0.1 mg of sample, and detection limits of less than 10 ng of sample dispersed in KCl have been observed [33].

These astonishingly low detection limits, even compared to transmission spectroscopy, are observed because of the phenomena that govern transmission and diffuse reflectance spectroscopy. In transmission spectroscopy the S/N ratio of a spectrum is directly proportional to sample concentration, which can be shown as a consequence of Beer's law. In diffuse reflectance spectroscopy the S/N ratio is proportional to the square root of the sample concentration. So if the FT-IR and accessory are configured such that an adequate signal is reaching the detector, which is easily attainable, DRIFTS is actually a much more sensitive technique than transmission spectroscopy.

DRIFTS as a Chromatographic Detector

Although it is advantageous to dilute a sample in KCl before obtaining a DRIFT spectrum, in some cases this may not be desirable. One such example is the use of DRIFTS as a detector for species separated by thin-layer chromatography (TLC-IR). While this approach has met with some success as demonstrated by a review article on the subject [34], the strong absorption bands from silica below $\sim 1500\,cm^{-1}$ results in considerable spectral distortion. Many methods have been tried to eliminate the strong absorbance from silica. Most of these methods involve analyte transfer from the TLC plate [35, 36] which, of course, requires significant sample preparation. Danielson et al. [37] have gone so far as to perform TLC on a zirconium oxide stationary phase. These authors' reasoning is that since zirconium oxide does not exhibit intense absorptions until $\sim 1100\,cm^{-1}$, the useful spectroscopic range could be extended without the need for sample transfer. It is probably safe to predict that work will continue in this area before DRIFTS detection of separated species on TLC plates is widely applicable.

DRIFTS in Surface Science and Heterogeneous Catalysis

For a variety of reasons, DRIFTS is becoming the single most important technique for studying the surface chemistry of high surface area powders. A diffuse reflectance spectrum is much more sensitive to surface species than the bulk because the beam is reflected off the surface of particles. This is unlike transmission spectroscopy where a beam passes through the entire sample. Because a scattering sample is mandatory, spectra show no energy loss at high wavenumbers characteristic of transmittance spectra of samples like silica or alumina. Pressing pellets of pure silica or alumina can be very difficult, along with some drawbacks inherent in pellet pressing of these types of samples [2, 3]. Also, in catalytic studies, making a KBr pellet can inhibit adsorption of reactants onto catalyst active sites.

Figure 5.11 illustrates the surface selectivity afforded by DRIFTS as compared to transmission spectroscopy. These spectra are of organosilane-modified Cab-O-Sil^{R} fumed silica. The transmission spectrum yields no useful spectroscopic information below $\sim 1400\,cm^{-1}$, whereas the DRIFTS spectrum exhibits peaks to $400\,cm^{-1}$ which is the detector cutoff. The surface selectivity of the DRIFTS technique has allowed the investigation of surface reactions on silica in the $1000\,cm^{-1}$ region underneath the strong silica matrix absorption using spectral subtraction [38]. Many applications of DRIFTS to study surface reactions and interactions both qualitatively [39, 40] and quantitatively [41] have been reported.

The idea that DRIFTS is a highly sensitive technique has been shown to be true in surface analysis as well as in conventional microsampling. Van Every et al. [42] investigated the detection limit of CO adsorbed on a heterogeneous catalyst. These workers showed that adsorbed CO on a 3% Rh/Al_2O_3 catalyst can be detected at 10^{-6} of a monolayer. This detection limit is comparable to that of some ultrahigh vacuum (UHV) methods often

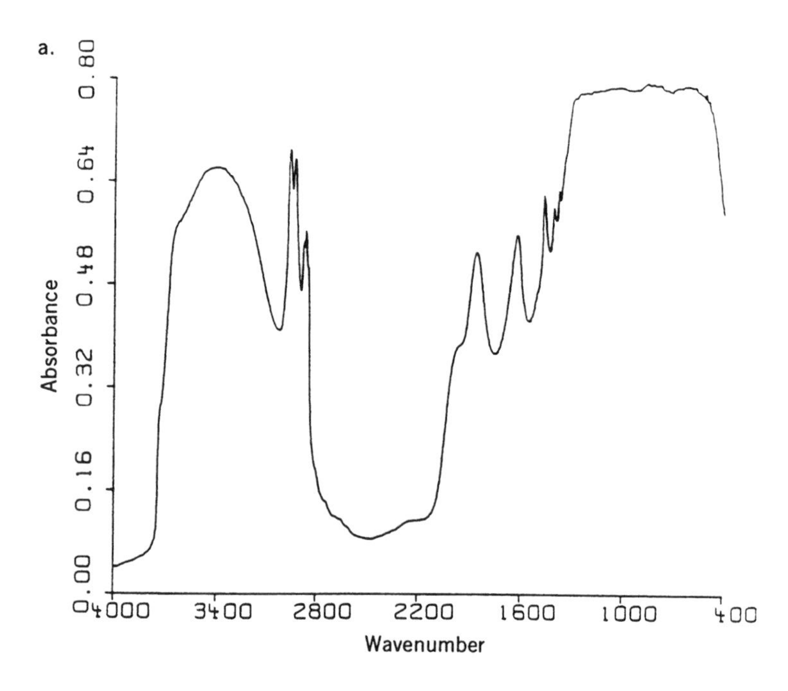

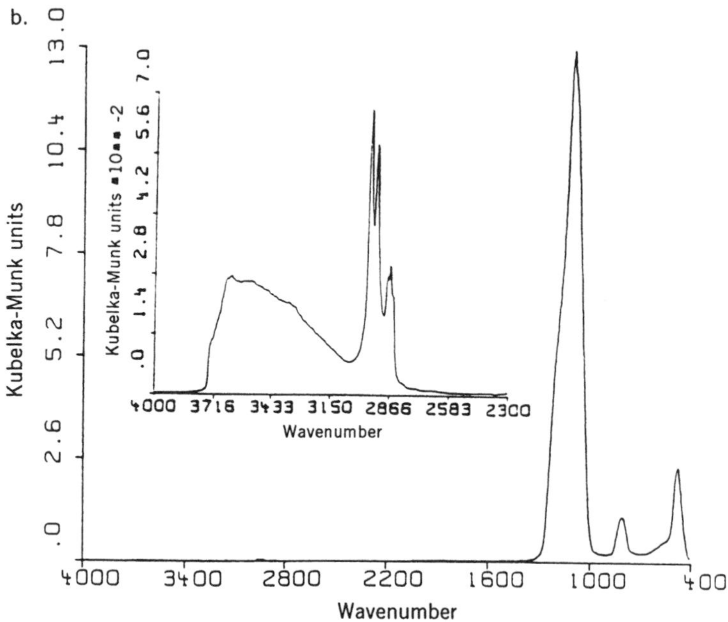

Fig. 5.11 Infrared spectra of a modified silica. (*a*) Transmission spectrum, and (*b*) diffuse reflectance spectrum.

used in heterogeneous catalyst characterization without the UHV constraints.

When using DRIFTS for catalytic or degradations studies, it is necessary to perform variable temperature/controlled atmosphere work. These studies require special cells for this work, and extra precautions must be taken to avoid spectral artifacts. These variable temperature/controlled atmosphere cells can be obtained commercially from the same vendors who cell the optical accessories. Since vendors' accessories differ, variable temperature cells are generally not transferrable between different companies' optical accessories.

Murthy et al. [14] were the first to report the fact that sample position varies as a function of temperature in variable temperature DRIFTS work. These workers found that thermal expansion of the sample post as temperature is raised causes significant spectral changes. Thus precise control of sample height is a necessity in performing careful variable temperature DRIFTS work. Venter and Vannice [43] subsequently provided some excellent suggestions for modifying DRIFTS accessories for variable temperature/controlled atmosphere work.

At least one additional factor must be considered when performing variable temperature DRIFTS work. At high temperatures the sample itself becomes an infrared emitter [44] which can saturate the signal reaching the detector. MCT detectors (unfortunately but nor surprisingly) are more susceptible to this effect than DTGS detectors [45]. The result of this is that when one performs variable temperature work, it is important that the background spectrum be obtained under identical conditions to the sample spectrum. The spectra should be collected when the sample is at the optimum sample position and at the same temperature to avoid potentially severe spectral artifacts.

In performing catalytic or degradation studies using variable temperature DRIFTS, some workers have found that interfacing the DRIFTS accessory to a mass spectrometer can be very useful. This allows one to simultaneously monitor reactions occurring at the surface of the particle and desorption reaction products. One such apparatus is detailed by White [46] to study polymer degradation. Augustine et al. [47] used DRIFTS/MS to study surface reactions and kinetics of a heterogeneous catalyst for vinyl acetate synthesis. An excellent DRIFTS/MS system for catalyst research has also been described by Highfield and coworkers [48]. One DRIFTS/MS apparatus is schematically shown in Fig. 5.12.

Miscellaneous Applications of DRIFTS

The literature is too diverse to provide a comprehensive review of DRIFTS applications. In this final section a few more applications will be briefly discussed, and the reader can refer to the references provided herein.

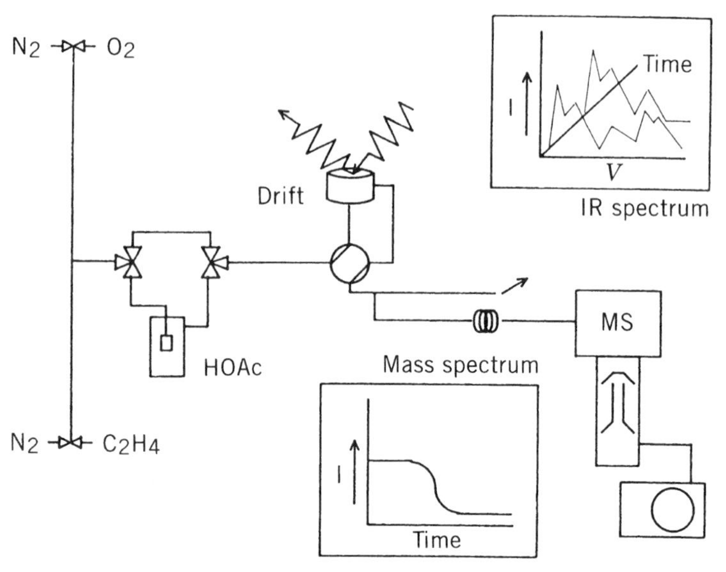

Fig. 5.12 Schematic of an FTIR diffuse reflectance accessory interfaced to a mass spectrometer. The DRIFTS cell, filled with catalyst, is used as a reactor, while the mass spectrometer detects vapor phase reaction products. Reprinted with permission from *Appl. Spectrosc.* 45(1991): 1746.

Polymer studies (powder, composites, foams, fibers) have been particularly amenable to DRIFTS studies. Recently Mitchell published an excellent review on this subject [49] which is highly recommended for anyone new to the field of DRIFTS. Nguyen et al. [50] published a fairly comprehensive treatise on the use of DRIFTS in soil studies. These workers discuss both advantages and disadvantages of the DRIFTS technique for characterizing a range of soil samples. They found that DRIFTS is most useful for the determination of clay mineralogy and organic matter. Suzuki et al. have published a series of papers concerning the application of DRIFTS to forensic science. Their first paper [51] discusses the nuts and bolts of applying DRIFTS to these types of samples. Recently Gilber et al. [52] discuss the application of DRIFTS and chemometrics to distinguish different cellulosic fabric types and the effects of sequential stages of the textile process.

5.5 CONCLUSIONS

Diffuse reflectance spectroscopy is a technique that is widely applied from the UV to the mid-infrared spectral regions. Whereas diffuse reflectance spectroscopy is most readily applicable to powders, any sample that scatters radiation can be characterized using this technique. To obtain reliable results, one should have some understanding of the fundamentals. The need to dilute samples to minimize specular reflectance, especially in the mid-infrared, should be understood. Each spectral region contains inherent information and also different considerations in performing diffuse reflectance work. The UV-visible and NIR regions primarily utilize integrating spheres to collect diffuse reflectance spectra. In the mid-infrared, biconical optical devices are used. The mid-infrared region contains the most specific chemical information, but in many ways it is also the most demanding region to perform diffuse reflectance work. Many applied studies have been carried out in all of the spectral regions, with the NIR and mid-IR exhibiting the most growth in the past 10 years.

ACKNOWLEDGEMENT

Thanks to Dr. Art Springsteen for generously providing valuable materials on integrating sphere technology.

REFERENCES

1. G. Kortüm, *Reflectance Spectroscopy*, Springer-Verlag, Germany, 1969.
2. P. W. Yang and H. L. Casal, *Appl. Spectrosc.*, **40**, 1070 (1986).
3. J. P. Blitz, R. S. S. Murthy, and D. E. Leyden, *Appl. Spectrosc.*, **40**, 829 (1986).
4. J. W. Childers, R. Rohl, and R. A. Palmer, *Anal. Chem.*, **58**, 2629 (1986).
5. J. W. Childers and R. A. Palmer, *Am. Lab.*, **3**, 22–28 (1986).
6. P. J. Brimmer, P. R. Griffiths, and N. J. Harrick, *Appl. Spectrosc.*, **40**, 258 (1986).
7. P. J. Brimmer and P. R. Griffiths, *Appl. Spectrosc.*, **41**, 791 (1987).
8. P. J. Brimmer and P. R. Griffiths, *Appl. Spectrosc.*, **42**, 242 (1988).
9. More rigorously, every scattering event in the sample must scatter the radiation isotropically. The requirement is not possible to achieve given the refraction/refraction/diffraction mechanisms proposed for diffuse reflectance.
10. B. Felder, *Helv. Chim. Acta*, **47**, 488 (1964).
11. G. H. C. Freeman, In *Making Light Work: Advances in Near Infrared Spectroscopy*, I. Murray and I. H. Cowe, eds. VCH, New York, 1992, p. 115.
12. F. J. J. Clarke and J. A. Compton, *Colour Res. Appl.*, **11**, 253, 1986.
13. M. P. Fuller and P. R. Griffiths, *Anal. Chem.*, **50**, 1906 (1978).
14. R. S. S. Murthy, J. P. Blitz, and D. E. Leyden, *Anal. Chem.*, **58**, 3167 (1986).
15. W. W. Wendlandt and H. G. Hecht, *Reflectance Spectroscopy*. Wiley, New York, 1966.

16. A. Springsteen, *A Guide to Reflectance Spectroscopy.* Labsphere Tech Guide, North Sutton, NH, 1994.

17. J. A. Schmidt and C. Heitner, *Tappi J.*, **76**, 117 (1993).

18. W. Zou and R. D. Gonzalez, *J. Catal.*, **133**, 202 (1992).

19. D. L. Wetzel, *Anal. Chem.*, **55**, 1165A (1983).

20. W. F. McClure, *Anal. Chem.*, **66**, 43A (1994).

21. P. J. Gemperline, L. D. Webber, and F. O. Cox, *Anal. Chem.*, **61**, 138 (1989).

22. H. Mark, *Anal. Chem.*, **58**, 2814 (1986).

23. H. Mark, *Anal. Chem.*, **59**, 790 (1987).

24. F. E. Barton, D. S. Himmelsbach, J. J. Duckworth, and M. J. Smith, *Appl. Spectrosc.*, **46**, 420 (1992).

25. I. Ben-Gara and K. H. Norris, *Israel J. Agr. Res.*, **18**, 125 (1968).

26. J. M. Olinger and P. R. Griffiths, *Anal. Chem.*, **60**, 2427 (1988).

27. I. Murray and I. A. Cowe, eds., *Making Light Work: Advances in Near Infrared Spectroscopy.* VCH, New York, 1992.

28. R. A. Lodder, M. Selby, and G. M. Hieftje, *Anal. Chem.*, **59**, 1921 (1987).

29. M. Tamura, *Jpn. Circ. J.*, **56**, 366 (1992).

30. R. Marbach and H. M. Heise, *Appl. Optics*, **34**, 610 (1995).

31. R. Hogue, *Fresenius' J. Anal. Chem.*, **339**, 68 (1991).

32. P. Robert, D. Bertrand, M. F. Devaux, and A. Sire, *Anal. Chem.*, **64**, 664 (1992).

33. M. P. Fuller and P. R. Griffiths, *Appl. Spectrosc.*, **34**, 533 (1980).

34. P. R. Brown and B. T. Beauchemin, *J. Liq. Chromatogr.*, **11**, 1001 (1988).

35. J. M. Chalmers, M. W. Mackenzie, J. L. Sharp, and R. N. Ibbett, *Anal. Chem.*, **59**, 415 (1987).

36. K. M. Shafer, P. R. Griffiths, and W. Shu-Quin, *Anal. Chem.*, **58**, 2708 (1986).

37. N. D. Danielson, J. E. Katon, S. P. Bouffard, and Z. Zhu, *Anal. Chem.*, **64**, 2183 (1992).

38. J. P. Blitz, *Colloids Surfaces*, **63**, 11 (1992).

39. R. W. Snyder, *Appl. Spectrosc.*, **41**, 460 (1987).

40. R. W. Fries and F. M. Mirabella, In *Transition Metal Catalyzed Polymerizations*, R. P. Quirk, ed. Cambridge University Press, Cambridge, 1988, pp. 314–326.

41. R. S. S. Murthy and D. E. Leyden, *Anal. Chem.*, **58**, 1228 (1986).

42. K. W. Van Every and P. R. Griffiths, *Appl. Spectrosc.*, **45**, 347 (1991).

43. J. J. Venter and M. A. Vannice, *Appl. Spectrosc.*, **42**, 1096 (1988).

44. I. M. Hamadeh, D. King, and P. R. Griffiths, *J. Catal.*, **88**, 264 (1984).

45. R. Lin and R. L. White, *Anal. Chem.*, **66**, 2976 (1994).

46. R. L. White, *J. Anal. Appl. Pyr.*, **18**, 325 (1991).

47. S. M. Augustine and J. P. Blitz, *J. Catal.*, **142**, 312 (1993).

48. J. G. Highfield, J. Prairie, and A. Renken, *Catalysis Today*, **9**, 39 (1991).

49. M. B. Mitchell, In *Structure-Property Relations in Polymers,* Advances in Chemistry Series 236, M. W. Urban and C. D. Craver, eds. ACS, Washington DC, 1993.

50. T. T. Nguyen, L. J. Janik, and M. Raupach, *Aust. J. Soil Res.*, **29**, 49 (1991).

51. E. M. Suzuki and W. R. Gresham, *J. Forensic Sci.,* **31**, 931 (1986).

52. C. Gilbert, S. Kokot, and U. Meyer, *Appl. Spectrosc.*, **47**, 741 (1993).

6 Photoacoustic Spectroscopy

J. F. McClelland, S. J. Bajic, R. W. Jones, and L. M. Seaverson

6.1. INTRODUCTION

Photoacoustic spectroscopy (PAS) measures a sample's absorbance spectrum directly with a controllable sampling depth and with little or no sample preparation. This rapid direct analysis capability is applicable to nearly all samples encompassing a wide range of absorbance strengths and physical forms. Among the other key features of PAS are that it is nondestructive, noncontact, applicable to macrosamples and microsamples, insensitive to surface morphology. It has a spectral range from the UV to far-infrared; and is operable in photoacoustic absorbance, diffuse reflectance, and transmission modes; and is capable of measuring spectra of all types of solids (sheet, chunk, pellet, powder, semisolid) without exposure to air or moisture, as well as liquids and gases. This high level of versatility provided by PAS is the primary reason that it is used in industrial and other research laboratories.

Modern Techniques in Applied Molecular Spectroscopy, Edited by Francis M. Mirabella.
Techniques in Analytical Chemistry Series.
ISBN 0-471-12359-5

History and Basic Idea of PAS

Alexander Graham Bell discovered PAS in 1880 while he was investigating possible means for optical communications [1] as shown in Fig. 6.1. Bell observed that a beam of light focused on a sample produced an audible sound in the air around the sample if the beam was turned on and off at an acoustic frequency. He also found that strongly absorbing samples produced a louder sound or photoacoustic signal than weakly absorbing materials. The photoacoustic signal variation with sample absorbance was recognized by Bell as a way to measure absorbance spectra of materials, and he was

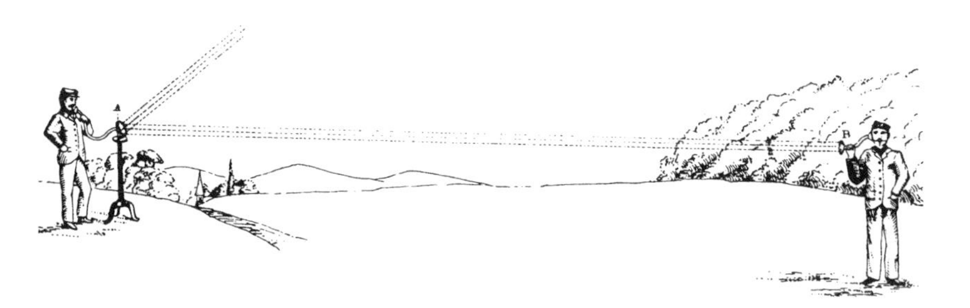

Fig. 6.1. In his 1881 publication [1] A. G. Bell shows a voice message being communicated by intensity modulated sunlight utilizing his recently discovered photoacoustic effect.

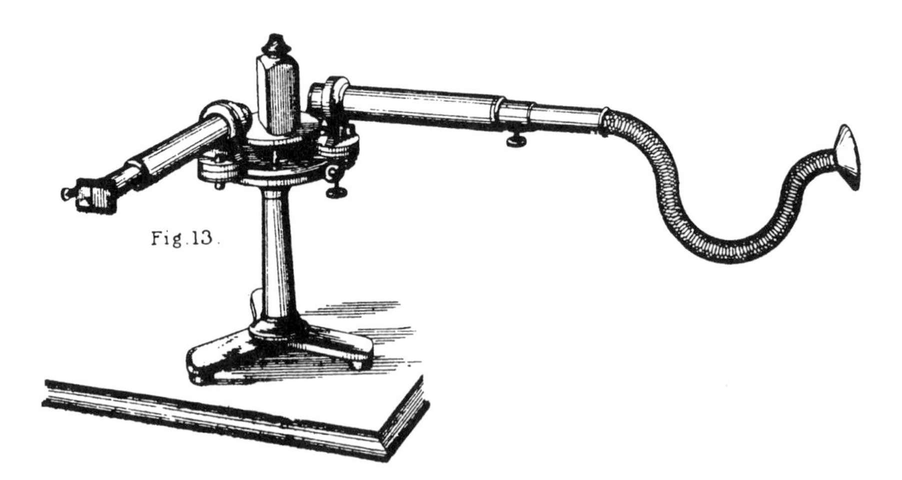

Fig. 6.2. Bell recognized the possibility of using the photoacoustic effect in spectroscopic applications and developed a prism spectrometer including a hearing tube to facilitate listening to the photoacoustic signal response form various samples. Reproduced from [1].

able to demonstrate a primitive form of PAS. His spectroscopic apparatus appears in Fig. 6.2. Practical exploitation of Bell's discovery was not initially achievable because further advances were needed in photoacoustic signal generation theory, in photoacoustic detector and spectrometer instrumentation, and in photoacoustic methods for specific analyses.

Since the 1980s PAS has become a valuable analytical technique due to the development of the Fourier transform infrared (FTIR) spectrometer, low-noise electronics, high-sensitivity microphones, computerized data handling, and methods for specific analyses. Research and development continues in all the areas mentioned. For example, new capabilities of instrumentation such as step-scan FTIR are enabling spectroscopists to use PAS to obtain results not previously possible. This chapter will discuss new advances since the publication of the authors' last chapter on the subject entitled "A Practical Guide to FTIR Photoacoustic Spectroscopy" [2].

Controllable Sampling Depth

The controllable sampling-depth feature of PAS is a result of the photoacoustic signal generation process, which involves absorption of light in the sample and production of heat followed by propagation of heat-generated thermal waves to the sample surface. Heat is then transferred into the adjacent gas, changing its pressure, which is sensed by a microphone as the photoacoustic signal. As the thermal waves propagate from the region where absorption occurred to the sample's light-irradiated surface, they decay rapidly, much like light does as it propagates through an absorbing medium. This decay process limits the depth beneath the sample surface from which signal generation occurs. Thus there is a mechanism for controlling sampling depth. Thermal waves have a decay coefficient that is proportional to the square root of the frequency at which the light is modulated. Consequently the sampling depth L is decreased when the modulation frequency f is increased. The sampling depth is given by

$$L = \left(\frac{D}{\pi f}\right)^{1/2}, \tag{6.1}$$

where D is the sample's thermal diffusivity [3]. The sampling depth in photoacoustic measurements is usually stated to be the length L, although due to the decay process there is not a sharp cutoff at a depth of L. For typical polymer materials where D is approximately $10^{-3}\ \mathrm{cm^2/s}$, L can be varied in FTIR photoacoustic measurements from a few micrometers to 100 or so micrometers by adjusting the frequency f.

Information regarding the depth at which light is absorbed in a sample can also be gained from the phase shift between the intensity modulation of the light beam that is incident on the sample and the photoacoustic signal response. The phase shift between excitation and response is due to the finite time required for thermal waves to propagate from the site within the sample where absorption occurs to the surface of the sample where heat is transferred to the detector's gas atmosphere where gas expansion and pressurization results.

Practical Uses of Controllable Sampling Depth

The controllable sampling-depth of PAS is the key feature of the technique and has a number of uses:

1. It allows spectra of totally opaque samples to be measured without the usual thinning or dilution common to transmission and diffuse reflectance spectroscopies. In PAS the effective sample thickness or optical path length for a sample is not determined by physical thickness but rather by the sampling depth given by equation (6.1). It can be adjusted to an appropriate value for the absorbance values to be measured by properly setting the modulation frequency produced by the FTIR spectrometer. Thus the laborious and error prone task of reproducible sample thinning or dilution required for quantitative analyses in transmission and diffuse reflectance spectroscopies is replaced by simply setting an FTIR mirror velocity parameter for most PAS analyses.
2. It facilitates analysis of layered samples or samples with concentration gradients. Both the magnitude and the phase of the photoacoustic signal can provide depth-distribution information. Signal magnitudes change with modulation frequency, in part because of the change in sampling depth given by equation (6.1). The magnitudes of bands from components having different depth distributions will therefore change with modulation frequency differently, so the comparison of spectra taken at various frequencies can provide qualitative depth distributions.

The phase shift approach is especially useful in determining the ordering and thickness of layered polymer materials. With this approach the ordering sequence of layers is determined by the relative phase shifts of specific absorbance bands associated with the different layers. Bands with larger phase shifts are assigned to deeper layers. The thickness of a layer on top of a substrate can be determined from the amount of phase shift caused to substrate bands due to the finite transit time for thermal waves transiting across the layer.

The various uses of controlled sampling depth and its limiting factors will be discussed in more detail in the theory and applications sections of the chapter.

6.2. SIGNAL GENERATION THEORY AND DATA ANALYSIS TREATMENTS

Photoacoustic signal generation theory has been based on various models since the discovery of PAS by Bell. The modern theory uses a model with elements from the works of Parker [4], Rosenscwaig and Gersho [3], and McDonald and Wetsel [5]. A one-dimensional model is shown in Fig. 6.3 with the light beam of average intensity I_0 assumed to oscillate in intensity at a modulation frequency f and to be incident normal to the sample's surface. After a reflection loss RI_0 at the sample's surface, the light beam intensity decays exponentially as it propagates into the sample at a rate according to the sample's optical absorption coefficient α. Heat is produced proportional to the amount of light absorption occurring in each layer of the sample. This process results in each light absorbing layer of the sample oscillating in temperature at the frequency f and becoming a source for thermal waves that propagate in the one-dimensional model back toward the light-irradiated surface of the sample. The thermal waves decay during propagation with a decay coefficient,

$$a_s = \left(\frac{\pi f}{D}\right)^{1/2}, \tag{6.2}$$

where D is the sample's thermal diffusivity [3]. Thermal waves from layers within one decay length given by $1/a_s = L$ of the irradiated sample surface are responsible for transferring most of the heat into the gas and for thermal expansion driven photoacoustic signal generation. L is usually stated to be the sampling depth of a photoacoustic measurement as described in the introduction, although some signal contribution comes from as deep as $2\pi L$.

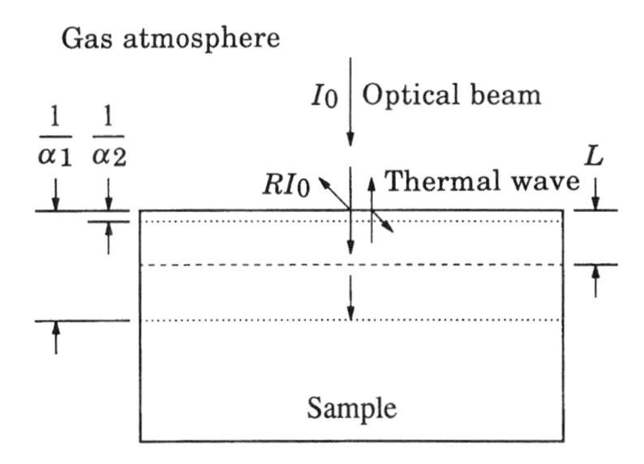

Fig. 6.3. One-dimensional model schematic of photoacoustic signal generation as discussed in the text.

When a thermal wave reaches the sample/gas interface, a small fraction of the thermal wave crosses the interface and causes a temperature oscillation in the gas, which generates the photoacoustic signal in the form of a gas pressure fluctuation. Unfortunately, a large fraction of the thermal wave is reflected back into the sample at the interface. Most samples are physically much thicker than L, and the reflected thermal wave decays to zero within the sample. The thermal-wave reflection loss is an unfortunate fact of the physics of PAS signal generation and is a costly loss in terms of the signal-to-noise ratio of PAS measurements. Consequently PAS detectors and spectrometers must be designed with care in order to optimize performance and to produce a signal-to-noise ratio that is acceptable relative to other sampling techniques. On the other hand, if the sample is much thinner than L, as with microsamples, the signal is enhanced by multiple thermal energy transfers as the thermal wave reflects back and forth from various faces of the sample.

Magnitude of the Photoacoustic Signal

The schematic diagram of Fig. 6.3 shows the one-dimensional signal-generation model, where L is the nominal sampling depth [3, 5]. As the beam penetrates the sample, its intensity decays exponentially, with a decay length equal to the reciprocal of the optical absorption coefficient α. If α increases, the signal magnitude increases because a larger fraction of the beam energy is deposited within a distance L of the sample surface. When $1/\alpha \gg L$ (as for α_1 in Fig. 6. 3), the signal increases linearly with α, as shown by the Thermal Only line in Fig. 6.4. As $1/\alpha$ approaches L, the linear relation is lost (onset of saturation) because most of the beam energy is now absorbed within depth L. Eventually, as for α_2 in Fig. 6.3, $1/\alpha \ll L$ and full saturation is achieved. The signal no longer increases after full saturation with increasing α because virtually all of the beam is absorbed well within L. Beyond the full saturation point, only the sample reflectivity R, which is related to α, affects the signal magnitude because the reflectivity affects the fraction of the incident beam that actually enters the sample.

Another signal generation mechanism, quite different from thermal waves, is observable from mid-range to low values of α. In this instance the photoacoustic signal has a thermal wave component, as discussed above, and a small acoustic component due to thermal expansion and contraction of the sample itself, excited by the temperature oscillations within the sample [5]. This acoustic contribution causes a small increase as shown in Fig. 6.4 in the thermally driven gas signal in the mid to low range of α, which is negligible at higher α.

Figure 6.4 also shows a third contribution from a background signal which is sometimes observed as α decreases. This contribution is due to

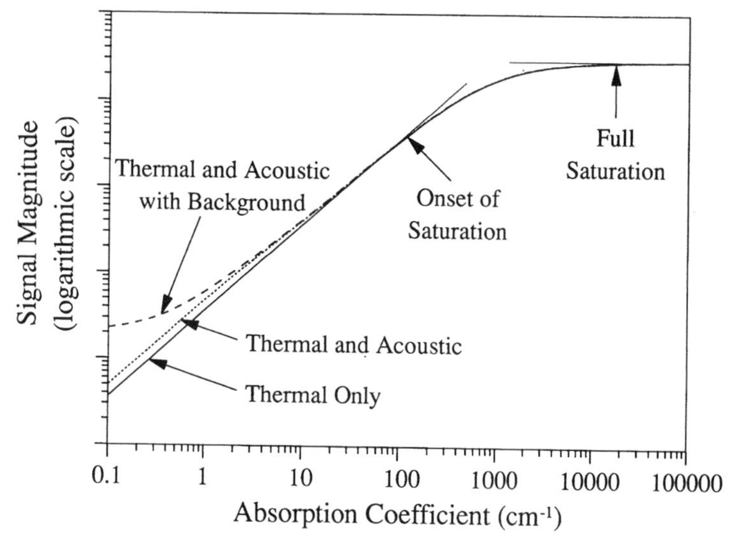

Fig. 6.4. Photoacoustic signal magnitude versus absorption coefficient as calculated from models based on thermal, and thermal and acoustic signal generation mechanisms. The background contribution was estimated from experimental data and added to the thermal and acoustic curve because this contribution would be difficult to calculate from first principles.

scattered light being absorbed by the walls of the sample chamber within the photoacoustic detector. The background contribution shown is estimated from typical experimental data.

Phase of the Photoacoustic Signal

The discussion of signal generation so far has focused on the behavior of the magnitude of the photoacoustic signal. The signal's phase also has valuable information. The phase of a photoacoustic signal is defined as the phase difference between the modulation of the incident light beam absorbed by the sample and the photoacoustic signal response. For a thermally and optically homogeneous sample, the phase is predicted to approach a value of 45° at high values of α and 90° at low values of α provided that the sample is thick relative to L and that the thermal model is used without the acoustic and background signal contributions [3]. This phase behavior is shown as the Thermal Only line in Fig. 6.5, which also shows the phase dependence when the acoustic and background signal contributions are included, both of which cause smaller phase shifts or faster signal evolution than the thermal model alone in regions of low α. The phase of the background signal is estimated from typical experimental data. The background contribution

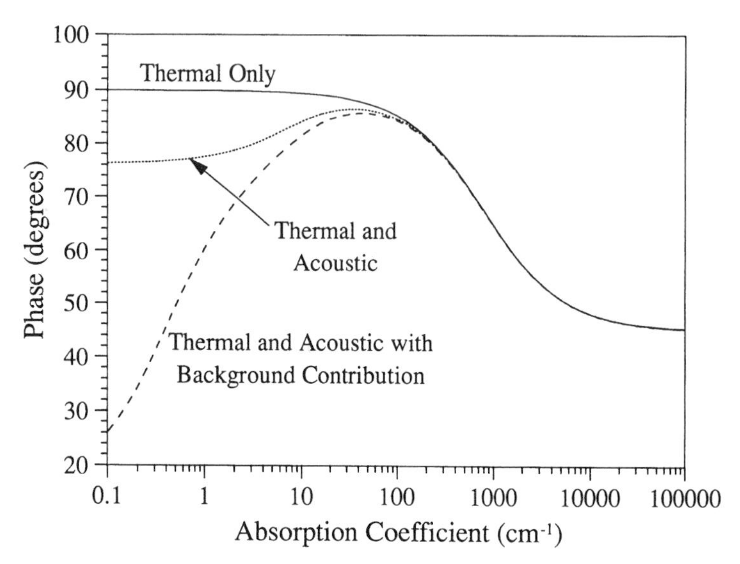

Fig. 6.5. Photoacoustic signal phase versus absorption coefficient as calculated from thermal, and thermal and acoustic signal generation models. The background contribution was estimated from experimental data due to the impracticality of calculating it from first principles.

produces a smaller phase shift or faster signal response than otherwise expected. This signal behavior can be explained, as in the magnitude case, by scattered light generating a signal at the interior walls of the photoacoustic detector, which are usually anodized aluminum comprising a layered material (thin lower thermal conductivity layer on a higher thermal conductivity substrate). This type of sample is not homogeneous as assumed in the signal generation model results plotted without background contribution in Fig. 6.5. Theoretical calculations [6] and the authors' experimental results show that a layered signal generator of the type described can produce signals with phase shifts smaller than 45°, whereas bare metal such as stainless steel produces a background signal at 45° in agreement with the homogeneous sample model.

Photoacoustic Signal Saturation

As shown in Fig. 6.4, the photoacoustic signal ceases to increase with absorption at high values of α, and therefore the PAS signal may saturate, as concentrations of analytes in a sample increase, depending on an analyte's absorption coefficient. In nearly all PAS analysis applications, however, full saturation does not occur. Measurements are made in the linear range

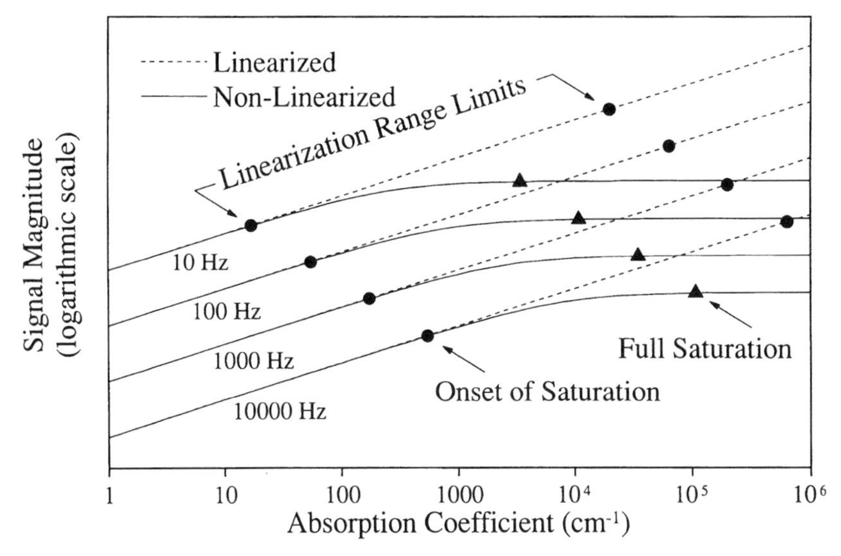

Fig. 6.6. Photoacoustic signal behavior for different modulation frequencies as a function of sample absorption coefficient. Onset and full saturation points are shown as well as linearization extensions discussed in the text.

before the onset of saturation and the nonlinear range after onset. Consequently the strong absorbance bands in PAS spectra are truncated relative to the weak bands due to saturation. Thus weak bands appear more prominently in plots of PAS spectra as a result of the nonlinearity of PAS spectra relative to transmission spectra where saturation is not present due to thinning or dilution. Figure 6.6 shows typical PAS magnitude signal behavior versus α for several different modulation frequencies. These curves are calculated from the thermal model [3]. Note that the onset and full saturation points move to higher α as the modulation frequency is increased. This frequency dependence implies that saturation of strong bands will be less in high modulation frequency spectra than in low, as is expected since the sampling depth L decreases with frequency.

The signal magnitudes shown in Fig. 6.6 decrease with frequency as $1/f^{3/2}$ before onset of saturation and as $1/f$ after full saturation. The decrease in signal magnitude with f reduces the signal-to-noise ratio of PAS spectra if f is increased to reduce saturation and equal signal averaging times are used. Thus there is a trade-off situation between signal-to-noise ratio and reduction of spectral saturation to be addressed in both quantitative and qualitative analysis applications. In quantitative PAS applications, factor analysis treatment of spectra by either principal components regression (PCR) or partial least squares (PLS) is usually able to accommodate considerable spectral nonlinearity while still producing standard errors of prediction (SEP) less than 1%. Therefore the higher signal-to-noise ratio and more rapid spectral acquisition of lower frequency PAS measurements are the trade-off choice. This is also usually the choice for qualitative analyses with computer searching, because the presence of nonlinearity does not strongly influence peak positions and most search software can perform searches based only on peak positions.

Sampling Depth

PAS measurements over a range of modulation frequencies, including high frequencies, are important when samples with depth varying compositions are analyzed. Figure 6.7 shows the variation in sampling depth L as a function of FTIR spectrometer optical path difference (OPD) velocity and phase modulation frequency and the spectrum wavenumber for samples with a thermal diffusivity, $D = 10^{-3}\ \mathrm{cm^2/s}$, which is typical of many polymers. The diagonal lines in the plot each have a wavenumber label because an FTIR spectrometer, operating in constant velocity mode, modulates each wavenumber across the spectrum at a different frequency that depends on the mirror velocity of the spectrometer's interferometer according to the formula,

$$f = V\mathbf{v}, \tag{6.3}$$

where V is the optical path difference (OPD) velocity of the interferometer mirror and $\mathbf{v}$ is the wavenumber. The mirror velocity is usually expressed either as the OPD velocity (right-hand vertical axis of Fig. 6.7) or as the helium-neon laser-fringe frequency (left-hand vertical axis of Fig. 6.7) which is the modulation frequency occurring at the laser wavelength of 632.8 nm corresponding to a laser wavenumber of 15,800 $\mathrm{cm^{-1}}$.

When L is used to estimate sampling depth, it is important to note that this is valid only if there is adequate penetration of light to the depth L to generate a photoacoustic signal. If the optical decay length, $1/\alpha$, is shorter than L at a particular wavenumber in a spectrum, then the sampling depth is determined optically and the signal will have substantial saturation at that wavenumber. Due to the exponential decay of both light waves and thermal waves, the information obtained at a particular sampling depth is always skewed toward the sample surface.

The very deep sampling depths indicated in the lower right hand corner of Fig. 6.7 are achieved only for α values in the 10 $\mathrm{cm^{-1}}$ and below range. Due to the low α values common to near-infrared spectra relative to the mid-infrared region, the near-infrared region often will offer the best opportunity for very deep sampling.

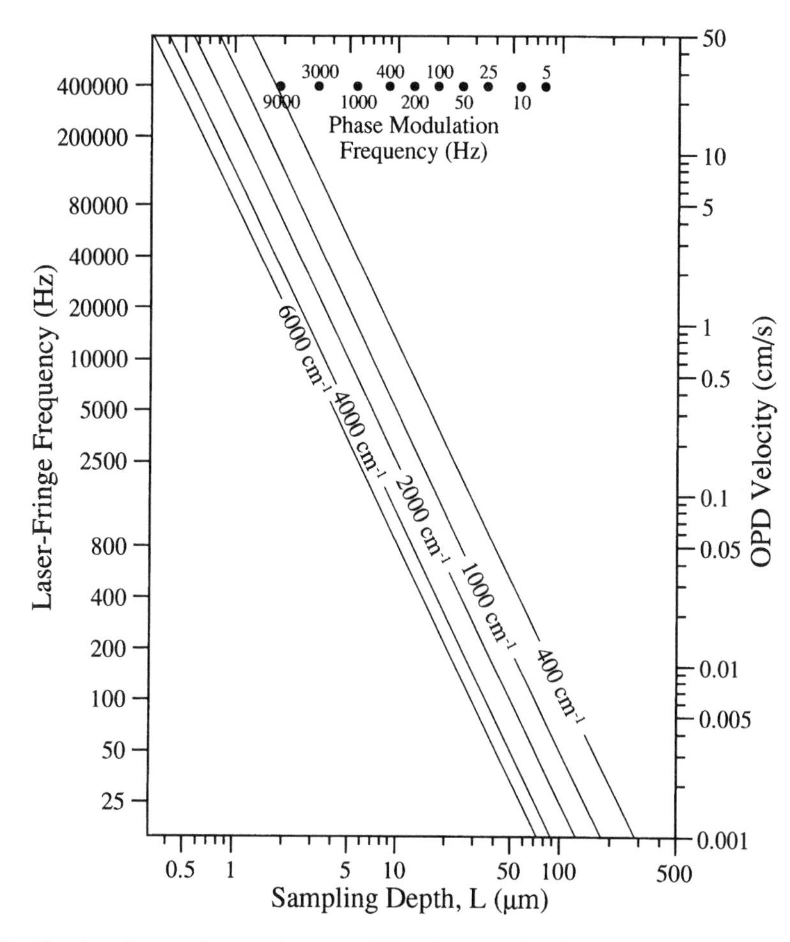

Fig. 6.7. Sampling depth dependence on FTIR mirror velocity given in terms of optical path difference (OPD) velocity and helium-neon laser fringe frequency. A thermal diffusion coefficient D of 1×10^{-3} cm^2/s, typical of many polymers, is assumed. An increase in D would cause deeper sampling, and vice versa.

Very shallow sampling also presents some special considerations because PAS signal generation becomes weaker at high frequency, the acoustic response of the typical photoacoustic detector falls off above approximately 16 kHz, and FTIR interferometers have mirror velocity limits. In view of these limitations, it is important to note that it is possible to compute spectra using magnitude and phase data that have an effective sampling depth approximately a factor of 3 shallower than calculated from the actual modulation frequency. Consequently the vertical axes of Fig. 6.7 are extended above the 80 kHz and 5 cm/s OPD velocity points, which approximately mark the instrumental limit, to indicate the range extended by calculation. The computed spectra are called linearized spectra because the shallower effective sampling depth results in linear signal dependence on α for α values above the onset of saturation as shown in Figure 6.6. The linearization process involves synchronizing the interferograms of the sample and a surface absorber such as glassy carbon. The interferograms are transformed using the surface absorber as a phase guide, and the real and imaginary components are saved. The linearized spectrum, S_l, is calculated from

$$S_l = \frac{S_R^2 + S_I^2}{1.414(S_I R_R - S_R R_I)} \tag{6.4}$$

where the subscripts R and I denote the real and imaginary parts of the sample (S) and reference (R) spectra [7]. The linearization calculation is based on the thermal signal generation model. The linearization calculation recently has been incorporated into the standard software of one FTIR manufacturer (Win-IR Pro, BioRad, Inc.) and may be available from other FTIR firms in the future.

Phase modulation frequency points are shown at the top of Figure 6.7. Phase modulation is generated by operating the FTIR in a step-scan mode and oscillating the position of one of the interferometer mirrors at an amplitude corresponding to some number of helium-neon laser wavelengths. In this way an intensity modulation is imposed on all wavenumbers of the infrared spectrum at the same phase modulation frequency, so only one value of L applies to all wavenumbers. If the waveform of the mirror oscillation is a square wave rather than sinusoidal, measurable photoacoustic signals are also generated at higher odd harmonics, such as the 3rd and the 9th, of the fundamental modulation frequency, which are shown in Fig. 6.7 for a 1 kHz fundamental. State-of-the-art digital signal processing makes it possible to simultaneously measure the magnitude and phase of photoacoustic signals at the fundamental and the odd harmonics of the phase modulation frequency. Thus spectra are obtained, for instance, with three different sampling depths simultaneously.

Interpretation of Photoacoustic Signal Phase Data

If the waveforms of the infrared beam's intensity modulation and of the photoacoustic signal excited by that beam are plotted together, one observes that the photoacoustic response lags the excitation in time. This time lag between the two waveforms is due to the finite time for thermal waves to propagate first from where light is absorbed in the sample to its irradiated surface and then into the gas adjacent to the sample surface.

The time lag of the photoacoustic signal can also be viewed as a phase angle shift between the excitation and response waveforms. The phase shift due to propagation only in the gas is always 45° as can be seen for very large values of α in Fig. 6.5. When α is very large, all of the incident beam is absorbed right at the irradiated surface and thermal wave propagation within the sample is negligible. Note that for the phase to be 45° it is also necessary that the sample thickness be at least half the maximum distance that a thermal wave can travel in one modulation cycle. Otherwise, the phase is shifted by thermal waves reflecting off the back face of the sample and then propagating back to the irradiated surface. It is also necessary that the back face be thermally in contact with a backing material so that signal generation occurs only at the irradiated front surface of the sample.

As the sample's absorption coefficient decreases from high values to low, light is absorbed at increasingly deeper depths within the sample. Figure 6.5 shows the increasing phase shift for an optically and thermally homogeneous sample as α decreases to the low-absorption range where the acoustic and scattered light background signal contributions become evident.

Layered polymer samples present a case where the sample is often highly inhomogeneous optically but usually fairly homogeneous thermally. A simple example is a thin coating on a thick substrate. If the substrate has a strong absorbance band where the coating is transparent, then absorbance by that band results in thermal waves being generated very close to the interface between the coating and the substrate. A phase shift is caused by the finite transit time for the thermal wave to propagate across the coating and into the gas. The phase shift θ_t can be calculated for thermal-wave propagation across a coating of thickness t from the thermal-wave velocity,

$$V = (4\pi D f)^{1/2}, \tag{6.5}$$

as given in the classic text by Carslaw and Jaeger [8]. If T is the oscillation period ($T = 1/f$), then the product $VT = 2\pi L$ is the distance that a thermal wave travels in one cycle or 360°. This result is obtained by the substitution $L = (D/\pi f)^{1/2}$, equation (6.1). The phase shift θ_t is then given by

$$\theta_t = \left(\frac{t}{2\pi L}\right) 360^\circ \qquad \text{in degrees,} \tag{6.6}$$

or

$$\theta_t = \frac{t}{L} \qquad \text{in radians.} \tag{6.7}$$

The propagation distance of a thermal wave per unit angle of phase shift is

$$\frac{t}{\theta_t} = \frac{2\pi L}{360^\circ} \quad \text{in degrees,} \tag{6.8}$$

or

$$\frac{t}{\theta_t} = L \quad \text{in radians.} \tag{6.9}$$

Table 6.1 shows the ratio of propagation distance to angle for a range of modulation frequencies using a value of $D = 10^{-3}\ \text{cm}^2/\text{s}$, which is typical of polymers. The table also includes for comparison the corresponding values for the sampling depth L and for $VT = 2\pi L$, which is the distance a thermal wave can travel in one modulation cycle.

Table 6.1 indicates that very small changes in depth cause degree-sized changes in phase with increasing sensitivity at higher frequencies. It is advantageous to make step-scan, phase-modulation measurements that readily yield phase information when using photoacoustic spectroscopy to study layered samples. This approach also results in constant modulation frequency across the entire spectrum rather than a linear frequency variation with wavenumber as in continuous-scan measurements.

In a typical phase-modulation measurement orthogonal interferograms are measured simultaneously at the phase-modulation frequency. More advanced FTIR systems with digital signal processing (DSP) capability and square-wave phase modulation are also able to simultaneously acquire data at the odd harmonic frequencies of the fundamental modulation frequency. The higher-frequency data provide higher depth/degree resolution near the sample surface as indicated in Table 6.1.

TABLE 6.1 Thermal-wave Propagation Parameters for Different Modulation Frequencies

f(Hz)	L(μm)	VT(μm)	t/θ_t(μm/deg)
1	178	1120	3.1
10	56.4	354	0.98
50	25.2	159	0.44
100	17.8	112	0.31
400	8.92	56.0	0.16
1000	5.64	35.4	0.10
10,000	1.78	11.2	0.031

A major interest in phase-modulation measurements is the relative phase shift differences between different absorbance bands in a sample's absorbance spectrum. As discussed earlier, the larger phase shifts are associated with absorption occurring deeper in the sample.

The phase shifts associated with different absorbance bands are best observed by plotting the amplitudes of absorbance peaks versus phase angle. Such plots are generated by measuring in-phase I_0 and quadrature I_{90} interferograms simultaneously and calculating the interferogram I_φ at detection angle φ from

$$I_\varphi = I_0 \cos\varphi + I_{90} \sin\varphi. \tag{6.10}$$

The result of equation (6.10) is Fourier transformed to get the spectrum S_φ at a specific detection angle φ. Repeating this process for different values of φ allows a plot to be made of S_φ versus φ. S_φ is plotted versus φ for typical strong and weak absorbance bands in Fig. 6.8. The plot shows a 35° phase shift between the strong (band 1) and weak (band 2) bands. Figure 6.8 could also represent a substrate band before (band 1) and after (band 2) a transparent coating is applied to the substrate. In this case the coating thickness t, which produces a phase shift θ_t, can be calculated from the expression

$$t = \frac{2\pi L\theta_t}{360^\circ}. \tag{6.11}$$

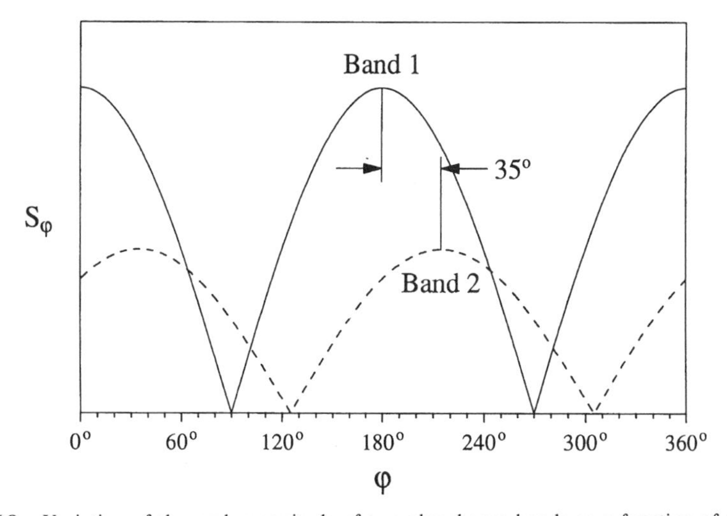

Fig. 6.8. Variation of the peak magnitude of two absorbance bands as a function of phase angle. The origin of the phase angle axis is arbitrary.

For a 35° phase shift, equation (6.11) predicts a thickness of 10.9 μm using a thermal diffusivity of 10^{-3} cm^2/s and modulation frequency of 100 Hz to calculate the $L = (D/\pi f)^{1/2}$ term.

In many practical measurements the coating is not completely transparent at the peak wavenumber position of the substrate band being used to measure coating thickness versus phase shift. This results in a smaller phase shift being measured than would be the case if the coating were completely transparent. The smaller shift is due to the signal contribution from the coating which involves a shorter thermal-wave propagation distance.

The reduced phase shift can be corrected for by a calculation based on the fact that the measured magnitude and phase are the vector sum of the signals generated in the coating and substrate [9]. In order to determine coating thickness, it is necessary to obtain the phase shift θ_S, for the substrate contribution alone. This can be done from the law of cosines formulas in trigonometry:

$$S_S = (S_T^2 + S_C^2 - 2S_T S_C \cos(\theta_T - \theta_C))^{1/2} \tag{6.12}$$

and

$$\theta_S = \theta_T + \cos^{-1}\left(\frac{S_T^2 + S_S^2 - S_C^2}{2S_T S_S}\right), \tag{6.13}$$

where S_S, S_T, and S_C are the magnitudes of substrate, total, and coating spectra at the peak position of the substrate band used in the analysis. S_T and S_C and the corresponding phases θ_T and θ_C are measured from the spectra of the coated sample and a thick slab of the coating material. The two spectra are scaled so that the closest coating band to the substrate band is of equal magnitude prior to measuring S_T and S_C. S_S is calculated from equation (6.12) and then θ_S is calculated from equation (6.13). An example of this method is given in Section 6.4.

6.3. INSTRUMENTATION

Photoacoustic Detector

Photoacoustic detectors for FTIR spectrometers are available from the major FTIR spectrometer firms and FTIR accessory companies throughout the world and directly from the sole detector manufacturer (MTEC Photoacoustics, Inc., PO Box 1095, Ames, IA 50014, USA) as of this writing. The MTEC model 300 photoacoustic detector is the instrument currently being

Fig. 6.9. Photograph of the Model 300 MTEC photoacoustic detector mounted in a typical FTIR slide mount fixture (Nicolet Instruments). The helium flowmeter and desktop power supply are shown with their connections to the detector. The black cable with the BNC connector carries the detector signal to the FTIR. The rear control lever is visible between the flowmeter and detector. The black cylinder adjacent to the power supply is the purge coupler for the infrared beam path.

manufactured and will be discussed in this section. An earlier book [2] treated the MTEC model 200, the predecessor of the model 300. The model 300 is pictured in Fig. 6.9.

The model 300 is designed primarily for FTIR analysis of solid samples, but it can also be used for liquids and gases. The prefocused unit mounts in the FTIR sample slide fixture that is common to all FTIR spectrometers. A single lever controls the opening and closing (sealing) of the sample chamber and the flow of helium purge gas through the sample chamber. The photoacoustic signal preamplifier gains are controlled with a 12-step switch on the detector, which is powered by a desktop power supply that operates on 115 V (60 Hz) or 230 V (50 Hz) line voltages.

The spectral range of a photoacoustic detector depends on the transmission range of its window (see Table 6.2) and on the availability of suitable modulation frequencies for the spectral range of interest. Because the modulation frequency of a spectrometer beam is proportional to its wavenumber, as given by equation (6.3), FTIR mirror velocities that are appropriate for the mid-infrared can lead to modulation frequencies that are too high at shorter wavelengths. At 50,000 cm^{-1} in the ultraviolet, a

TABLE 6.2 Transmission Ranges of Commonly Used Photoacoustic Detector Window Materials

Material	Transmission Range
KBr[a]	40,000 cm^{-1}–385 cm^{-1}
Quartz (UV grade)	54,000 cm^{-1}–3700 cm^{-1}
CsI	33,000 cm^{-1}–200 cm^{-1}
ZnSe	10,000 cm^{-1}–515 cm^{-1}
Polyethylene	625 cm^{-1}–10 cm^{-1}

[a]KBr is the standard window material used in the MTEC model 300 detector.

practical modulation frequency of 316 Hz occurs at a mirror velocity of 100 Hz on the helium-neon laser fringe. In the near-infrared the combination of the wavenumber and low absorption coefficients leads also to a need for low velocities such as 100 Hz. These considerations are summarized in the next section. In step-scan with phase modulation, the modulation frequency and amplitude are important. In this instance, the modulation frequency remains constant across all wavenumbers, but the efficiency of modulation varies with wavenumber, depending primarily on the modulation amplitude as discussed in the section on phase-modulation.

The optics of the photoacoustic detector collect the light from the focal point at the FTIR's slide mount fixture and refocus the light downward into the detector's sample cup. A single 2X off-axis ellipsoidal mirror performs this function producing approximately a 5-mm diameter focal spot about 1.5 mm below the detector window in the sample cup.

The model 300 is prefocused and usually requires no optical adjustment when it is placed in the FTIR slide mount fixture. A liquid crystal infrared imaging disk is supplied by MTEC to visualize the position of the infrared beam focal spot in the sample cup and guide lateral and vertical adjustments, if necessary, to center the beam.

The photoacoustic detector is positioned in the sample compartment of the FTIR by sliding it into the slide mount fixture common to all FTIR instruments as shown in Fig. 6.9. The slide mount fixture should be located in the focal plane of the FTIR and centered on the beam axis.

Electric power ($\pm$15 VDC) for the detector's preamplifier is provided by a desktop power supply (115/230 V) with a switch to select the appropriate line voltage. Some FTIR spectrometers power the detector via a connector in the sample compartment making the desktop power supply unnecessary. The photoacoustic detector signal is connected to the input of the FTIR electronics via a cable and connector specific for each instrument.

Helium gas of zero grade (99.995%, total hydrocarbons <0.5 ppm) or better should be used for purging the detector sample chamber. Purge rates from 5 to 20 cc/s are typical. There should be no valve between the gas cylinder pressure regulator valve and the detector's flow meter so that pressures only high enough to sustain the desired flow rate are used. Consequently the flow rate is to be controlled by the pressure regulator valve alone. This precaution reduces the chance of exposing the detector inadvertently to pressure surges. To further protect the detector, it is advisable to turn the gas on and off and to adjust pressure only when the detector's rear lever is in the "seal" position. In this position gas bypasses the sample chamber and the microphone cannot be damaged by pressure surges.

It is advantageous to maintain a water vapor and CO_2 free atmosphere not only in the FTIR but also in the optical path leading to the detector. Water vapor and CO_2 produce interfering absorbance bands, and water vapor can damage moisture-sensitive optical components. The model 300 can be coupled directly to most FTIRs by a purge coupling tube, as shown in Fig. 6.9. Use of the purge coupling allows samples to be changed rapidly without having to wait for the FTIR sample compartment to be purged. The sample compartment door is normally left open at all times when the purge coupling is used. The purge coupling can also be used in sealed and desiccated FTIR spectrometers if it is connected to a dry air source.

The main sample cup of MTEC detectors accommodates samples up to 10 mm in diameter and 6 mm in thickness. Normally sample specimens are considerably smaller. If samples are not already in the form of small specimens, they can be cut, punched, or abraded to obtain small specimens using a razor blade, scissors, hacksaw, cork borer (#6 and smaller), bench punch, or abrasives. Samples are usually placed in stainless steel inner cups that fit concentrically into the main cup. Different inner cup sizes are used with combinations of spacers to displace surplus gas volume in the main cup in order to enhance the signal, which is inversely proportional to the gas volume. There should, however, be an empty gas volume between the top of the sample and the main cup rim for generation of the photoacoustic signal. Usually the top of the sample should be approximately 1 to 2 mm below the main cup rim. This positioning is a reasonable compromise between signal generation efficiency and the main cup dimensions. Additional sample handling information is given in [2].

H_2O and CO_2 often evolve from the sample into the helium gas atmosphere of the sample chamber. Such vapors and gases produce absorbance bands in the spectra that can cause spectral interferences. There are several ways to reduce this problem. Samples can be predried in an oven or under a heat lamp prior to analysis. Very small amounts of powders can be put in the sample cup to lower the amount of contamination that would evolve from a full cup. Desiccants, such as magnesium perchlorate or molecular sieve spheres, can be placed in a second inner cup that is placed beneath the inner sample cup. Finally the sample chamber purge time can be increased to dry the sample.

The detector is purged with helium to increase the signal magnitude by two to three times compared to air and to remove water vapor and CO_2 gas from the sample chamber. Dry, clean helium gas should be used with a "zero-grade" rating or better. When the photoacoustic detector is in use, the flow rate is usually set between 5 and 20 cm^3/s and left at that setting throughout the analysis run for the sake of reproducibility and convenience. The detector purge valve is controlled by the rear lever which also controls the opening and closing of the sample chamber. The lever has four positions: open, open purge, sealed purge, and seal. When the lever is in the open purge position, helium gas flows over the sample. When the lever is in the sealed purge position, gas flows past but not over the sample. Fine powders can be blown out of the sample cup when the lever is in the open purge position. Therefore it is recommended that the plastic tubing carrying the helium gas to the detector be momentarily finger pinched closed as the lever is moved between the open and sealed purge positions when fine powders are being measured. Other samples do not require this precaution. Many samples may be loaded without any pause for purging when the detector is sealed. If the sample has moisture associated with it, a pause for purging may be necessary. In quantitative measurements it may be important to perform purging in a reproducible manner. Reproducibility can be checked by acquiring several spectra with a particular gas flow rate and purge time and then overlaying the spectra to compare them. It is important to leave the detector's rear lever in the seal position when the helium tank is first turned on and the gas line leading to the detector is being purged prior to purging the detector itself. In the seal position the sample chamber is bypassed and the microphone is protected from inadvertent pressurization due to an excessively high gas flow rate.

It is usually desirable to ratio the sample spectrum with the spectrum of a black absorber to remove spectral variations due to the source, optics, and detector responses. Carbon black powder has traditionally been used for this purpose, but it is a difficult substance to handle and is easily blown out of the sample cup. Consequently a special carbon black coated membrane reference sample is supplied with MTEC detectors; it remains permanently in a dedicated sample holder. This reference eliminates the need to handle carbon black powder and generates a very strong signal due to its low thermal mass. At very low modulation frequencies the MTEC reference is unsuitable because the high-signal level and low frequency cause vibration of the membrane. Therefore a carbon black powder, glassy carbon, graphite, or 60% carbon black filled rubber sample should be used.

Photoacoustic spectra have arbitrary units in that the gain settings used when sample and reference spectra are measured will influence the scale of ordinate axis units. The thermal response of the sample, reference, and cell atmosphere also introduce an ordinate scale factor. As long as gain settings and thermal responses are kept consistent, the arbitrary nature of photoacoustic units does not interfere with quantitative analyses.

When very strongly absorbing samples such as a rubber tire are being analyzed, it is not uncommon to find that the baseline is not flat and must be corrected via the FTIR software. If a series of measurements are being done, for instance, on tire rubber as a function of thermal/oxidation exposure, spectrum normalization can be done as follows: Single-beam sample spectra are acquired and copied. The copied spectrum is very strongly smoothed to remove most of the sharp spectral peaks while leaving the broad spectral envelope unchanged. This smoothed spectrum is then used as the reference spectrum for normalization. The resulting normalized spectrum will have a very flat baseline and somewhat derivatized spectral band shapes. Spectra of this type can be subtracted with no scaling. In this way very small spectral changes can be observed in the spectra of strongly absorbing materials. One caution is that prior to spectral subtraction sharp features due to the optics or gas contamination may be present in the normalized spectra. Such features are removed when subtractions are done if their source remains fixed during analyses.

It is important that the detector gain switch be consistently set when sample and reference spectra are collected. Otherwise, scale changes will be present in spectra. Initially settings should be determined by checking the magnitude of the interferogram to see that it is of a reasonable level for a given spectrometer. Usually several tenths to several volts, peak to peak, is suitable. Signal levels may be lower at high modulation frequencies.

The optical alignment of the detector should be checked using the MTEC liquid crystal infrared imager. A 100% line signal-to-noise test should produce a peak-to-peak noise of less than 0.2% at $2000\,cm^{-1}$ ($\pm 50\,cm^{-1}$) for the conditions in Table 6.3.

Some FTIR systems will not allow exact duplication of the settings given in Table 6.3, and then the closest available settings should be used. If the FTIR does not meet the noise level test, the beam-splitter alignment should be checked using the FTIR's detector, and the FTIR's mirror velocity control servo loop electronics should be tested.

The MTEC model 300 and earlier MTEC detectors can be used with special options for diffuse reflectance (DRIFTS) and transmittance sampling (SH003) or for microsampling (SH004) [2]. Option SH003 allows the

TABLE 6.3 Test Parameters for Measuring the FTIR-PAS System Noise

Mirror velocity (OPD) = 0.1 cm/s
Resolution = 8 cm^{-1}
Source aperture = maximum
Spectral range = 400–4000 cm^{-1}
Number of scans = 8
Apodization = medium Beer-Norton
Sample = MTEC carbon black reference
Detector gas atmosphere = helium

spectroscopist to switch between PAS, DRIFTS, and transmittance sampling by changing sampling heads on the standard sample holder. This is a useful option because the three sampling methods often produce complimentary data and can be used from the ultraviolet though the mid-infrared spectral regions. In other instances SH003 allows the spectroscopist to rapidly determine which sampling method works best for a given sample. Option SH004 allows measurement of single particle or fiber spectra without pressing to thin samples as is commonly done with infrared microscopes. Fibers with diameters as small as 10 μm and particles with diameters of 50 μm have been measured with SH004. Significantly higher sensitivity is possible by replacing the current 2X mirror with one of higher magnification.

FTIR Spectrometer

The performance level and the utility of PAS measurements are strongly influenced by the performance quality of the FTIR and the FTIR operating parameter selections available.

Proper selection of FTIR operating parameters is important in photoacoustic measurements in order to optimize the signal-to-noise ratio and to control sampling depth as appropriate for a given investigation. Table 6.4 summarizes the parameter considerations to be made which are of particular importance in photoacoustic measurements.

A more comprehensive discussion of modulation frequency than can be accommodated in Table 6.4 is necessary due to its importance in varying the sampling depth of photoacoustic measurements. FTIR spectometers modulate the intensity of their light beam via changes in the optical path difference of the interferometer. Continuous motion of one of the interferometer mirrors at constant velocity has been the traditional method of beam modulation in commercial instruments. More advanced FTIR spectrometers allow selection of either continuous motion or step-scan

TABLE 6.4. FTIR Parameter Selection Considerations for Photoacoustic Measurements

Parameter	Considerations
Resolution	When resolution is reduced by a given factor, signal-to-noise is improved by the square root of that factor for the same measurement time. Consequently, if an analysis can be done at 16- or 32-cm^{-1} resolution rather than at 8-cm^{-1} resolution, analysis time can be reduced significantly. Resolution higher than 8 cm^{-1} is rarely needed for analysis of solids.
Number of scans	Increasing the number of scans by a given factor increases signal-to-noise by the square root of that factor. 32 scans at 8-cm^{-1} resolution and a mirror velocity of 2.5 kHz (0.16 cm/s) is typical for survey work.
Continuous-scan mirror velocity	In studies where samples are homogeneous and variable sampling depth is not desired, a single velocity can be used. Signal-to-noise for a given spectrum acquisition time improves as mirror velocity is reduced, but photoacoustic signal saturation at high absorbance also increases. Mirror velocities from 400 Hz (0.25 cm/s) to 5 kHz (0.31 cm/s) are typical for both qualitative and quantitative analyses in the mid-infrared. In the near-infrared, absorption is lower, and a typical scan velocity of 100 Hz (0.063 cm/s) is more suitable. See Table 6.5 for information on modulation frequency ranges for spectral regions from the far-infrared to the ultraviolet.
Modulation frequency	See Section 6.3 on modulation frequency.
Beam splitter and source	Beam splitters and sources are optimized for specific spectral regions and should be chosen accordingly to optimize performance.
Aperture	The widest aperture should be chosen that is compatible with a given spectral resolution in order to maximize the incident beam intensity. An open aperture can usually be used for 8-cm^{-1} resolution and lower (16 cm^{-1}, 32 cm^{-1}, etc.).

operation. Step-scan operation can be used to simulate low-velocity continuous motion or in conjunction with phase modulation of the infrared beam intensity. In the step-scan mode the optical path difference of the interferometer is changed in a series of incremental steps until a suitable interferometer path difference has been generated appropriate for the desired spectral resolution. The stepping motion can be achieved either by moving one interferometer mirror in steps while the other remains stationary or by oscillating one mirror with a sawtooth waveform while moving the other mirror at a constant velocity. This second method avoids the difficulty of bringing the stepping mirror to a dead stop and having to wait for transient mechanical vibrations to dampen out. Phase modulation of the infrared beam intensity involves oscillating one of the mirrors about the step locations with an amplitude selected that corresponds to some number of helium-neon laser wavelengths. The phase-modulation frequency must be well above the stepping frequency. The phase-modulation frequency determines the sampling depth of the measurement and is constant across the whole spectrum being measured, in contrast to the variable sampling depth across the spectrum produced by continuous constant-velocity mirror motion. If the phase-modulation motion of the interferometer follows a square waveform, an ac signal can be measured that is generated by the beam intensity difference at the two extremes of the mirror oscillation. The infrared beam modulation frequencies generated by the square waveform are the fundamental frequency and the higher odd harmonics. Digital signal processing (DSP) allows simultaneous detection of orthogonal components of signals at the fundamental and odd harmonics. Data sets thus acquired allow infrared absorptions occurring in different sampling depth ranges to be studied on samples containing layers or concentration gradients.

In the continuous-scan mode the modulation frequency f is given by $f = V\mathbf{v}$, where $\mathbf{v}$ is the wavenumber location in the spectrum and V is the optical path difference (OPD) velocity. The OPD velocity is commonly given in units of cm/s but may also be given as the modulation frequency of the helium-neon laser beam f_l at a wavelength of 0.6328 μm (15,800 cm^{-1}). The expression $f = \mathbf{v}f_l/15{,}800\ cm^{-1}$ gives the modulation frequency f at a given wavenumber $\mathbf{v}$, corresponding to a velocity that generates a helium neon laser frequency of f_l. Table 6.5 summarizes the interferometer gener-

TABLE 6.5 FTIR Modulation Frequencies Generated by Different OPD Velocities

OPD Velocity		Far-IR	Mid-IR	Near-IR	UV-Visible	
(cm/s)	(Hz)	50 cm^{-1}	400 cm^{-1}	4000 cm^{-1}	10,000 cm^{-1}	50,000 cm^{-1}
0.0003	5			1.26 Hz	3.15 Hz	15.8 Hz
0.0006	10			2.53	6.30	31.6
0.0016	25			6.32	15.8	79.1
0.0063	100		2.53 Hz	25.3	63.2	316
0.025	400	1.27 Hz	10.1	101	253	1260
0.05	800	2.53	20.2	202	504	2530
0.16	2.5 k	7.91	63.2	632	1580	7910
0.31	5 k	15.8	127	1260	3150	15,800
0.63	10 k	31.6	253	2530	6300	
1.25	20 k	63.2	506	5060	12,600	
2.5	40 k	127	1010	10,100		
5.00	80 k	253	2020			
7.50	120 k	381	3030			

ated modulation frequencies as a function of mirror velocity at the wavenumber limits of spectral regions including the far-infrared, mid-infrared, near-infrared, and uv-visible. Frequencies above 16 kHz and below 1 Hz are not included in the table because they are beyond the frequency range for which the detector has good sensitivity operating with a helium gas atmosphere. Selection of a mirror velocity within the available ranges depends on the sampling depth, L, required, which is given by the expression $L = (D/\pi f)^{1/2}$. Sampling depths are plotted in Fig. 6.7 for different mirror velocities. The expression can also be written as $L = (D/\pi \nu V)^{1/2}$ or $L = (\nu_l D/\pi \nu f_l)^{1/2}$ where $\nu_l = 15{,}800\ \text{cm}^{-1}$. In measurements on polymers a value of $1 \times 10^{-3}\ \text{cm}^2/\text{s}$ is often used for D, since the thermal diffusivity of many common polymers is approximately equal to this value.

Phase Modulation

Phase modulation is used exclusively with step scanning. In phase modulation, one mirror oscillates at a selected frequency and amplitude about each step position. The signal from the photoacoustic detector is then processed by a phase-sensitive detector locked to the phase-modulation frequency to measure the two orthogonal components (in-phase and quadrature) of the signal, ultimately resulting in two interferograms. Both signal magnitude and phase can then be derived from these components. It is the phase-modulation frequency, not the spectrometer scanning speed, that controls the sampling depth and the depth resolution provided by phase, as illustrated in Fig. 6.7 and Table 6.1, respectively. If the phase-modulation oscillation produced by a spectrometer is not a sine wave (typically it is a square wave), then digital signal processing (DSP) can be substituted for a phase-sensitive detector to make measurements simultaneously at the fundamental phase-modulation frequency and certain of its harmonics. The amplitude of the phase-modulation oscillation affects how efficiently (or how completely) the spectrometer light beam is modulated [10]. Whenever the peak-to-peak amplitude of the oscillation equals the wavelength of the light or an integer multiple of it, the light is not modulated at all because the amount of interference produced by the interferometer is the same at the two extremes of the oscillation. Conversely, whenever the oscillation amplitude is equal to half the wavelength or an odd multiple of it, the modulation is maximized. Spectrometers are commonly set up so that the available peak-to-peak amplitudes of the phase-modulation oscillation are integer multiples of half the helium-neon laser wavelength. In that case the wavenumber positions of the modulation nodes, $\mathbf{v}_p^{\text{min}}$ and maxima, $\mathbf{v}_p^{\text{max}}$, are given by

$$\mathbf{v}_p^{\text{max}} = \frac{(p - 1/2)\mathbf{v}_{\text{HeNe}}}{n} \tag{6.14}$$

and

$$\mathbf{v}_p^{\min} = \frac{p\nu_{\mathrm{HeNe}}}{n} \tag{6.15}$$

where $\mathbf{v}_{\mathrm{HeNe}}$ is the helium-neon laser wavenumber (15,800 cm^{-1}), n is the peak-to-peak amplitude of the oscillation measured in helium-neon laser wavelengths ($n = 0.5, 1, 1.5, \ldots$), and p is an index ($p = 1, 2, 3, \ldots$). The first few nodes and maxima for selected values of n are given in Table 6.6. At present most spectrometers provide amplitudes only up to about $n = 2$. No photoacoustic signal is produced at the modulation nodes, and the highest instrumental signal-to-noise ratios occur at the modulation maxima. The phase-modulation amplitude should therefore be chosen to place all nodes outside the spectral range of interest and, when possible, a modulation maximum near the region of interest.

The higher n values given in Table 6.6 are not presently available on most step-scan spectrometers but were included to show how the extrema would vary with amplitude if such amplitudes are available. Selection of available n values on a particular spectrometer for the spectral region desired should be based primarily on the $\nu_1^{\max}$ and $\nu_1^{\min}$ values in Table 6.6, but the overall system response curve should be checked with the carbon black reference in the photoacoustic detector prior to running sample spectra, since it is sensitive to the other spectral variations of the source and optics.

Photoacoustic FTIR interferograms are transformed by the standard FTIR data system algorithms used with other sampling techniques and detectors. The resulting photoacoustic spectra are absorbance spectra and are normalized by dividing the sample spectrum by a background spectrum obtained with a carbon black or other reference sample. The ordinate label may be given as photoacoustic, absorbance, or transmittance units depending on the FTIR. Depending on how the particular FTIR software is set up, it may be necessary to change the label to absorbance, if it is not already so, in order to do qualitative or quantitative analyses. It is important to note that photoacoustic absorbance units are sensitive to amplifier gain settings, sample conditions (slab versus powder), and the gas atmosphere and volume of the detector. Qualitative analysis can be done with photoacoustic FTIR spectra using commercial and user-generated libraries. Libraries in the former case are often based on transmission measurements resulting in differences between ordinate values of library and photoacoustic spectra due to arbitrary PAS units and to different effective sampling thickness and consequent signal saturation variations. This situation is not unique to photoacoustic sampling, since many transmission- and reflectance-based unknown spectra do not perfectly match commercial library spectra on the ordinate axis. Photoacoustic spectra show good coincidence on the

TABLE 6.6. Interferometric Wavenumber Maxima and Minima for Different Phase Modulation Amplitudes Given in HeNe Laser Wavelentgths, *n*.

n	ν_1^{max} (cm^{-1})	ν_1^{min} (cm^{-1})	ν_2^{max} (cm^{-1})	ν_2^{min} (cm^{-1})	ν_3^{max} (cm^{-1})	ν_3^{min} (cm^{-1})	ν_4^{max} (cm^{-1})
0.25	31,610	63,210					
0.5	15,803	31,606	47,408	63,211			
1.0	7901	15,803	23,704	31,606	39,507	47,408	55,310
1.5	5268	10,535	15,803	21,070	26,338	31,606	36,873
2.0	3950	7901	11,852	15,803	19,753	23,704	27,655
2.5	3161	6321	9482	12,642	15,803	18,963	22,124
3.0	2634	5268	7901	10,535	13,169	15,803	18,437
5.0	1580	3161	4741	6321	7901	9482	11,062
10.0	790	1580	2370	3161	3951	4741	5531
20.0	395	790	1185	1580	1975	2370	2766

wavenumber axis with commercial libraries. User-generated photoacoustic spectral libraries produce good coincidence on the ordinate axis provided that amplifier gain is set consistently at the same levels for sample and background spectrum acquisitions and that the sample condition is consistent. If this condition varies, for instance, from a slab to a powder, the signal level will shift upward, but this does not significantly influence qualitative analyses. Coincidence between the photoacoustic library and the unknown spectra is excellent on the wavenumber axis. Standard FTIR factor analysis software packages for quantitative analysis work well with photoacoustic spectra, typically producing standard errors of prediction of less than 1%. Both PLS and PCR tolerate the nonlinearities (due to signal saturation) present in photoacoustic spectra well. Quantitation using a single band is also possible with FTIR-PAS spectra. Both approaches are demonstrated in the discussion of applications in Section 6.4.

6.4. APPLICATIONS

Applications discussed in this section were chosen from work in the authors' laboratory to illustrate important capabilities of PAS including rapid qualitative analysis, quantitative analysis, and variation of sampling depth to reveal depth varying composition. Readers are referred to the general scientific literature for a wider scope of applications [2, 11, 12, 13, 14, 15, 16].

Rapid Identification of Polymers for Recycling

Recycling of polymers is expanding due to both environmental and economic factors. There are two main classes of polymers. These are the less-valuable consumer polymers used, for instance, in plastic household containers and the more costly and more difficult to analyze engineering polymers used in motor vehicles, computers, and appliances. The value of recycled engineering polymers is increased if the material can be accurately sorted as to both general polymer type and subtype, which is related to filler and additive concentrations.

Polymer recycle operations require fast and modestly priced equipment given the economics involved. Hence a low-priced FTIR (Perkin-Elmer Paragon 1000) was used in this study [17] with an MTEC model 300 photoacoustic detector. Samples were punched (0.285 diameter) from typical polymer car components supplied by Ford and MBA Polymers Inc. Punching time was approximately two seconds per sample. Samples were positioned in a special sample holder so that the infrared beam impinged on the cross section, thus avoiding the need to remove surface coatings that are present on many car components. FTIR-PAS spectra (3000–500 cm^{-1})

were acquired in one scan (0.1-cm/s OPD velocity, 16-cm^{-1} resolution, helium-purge) and identified using the Perkin-Elmer "Compare" function in a total of five seconds. Minor modifications were made on the Paragon to increase the speed of data acquisition and analysis. A library of 19 spectra were used to identify 35 unknowns as to general polymer type (xenoy, polyurethane, ABS, and polypropylene) and as to talc filler level in polypropylene. The identification accuracy was 100% for both general polymer type and talc filler level [17].

Quantitative Analysis of Major and Minor Concentrations of Additives in Paper Products

Most paper products contain surface and bulk additives at different concentrations to produce various special properties. Figure 6.10 shows FTIR-PAS spectra of paper towels with varying concentrations of latex to produce wet strength. The spectra were measured at 8-cm^{-1} resolution, 128 scans, at a mirror velocity of 2.5-kHz laser fringe frequency. Magnesium perchlorate desiccant was used in the sample cup. The samples were predried overnight at 110°C in air. The starred cellulose band was used to scale the ordinate axis of the spectra. The latex constitutes a major bulk additive in the paper ranging from 12.3% to 29.8% by weight. Figure 6.11 shows the scaled peak height of the latex band at 1735 cm^{-1} as a function of latex concentration.

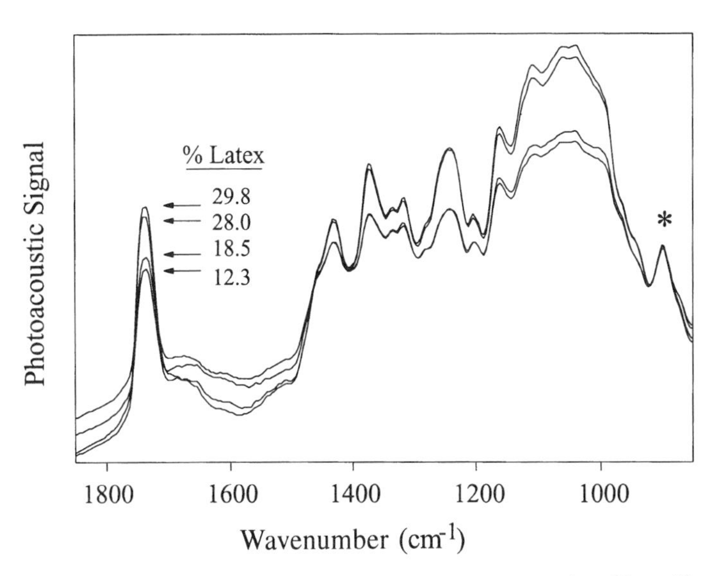

Fig. 6.10. FTIR-PAS spectra of paper towels with different concentration of latex. The spectra are normalized to the starred cellulose band.

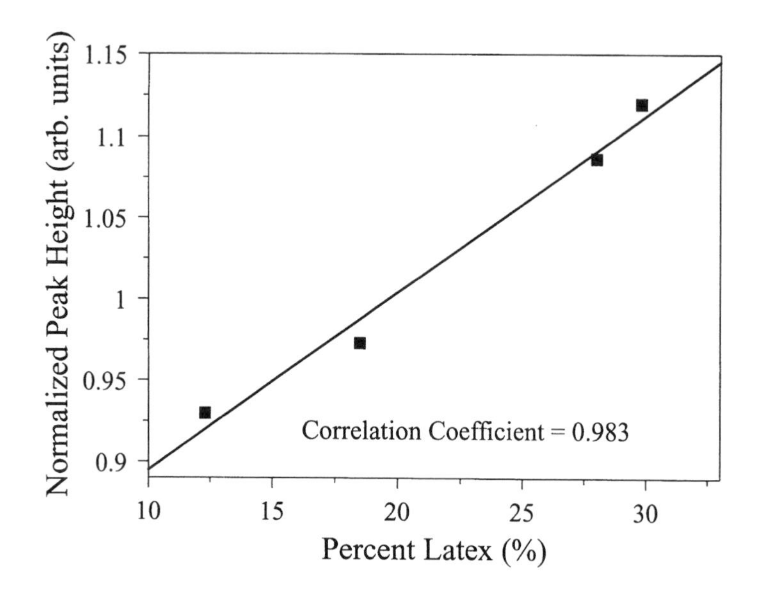

Fig. 6.11. Correlation plot for the latex in paper analysis.

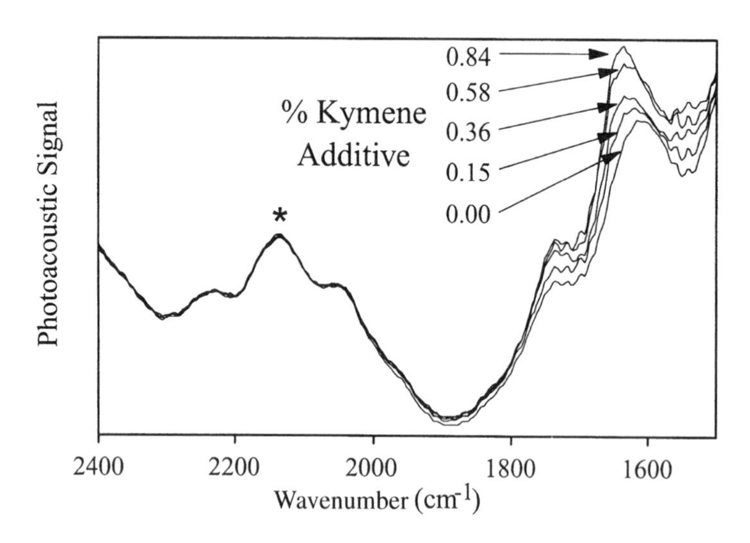

Fig. 6.12. FTIR-PAS spectra of Kymene in paper. The spectra are normalized to the starred cellulose band.

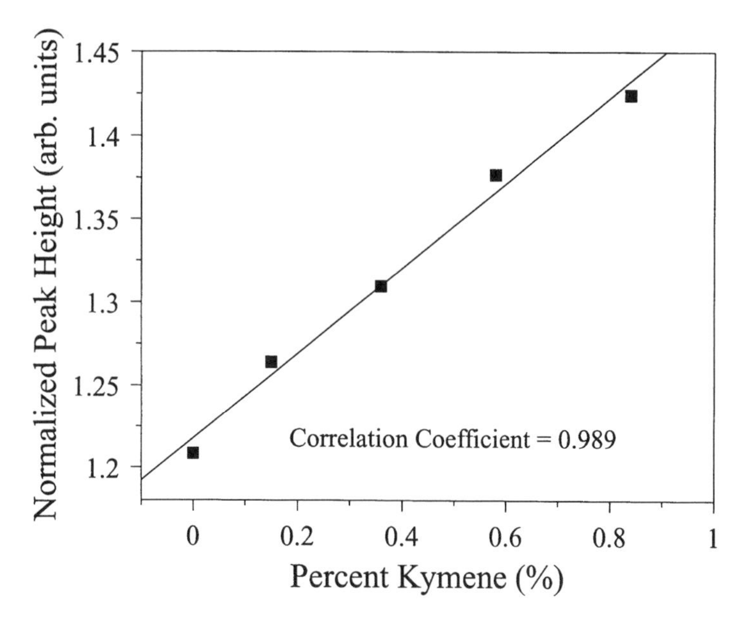

Fig. 6.13. Correlation plot for the Kymene in paper analysis.

The latex peak heights were measured relative to the baseline which varies significantly for these samples. A correlation coefficient of 0.983 was calculated for this analysis. Analysis of minor concentrations of additives in paper is also possible with FTIR-PAS. Paper spectra with a kymene bulk additive (concentration range 0 to 0.84% by weight) are shown in Fig. 6.12. The starred paper band at 2140 cm^{-1} was used to scale the spectra. Measurement and drying conditions were the same as for the latex analysis. Figure 6.13 shows the correlation between the kymene concentration and normalized kymene peak height near 1600 cm^{-1}. A correlation coefficient of 0.989 was calculated for this analysis. Factor analysis was not used in either paper analysis because of the limited number of samples.

Analysis of Aqueous Sludges with Soluble and Insoluble Species

Analysis of nuclear waste tank sludges is important in waste cleanup operations. FTIR-PAS is an ideal molecular analysis method because it allows small amounts of radioactive material to be analyzed with minimal handling and radiation exposure [18]. In this application approximately 1-mg samples of nonradioactive surrogate were analyzed. The samples contained both soluble and insoluble salts in an aqueous sludge. Soluble salts in the moist sludge tended to be partly in solution and partly

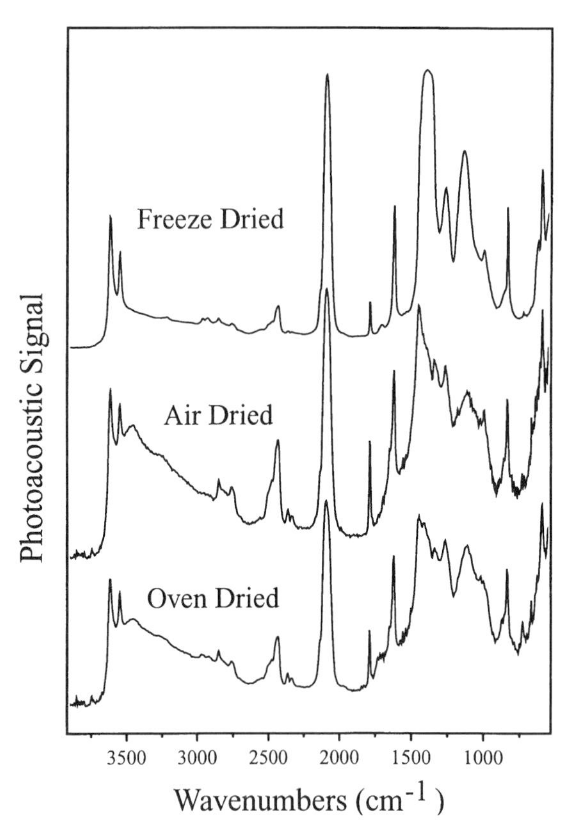

Fig. 6.14. FTIR-PAS spectra of a surrogate nuclear waste tank sludge which was dried by three different methods as indicated. Reproduced from [18].

crystalized, leading to variability in the infrared spectra as the moisture level varied. Therefore the samples were dried to obtain reproducible spectra. The drying procedure is critical for this type of sample because conventional air or oven drying causes the soluble components to migrate to the sample surface and form macrocrystals. This inhomogeneous condition results in an increase of soluble species' peak heights and spectral distortion from crystal refraction effects. Freeze-drying, on the other hand, produces microcrystals with no migration of soluble species because they are initially frozen in place and water is removed by sublimation. Figure 6.14 shows spectra of a surrogate sludge dried by the different methods. The presence of macrocrystal formation is especially apparent in the air-dried spectrum where the soluble nitrate band intensities (1798 cm^{-1} and 2433 cm^{-1}) are increased relative to the insoluble ferrocyanide band (2098 cm^{-1}). Figure 6.15 shows

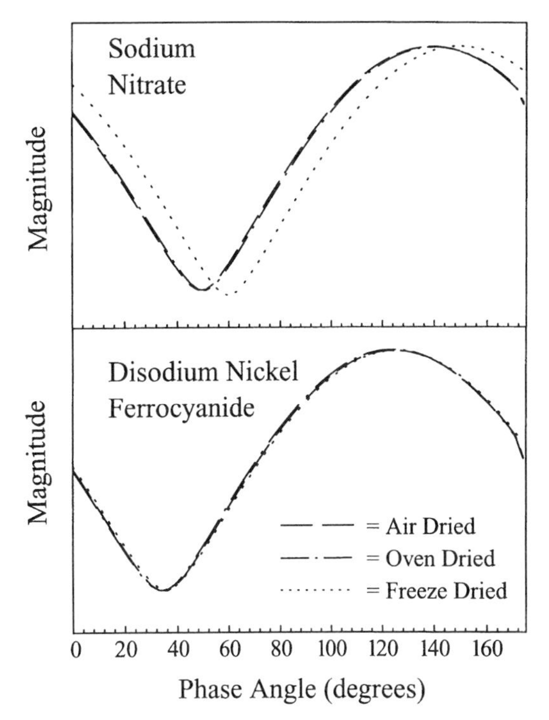

Fig. 6.15. Phase angle behavior for soluble sodium nitrate species and insoluble disodium nickel ferrocyanide species after drying by different methods. The phase zero is arbitrary. Reproduced from [18].

the phase behavior of the sodium nitrate band at 1789 cm^{-1} and of the ferrocyanide band at 2098 cm^{-1} for samples dried by the different methods. The strong ferrocyanide band's phase leads that of the weaker nitrate band as would be expected. The phase in the case of ferrocyanide is the same for all drying methods because there is no migration to the surface of insoluble species. On the other hand, the nitrate phase for air- and oven-dried samples leads that of the freeze-dried, indicating migration of soluble species to the surface for air- and oven-dried samples.

Figures 6.16, 6.17, and 6.18 are plots for nitrite, nitrate, and disodium nickel ferrocyanide, respectively, showing the correlations between the concentrations predicted from FTIR-PAS spectra by a PLS routine and the concentrations known from synthesis. Samples were prepared by adding known amounts of a single salt species to surrogate sludge supplied by the Westinghouse Hanford Company and then freeze-drying the sample for six hours at $-50°C$ using a Fisons Model 350 freeze-dryer. Spectra were measured at 8-cm^{-1} resolution with 256 scans and a mirror velocity of

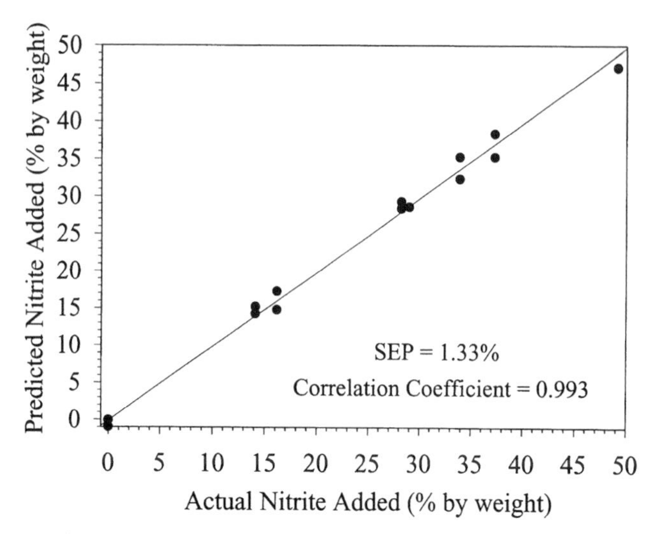

Fig. 6.16. Correlation plot for sodium nitrite in sludge analysis.

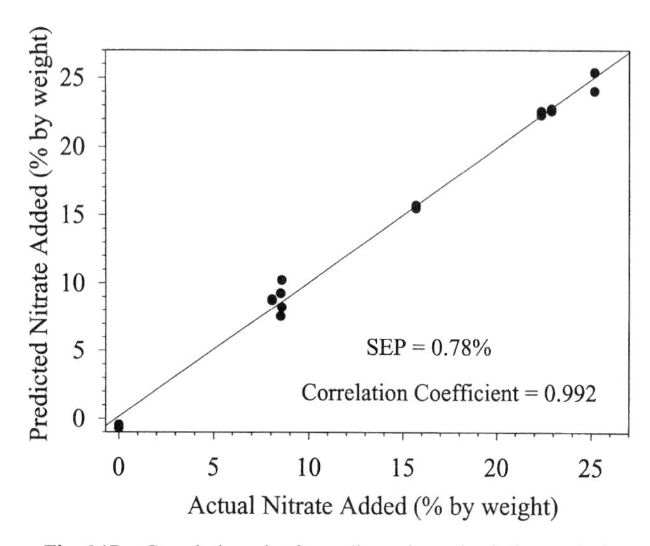

Fig. 6.17. Correlation plot for sodium nitrate in sludge analysis.

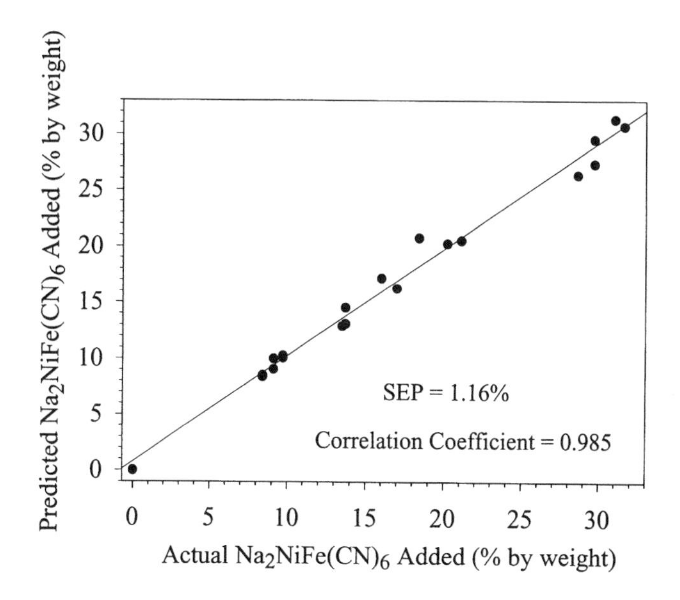

Fig. 6.18. Correlation plot for disodium nickel ferrocyanide in sludge analysis. Adapted from [18].

2.5 kHz. The standard errors of prediction (SEP) and correlation coefficients suffer to some degree from mixing difficulties during synthesis of these types of samples.

Quantitation of $CaCO_3$ Residual in Lime

Lime (CaO) is produced by heating limestone ($CaCO_3$) in a kiln. It is desirable to calcine as much $CaCO_3$ as possible into CaO without exposing the CaO to too much heating, which reduces its reactivity. Processing is considered optimal if the residual $CaCO_3$ concentration is approximately 1%. Figure 6.19 shows FTIR-PAS spectra of lime containing various concentrations of $CaCO_3$ (0–5%) measured at 8-cm^{-1} resolution, 2.5-kHz mirror velocity, and 256 scans. The spectra were normalized against an adjacent CaO band to remove scaling variations due to particle size or packing. The samples were synthesized in powder form from reagent grade $CaCO_3$ and CaO and were placed in stainless steel inner cups for analysis. They were dried overnight in a desiccator to remove moisture. This step would probably not be necessary for samples coming directly out of a kiln. The plot in Fig. 6.20 gives the correlation between the $CaCO_3$ peak at

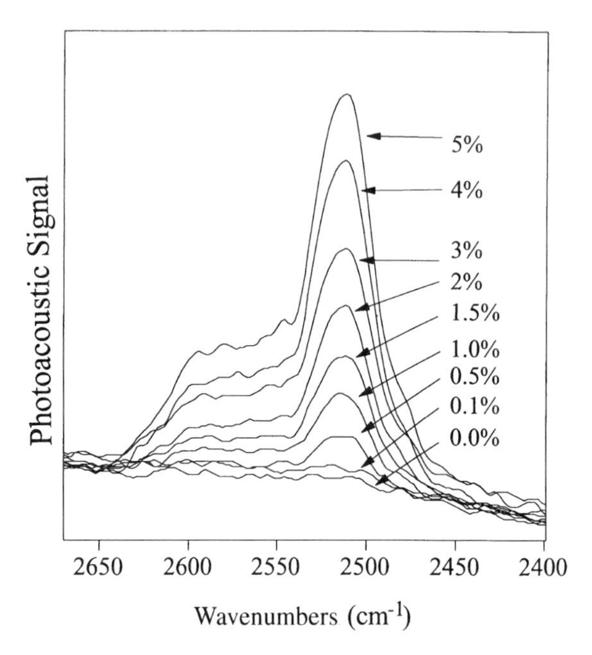

Fig. 6.19. FTIR-PAS spectra of lime with different concentrations of uncalcined $CaCO_3$.

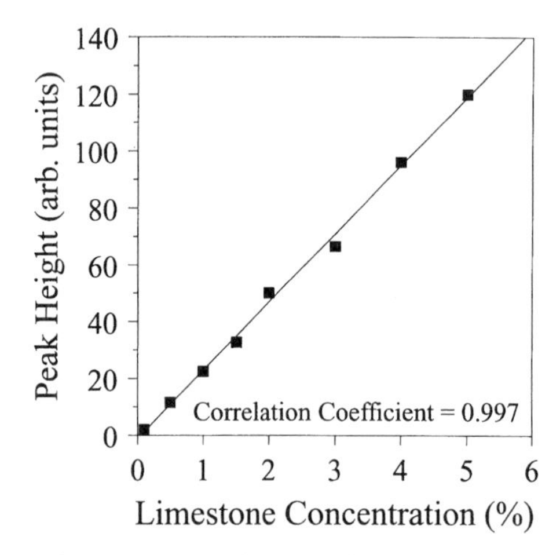

Fig. 6.20. Correlation plot of $CaCO_3$ in lime.

2513 cm^{-1} and the limestone concentration known from synthesis of the samples. The correlation coefficient for the analysis is 0.997.

Determining Surface or Bulk Character of Polymer Additives

Polymer additives are often specialized to concentrate at the surface or be homogeneously distributed in the bulk. Figure 6.21 shows FTIR-PAS spectra of polyethylene containing both surface (1645 cm^{-1}) and bulk (1019 cm^{-1}) additive bands measured at mirror velocities of 200 Hz (20 scans), 2.5 kHz (200 scans), and 20 kHz (10,000 scans) corresponding to sampling depths of approximately 50 μm, 10 μm, and 4 μm, respectively. The spectra have been scaled so that the polyethylene peak marked N is of constant height. Note that the surface additive peak height grows substantially as sampling depth is reduced, indicating a surface or near-surface additive concentration, whereas the bulk additive peak increases much more gradually due to reduction in signal saturation with increasing frequency. The strong off-scale polyethylene peak heights also increase due to reduced photoacoustic signal saturation at higher frequency as discussed previously in Section 6.2. The phase behavior of the surface S, the bulk B, and the strongest polyethylene (2920 cm^{-1}) bands are plotted in Fig. 6.22. Figure 6.23 shows a magnitude spectrum of the polyethylene sample. The data in

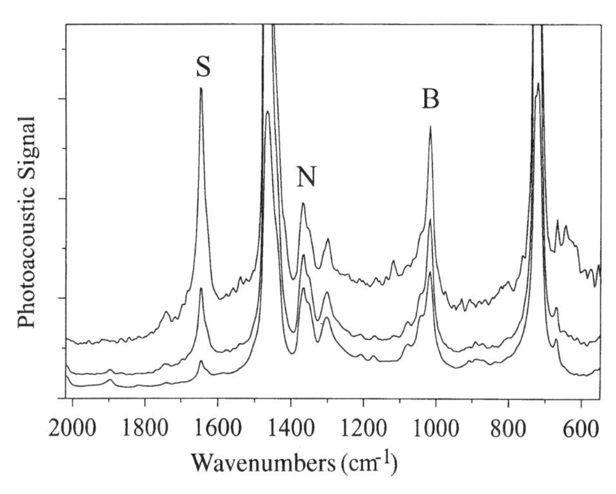

Fig. 6.21. FTIR-PAS spectra of polyethylene pellets with bands associated with a surface (S) segregating additive and a bulk (B) additive. The spectra were measured at mirror velocities of 200 Hz (*lower*), 2.5 kHz (*middle*), and 20 kHz (top) and are normalized to the band labeled N. The approximate sampling depths (top to bottom) are 4 μm, 10 μm, and 50 μm.

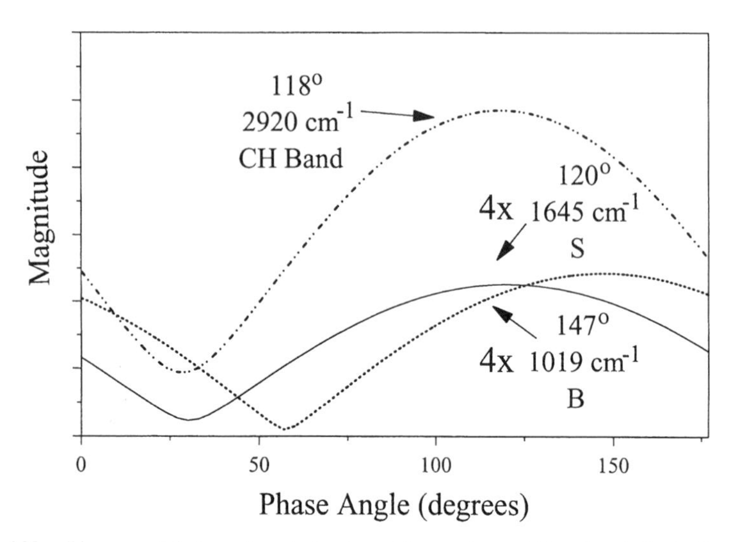

Fig. 6.22. Phase-modulation phase behavior of the surface additive band (*S*), bulk additive band (*B*), and the strong C–H polyethylene band. The phase zero is arbitrary.

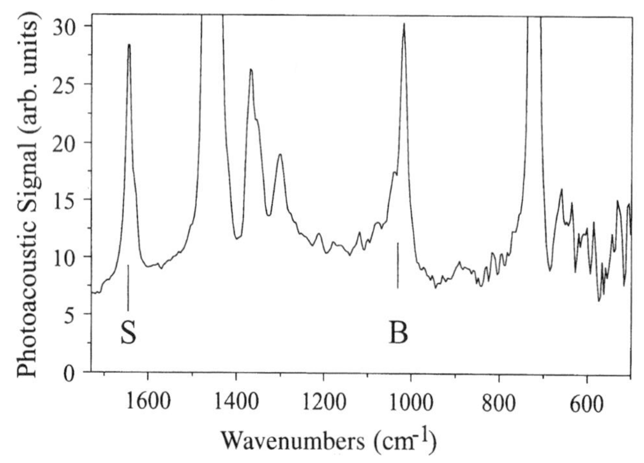

Fig. 6.23. FTIR-PAS magnitude spectrum calculated from the same data set used for Fig. 6.22. The data were measured by 400-Hz phase modulation.

Figs. 6.22 and 6.23 were measured using a phase-modulation optical path difference amplitude of 2 helium-neon laser wavelengths, 400-Hz modulation frequency, 8-cm^{-1} resolution, and 16 scans. The in-phase and quadrature photoacoustic signals were measured simultaneously, and the plots were calculated from these components. The magnitude of the surface additive band (*S*) is smaller than that of the bulk additive band (*B*) in Fig. 6.23, but the phase of the additive band (120°) is seen to lead that of the bulk band (147°). This magnitude and phase behavior indicates that the 1645 cm^{-1} band (*S*) is definitely associated with a surface-segregating additive. The phase data also allow determination as to if the additive consists of a layer on the surface or is localized near the surface. The fact that the phase of the strong polyethylene band at 2920 cm^{-1} leads the additive band in phase indicates that the additive is not a surface layer on top of the polyethylene but must extend to some degree below the polymer surface.

Fluorination of Polyethylene

Polyethylene containers and motor vehicle gas tanks are fluorinated to produce a surface layer that reduces fluid permeability. It is important to be able to measure the level of fluorination in order to achieve economical processing and to establish quality control standards. FTIR-PAS provides a method to satisfy both these production needs and to provide depth information to assist in process development. In Fig. 6.24, phase-modulation FTIR-PAS magnitude spectra are plotted for polyethylene with high and low levels of fluorination, as indicated by the broad band at 1100 cm^{-1}.

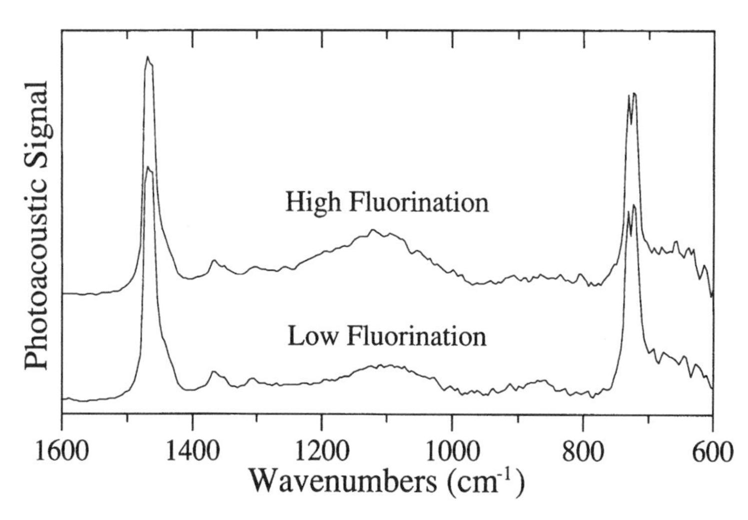

Fig. 6.24. Phase-modulation FTIR-PAS spectra of polyethylene with high and low levels of fluorination.

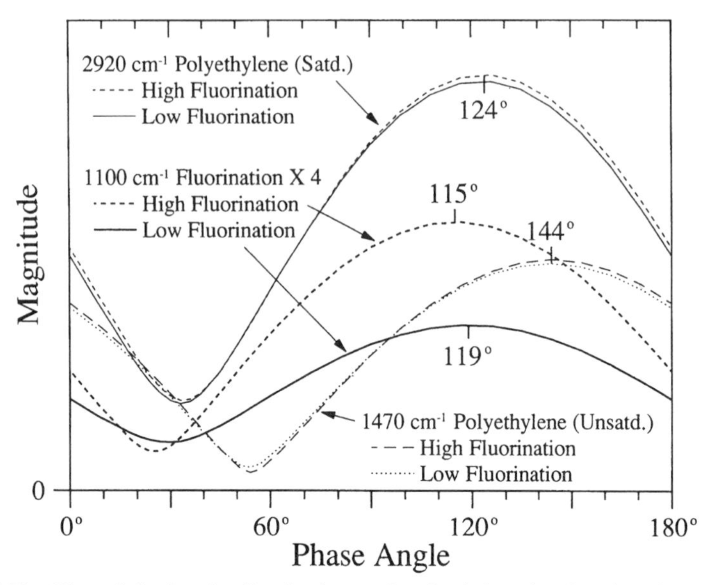

Fig. 6.25. Phase behavior for fluorination and polyethylene bands. The phase zero is arbitrary.

Figure 6.25 shows the phase behavior of the fluorination band at high and low levels, the strongest polyethylene band at $2920\,cm^{-1}$, and the next strongest at $1470\,cm^{-1}$. The phases of the high- and low-level fluorination band lead the phase of the strong polyethylene C–H band at $2920\,cm^{-1}$, indicating that the main fluorination region is within a few micrometers of the surface.

Chemically Treated Polystyrene Spheres

This application illustrates the use of phase information to indicate the layering order of a sample. Figure 6.26 shows the FTIR-PAS spectrum of polystyrene spheres that have a surface layer due to chemical treatment. The marked bands are associated with hydroxyl (OH), surface treatment (surface), and strong (SB) and weak (WB) bands of the polystyrene matrix. The phase behavior of these bands is also shown in Fig. 6.26. The hydroxyl and surface treatment bands are much weaker than the strong polystyrene band, but they still lead it in phase. The hydroxyl species must be on the surface of the chemical treatment species because the hydroxyl band has the smallest phase shift and has approximately the same peak magnitude as that of the chemical treatment species.

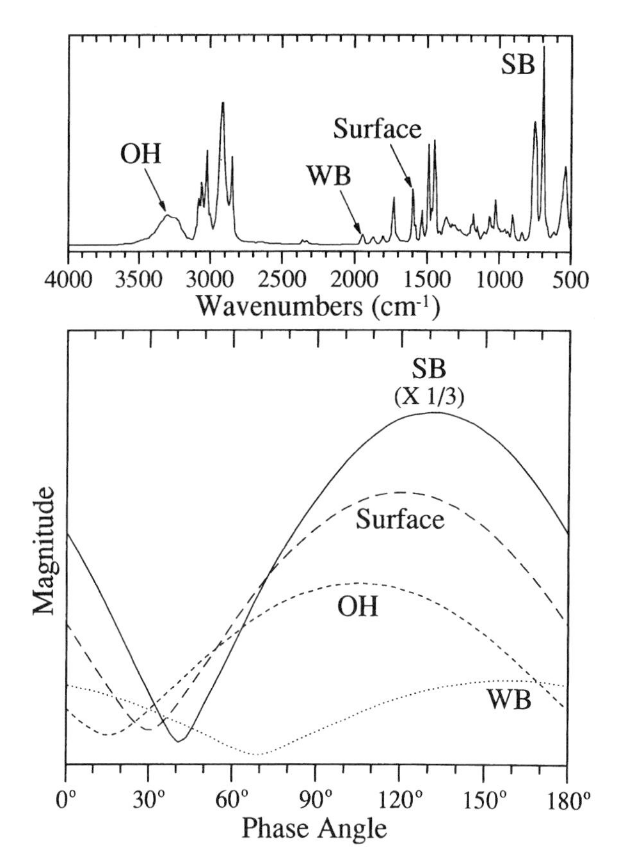

Fig. 6.26. Phase-modulation FTIR-PAS spectrum and phase behavior for polystyrene spheres with chemically treated surfaces. The phase zero is arbitrary.

Enhanced Surface Specificity by Linearization of FTIR-PAS Spectra

In our earlier discussion on sampling depth (Section 6.2), FTIR-PAS spectrum linearization using equation (6.4) was shown to decrease sampling depth beyond the limit imposed by the maximum modulation frequency that is practical for measurements [7]. Figure 6.27 gives the usual magnitude spectra (5 and 40 kHz) and the linearized spectrum (40 kHz) of a paper with a polyethylene inner coating and thin silicone outer coating. Note the growth with reduced sampling depth of the silicone bands (1261, 1022, and 798 cm^{-1}) relative to the polyethylene band at 1466 cm^{-1}. The sampling depths near 1000 cm^{-1} for the 5-, 40-, and 40-kHz-linearized spectra are 10, 3, and 1 μm. At these depths none of the paper bands are observed.

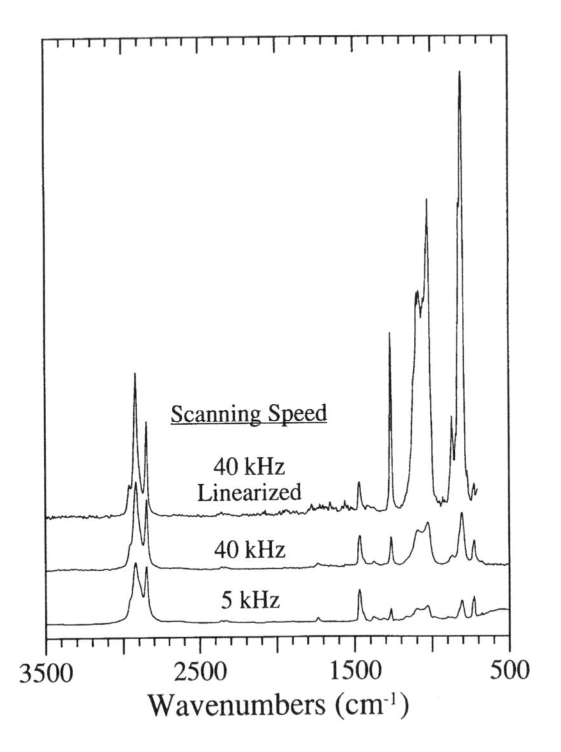

Fig. 6.27. Variable sampling depth FTIR-PAS spectra of a silicone-on-polyethylene coated paper. Spectra have been scaled to the 1466-cm^{-1} band.

Determination of Coating Thickness from Phase Data

In this application a calibration curve is produced for predicting the thickness of PET films on polycarbonate [9] based on equation (6.11). FTIR-PAS spectra are measured for five PET coating thicknesses (2.5, 3.6, 6.0, 12, and 23 μm), for the bare polycarbonate substrate, and for a thick specimen of PET. The latter measurement is needed to correct for the small absorption of the infrared beam by the thicker PET films at the 1778 cm^{-1} polycarbonate band, the band that is used for the phase versus coating thickness measurements [9]. Figure 6.28 shows the phase behavior for the 1730-cm^{-1} PET band and the 1778-cm^{-1} polycarbonate band that results for different coating thicknesses. The calibration curve is plotted in Fig. 6.29 using the phase data of Fig. 6.28. The solid circles are the phases observed in Fig. 6.28. The solid line is calculated from equation (6.11) using $L = 8.9\ \mu$m and plotted with an abscissa intercept of 132°. The first four

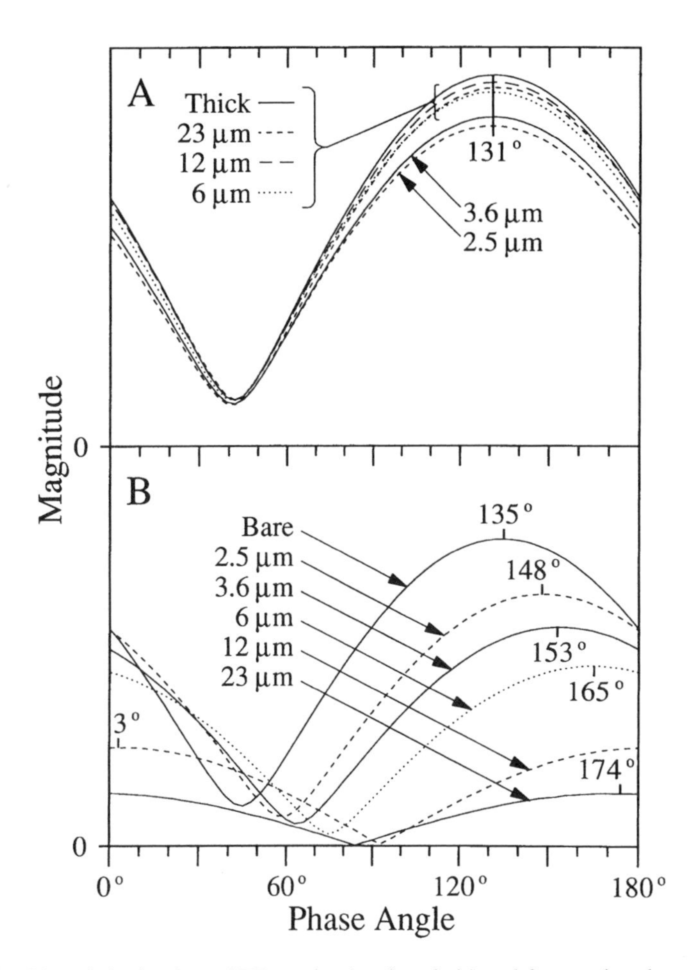

Fig. 6.28. Phase behavior for a PET coating band peak (*a*) and for a polycarbonate band peak (*b*) measured for samples having the indicated thickness of PET coated onto thick polycarbonate, as well as for "bare" polycarbonate and a "thick" piece of PET. The phase zero is arbitrary. Reproduced from [9].

points fit the calculation but the 12 and 23 μm points deviate substantially due to overlap by the wing of the 1730-cm^{-1} PET band. The absorption in the thick PET films contributes a fast signal component resulting in measurement of smaller phase angle values than would occur if the film were completely transparent. The measured phase values can be corrected using equations (6.12) and (6.13) [9]. The corrected points are shown as open circles in Fig. 6.29.

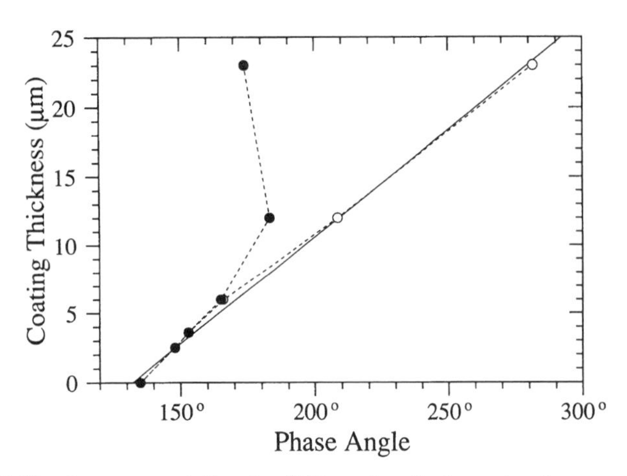

Fig. 6.29. Calibration curve relating the PET coating thickness to phase angle. The open circles represent corrected phase values. Reproduced from [9].

6.5. CONCLUSION

This chapter has considered advances in FTIR-PAS over the last four years since the publication in 1993 of [2]. Most significant are the reduction of FTIR-PAS analysis time to five seconds per sample discussed in Section 6.4 on the rapid identification of polymers for recycling and variable-depth probing using both magnitude and phase data. This latter area of advancement is closely tied to step-scan interferometry with phase modulation.

FTIR-PAS is expected to continue to grow in use at both ends of user sophistication. At the low end, users will continue to be attracted by the ease of dropping a sample in the cup and getting a spectrum for qualitative analysis. At the high end, researchers will develop more advanced variable depth analysis capabilities. Between these two user extremes, continued development and use of the quantitative analysis capabilities of FTIR-PAS is expected to occur.

ACKNOWLEDGMENTS

This work was funded by MTEC Photoacoustics, Inc. and the U.S. Department of Energy, Office of Science and Technology. Ames Laboratory is operated for the U.S. Department of Energy by Iowa State University under Contract No. W-7405-ENG-82.

The authors wish to thank the Digilab Division of Bio-Rad for the loan of an FTS 60A FTIR spectrometer used in some of the work and for technical support. Samples for application studies were supplied by Glenn

Norton (Ames Laboratory), Richard Kellner (Technical University of Vienna), Teo Rebagay (Westinghouse Hanford Co.), Mike Biddle (MBA Polymers), and Patsy Coleman (Ford Motor Co.).

REFERENCES

1. A. G. Bell, *Phil. Mag.*, **11**, 510 (1881).
2. J. F. McClelland, R. W. Jones, S. Luo, and L. M. Seaverson, In *Practical Sampling Techniques for Infrared Analysis*, P. B. Coleman, ed. CRC Press, Boca Raton, FL, 1993, ch. 5.
3. A. Rosencwaig and A. Gersho, *J. Appl. Phys.*, **47**, 64 (1976).
4. J. G. Parker, *Appl. Opt.*, **12**, 2974 (1973).
5. F. A. McDonald and G. C. Wetsel, Jr., *J. Appl. Phys.*, **49**, 2313 (1978).
6. A. Mandelis, Y. C. Teng, and B. S. H. Royce, *J. Appl. Phys.*, **50**, 7138 (1979).
7. R. O. Carter III, *Appl. Spectrosc.*, **46**, 219 (1992).
8. H. S. Carslaw and J. C. Jaeger, *Conduction of Heat in Solids*. Clarendon, Oxford, 1959.
9. R. W. Jones and J. F. McClelland, *Appl. Spectrosc.*, **50**, 1258 (1996).
10. R. A. Crocombe, J. C. Leonardi, and D. B. Johnson, In *Proc. 9th International Conference on Fourier Transform Spectroscopy*. SPIE Vol 2089, SPIE, Bellingham, WA, 1994, p. 244.
11. D. R. Bauer, M. C. P. Peck, and R. O. Carter III, *J. Coatings Tech.*, **59**, 103 (1987).
12. M.-L. Kuo, J. F. McClelland, S. Luo, P.-L. Chien, R. D. Walker, and C.-Y. Hse, *Wood Fib. Sci.*, **20**, 132 (1988).
13. C. Q. Yang, *Appl. Spectrosc.*, **45**, 102 (1991).
14. R. M. Dittmar, J. L. Chao, and R. A. Palmer, *Appl. Spectrosc.*, **45**, 1104 (1991).
15. S. Luo, C.-Y. F. Huang, J. F. McClelland, and D. J. Graves, *Anal. Biochem.*, **216**, 67 (1994).
16. A. O. Salnick and W. Faubel, *Appl. Spectrosc.*, **49**, 1516 (1995).
17. S. J. Bajic, J. F. McClelland, unpublished results.
18. S. J. Bajic, S. Luo, R. W. Jones, and J. F. McClelland, *Appl. Spectrosc.*, **49**, 1000 (1995).

7 Infrared Microspectroscopy

Jack E. Katon

7.1. HISTORICAL BACKGROUND

Once the power of infrared spectroscopy for the identification and quantitation of chemical compounds had been established after 1940, as discussed in detail in Chapter 2, the coupling of the infrared spectrometer and the microscope became a rather obvious good idea. Such a combination would allow the analyst to utilize infrared spectroscopy on smaller samples than would be possible using the spectrometer alone. The feasibility of this approach was demonstrated shortly after World War II and its first report appears to be the paper by Barer, Cole, and Thompson in 1949 [1]. There followed several other systems until Coates, Offner, and Siegler of the Perkin Elmer Company reported the first commercial infrared microscope designed to function in conjunction with an infrared spectrophotometer [2]. The microscope, designated the model 85, was interfaced to Perkin Elmer's most advanced infrared spectrophotometers of the time, the models 12, 112, or 13. The minimum sample size needed to obtain a satisfactory spectrum was

Modern Techniques in Applied Molecular Spectroscopy, Edited by Francis M. Mirabella.
Techniques in Analytical Chemistry Series.
ISBN 0-471-12359-5

100 ng. The time required for proper alignment and data acquisition was, however, so great that infrared microspectroscopy was essentially relegated to the position of a curiosity. It should be recognized, however, that the problems arose not so much from the microscope as from the inherent low-energy available in the infrared region and the rather poor utilization of this energy by the sodium chloride prism, dispersive infrared spectrometers that were used at that time.

In the 1970s the need of the semiconductor industry for an instrument to identify small foreign particles led to a renewal of interest in the problem. In 1978 Nanometrics, Inc. marketed their NanoSpec/20 infrared microspectrometer. This instrument utilized a new detector that was much more sensitive than the old thermocouple detector and a microprocessor by which multiple scans could be added to increase signal-to-noise ratio (S/N). In addition the infrared spectrophotometer used a filter system as a monochromator. This design allowed a considerably higher energy throughput but at a significant cost in spectral resolution. The NanoSpec/20 did not experience wide acceptance outside the semiconductor manufacturing industry.

In 1970 the first commercial Fourier transform infrared spectrometer (FT-IR) was delivered. Based on a fast scanning interferometer with on-line computing and plotting facilities, it gained rapid acceptance in the chemical infrared spectroscopy community. Somewhat later, the sensitive, liquid nitrogen cooled mercury-cadmium-telluride (MCT) detector became available. This combination resulted in a great improvement in S/N and allowed the development of many routine applications of heretofore very difficult infrared spectroscopic procedures.

In 1984 Analect Instruments (now KVB, Inc.) introduced a transmission-only microscope that was interfaced to their AQS FTIR. This system gained rapid acceptance in the forensic community and initiated the acceptance of infrared microspectroscopy in the overall chemistry community. There followed a succession of infrared microscopes for use with infrared spectrometers, introduced by various manufacturers. Notable among these was the Specta Tech, Inc. IR-PLAN, the first commercial infrared microscope to combine the high-quality visual imaging of a research light microscope with an infrared microscope. This system was introduced in 1986. Since that time infrared microspectroscopy has been widely accepted throughout industry. Nearly all organizations have problems with unidentified small particles in some phase of their operations. The ease of use of modern systems and the rapidity of obtaining spectra have led many infrared spectroscopists to use the microscope even when a large quantity of sample is available.

As an example of the performance of current systems, the spectrum of about 6 ng of isotactic polypropylene is shown in Fig. 7.1. The particle is roughly circular in cross section with a diameter of about 28 μm. This

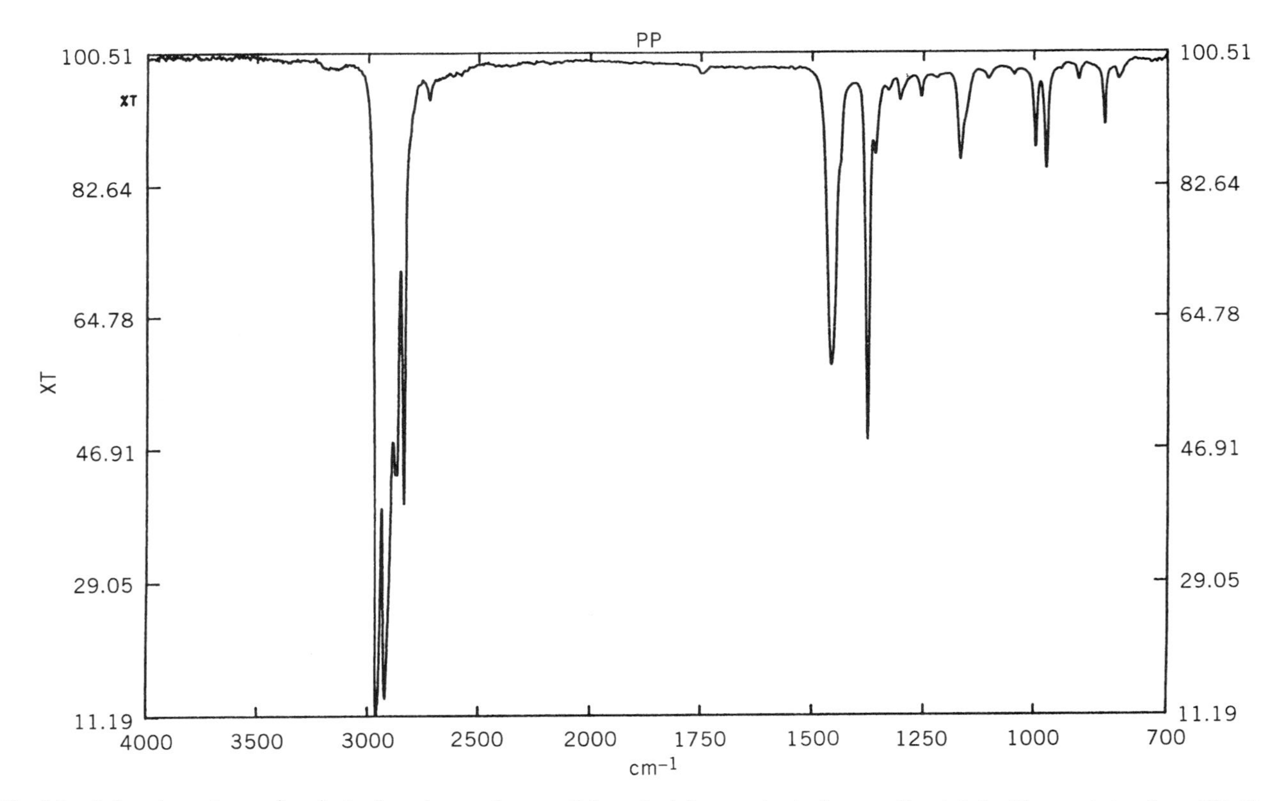

Fig. 7.1. Infrared spectrum of an isotactic polypropylene particle excised from a sheet of paper. Reprinted with permission from [9]. Copyright 1992 American Chemical Society.

spectrum was obtained with 64 scans at a resolution of 4 cm^{-1} with medium Beer-Norton apodization. The spectrum has *not* been flattened or smoothed and is clearly of reference spectrum quality. It was obtained with a Perkin Elmer Model 1800 FT-IR fitted with a nitrogen-purged Spectra Tech IR-PLAN.

7.2. INFRARED MICROSCOPE

It is useful, in discussing the infrared microscope used in spectroscopy, to review the principles of the optical microscope to make it more familiar to the reader. Comparisons can then be made. Initially we will consider transmission only. In an optical microscope, radiation from a white light source is first passed through a field diaphragm, or iris. This effectively controls the area of the illumination on the sample stage. The size of this focal spot can be continuously varied by the iris and is normally circular in shape. The light then passes through a lens, called the condenser, which focuses the radiation at the sample stage. The radiation is then collected by a second lens, called the objective, which forms a magnified image at its focal plane. This image is then focused, with additional magnification, on the observer's eye. The observer may differentiate two or more objects on the sample stage, if their size is sufficiently large and they are sufficiently far apart, by color or contrast. The various lenses are constructed of glass, and the magnification is brought about by the refraction of light by the lenses. The microscope is therefore said to be a refracting system.

In infrared microspectroscopy the source of radiation is the normal source of the infrared spectrometer, namely a heated body, often just a nichrome resistance wire. The visible radiation is removed by some type of filtering system, and the resulting infrared radiation is utilized by the microscope. Since glass is not transparent in the infrared region of the spectrum, refracting optics cannot be used. The optical components of the infrared microscope are therefore reflecting in nature. These may be simple shaped mirrors, such as paraboloidal, or they may be two mirrors working together to form a lens. The former is commonly called an off-axis element, the latter an on-axis element. Figure 7.2 is a cutaway diagram of such a lens. This is, strictly speaking, a Schwarzschild-type lens but is usually called a Cassegrainian lens or just a Cassegrain. The Cassegrain can be used as condenser, an objective, and as a lens to focus the radiation on the detector (the analogue to an eyepiece in an optical microscope). All infrared microscopes use Cassegrains as objective lenses, but there is variation between microscopes as to whether off-axis or on-axis elements are used for other purposes. The ray diagrams in Fig. 7.2 are bidirectional. The large

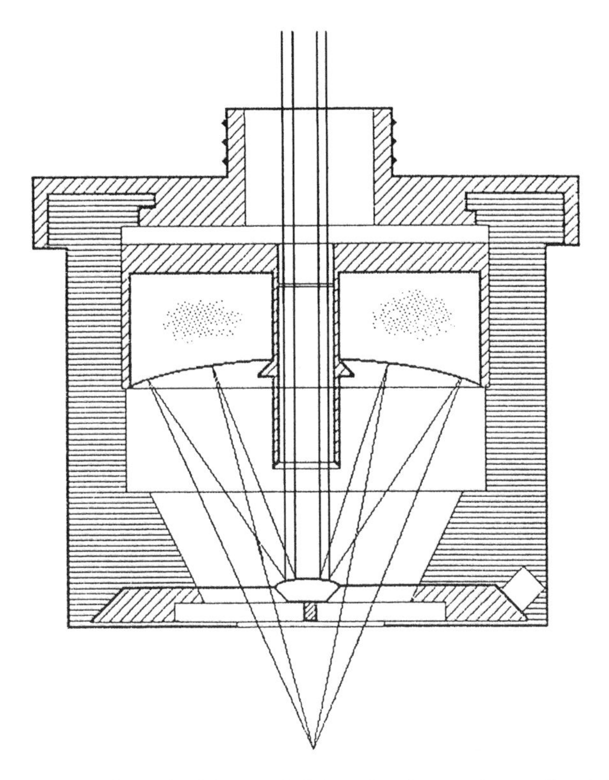

Fig. 7.2. Cutaway drawing of a "Cassegrain" lens showing ray traces. Reprinted from [5, p. 176] by courtesy of Marcel Dekker, Inc.

primary mirror works with the smaller secondary mirror so that the lens can be used to focus radiation (condenser) or collect it (objective). Note that the secondary mirror presents a partial barrier to the radiation. This is called the central obscuration, and although it can be minimized, it is always present. In addition to decreasing the energy throughput, it also affects the spatial resolution of the system (spatial resolution is discussed in detail later).

Two important characteristics of a microscope lens are its magnification and its numerical aperture (NA). Magnification is of course simply the ratio of the image size to the object size. NA is given by NA $= n \sin \theta$. Here n is the refractive index of the medium (the medium is always air with reflecting optics, so $n = 1$) and θ is the angle between the optic axis and the most extreme ray accepted by the lens. The magnification is important only to the visual operation of the microscope, since it does not affect energy throughput. However, the NA is directly proportional to the energy throughput and so should be as high as practical. The most widely used magnification factor of infrared lens is 15 ×, although higher magnification up to 36 × are readily available. Typical NA's of infrared microspectroscopic lenses are 0.5–0.7. Undesirable optical affects occur with higher NA's.

By far the most widely used detector in infrared microspectroscopy is the liquid nitrogen cooled MCT detector. Usually the microscope utilizes a dedicated detector rather than that of the infrared spectrometer. The difference between an infrared detector and the human eye requires that the infrared microscope have some additional features. The infrared detector essentially measures the total energy throughput of the system without reflecting the presence, or number, of areas in the field of view that may transmit differently. That is, differences in color and contrast are not detected by the infrared detector as they are by the human eye. Therefore it is required that some mechanism be incorporated to limit the field of view of the detector to that part of the image that is of interest. Such a device is called a field stop, and it is placed at the primary image plane of one of the microscope lenses (preferably the condenser). In infrared microspectroscopy it is commonly called an aperture. It consists of an adjustable opening (either circular or rectangular in shape) that can be used to mask the undesired part of the image so that no radiation strikes this part of the field (assuming the aperture is before the sample). The detector will then receive, in principle, only radiation transmitted by that portion of the field of view that is of interest. By placing the aperture at the image plane, it can be adjusted with respect to a magnified image, which is a much easier task than attempting to define the field at the sample. The aperture must be adjusted visually of course. This requires that an optical microscope be combined with the infrared microscope and that the two microscopes be parfocal with the visible and infrared radiation being collinear.

A final important consideration in the instrumentation required for high-quality infrared microspectroscopy is the area of the detector face that responds to the infrared radiation. For the best S/N this area should be matched to the area of the image of the sample. The system is then said to be throughput matched. Both the signal and the noise are proportional to the sensitive area of the detector. Thus, for a given size detector, the noise will be constant. If the area of the sample image is less than that of the detector, the noise will be greater than need be; if the area is larger, some of the signal energy will be wasted, since it will not be detected. Since samples to be studied are of various sizes, a compromise detector size is required. The standard detector is 250 × 250 μm, but other sizes are available. More detailed discussion of the coupling of the infrared microscope to the infrared spectrometer may be found in the literature [3–6].

7.3. SAMPLING

One of the more important concepts to consider in infrared microspectroscopy is that the sample is a part of the optical system. This statement has a number of significant implications with regard to the quality of the spectrum obtained from a given sample. Most samples dealt with in microspectroscopy are solid, although one can deal with liquids if they are not volatile. The sample characteristics of most importance are size, shape, refractive index, and homogeneity.

Since we are dealing with microspectroscopy, we expect the sample to be small. Whenever the size of a particle approaches that of the wavelength of light being used to study the particle we expect diffraction effects to appear. Thus, when the cross-sectional diameter of a sample becomes less than about 100 μm, we may begin to see the effects of diffraction in our resultant spectrum. These effects become increasingly more important as the size of the sample decreases. At sample sizes below about 8 μm, the energy throughput of the system becomes too low, and special techniques are required to obtain a reasonable spectrum. The diffraction effects arise from two different sources—the sample and the aperture edges that are being adjusted to define the sample. Diffraction by the edges of the aperture can be reasonably well compensated by recording the background spectrum (always necessary with an FT-IR) with an aperture identical to that employed for the sample. Failure to use this procedure will result in a sloping baseline in the observed spectrum (higher loss of transmission at longer wavelength). Dealing with the effects of diffraction by the sample is more difficult. In order to understand the observed result, it is useful to review some of the fundamentals of diffraction.

If one images a point source with a lens one finds that a simple image is not observed. Rather, a concentric series of rings of alternating maxima and minima are observed around a central bright image. This bull's-eye pattern is called the Airy disk, and it is due to diffraction. The bright central disk contains about 84% of the original energy. The remainder is spread into the outer rings. This implies that if one images two small particles, the radiation from the two will overlap to an extent that depends on their separation. If their separation is sufficiently small, the radiation overlap is so severe that one cannot distinguish the two; that is, they will not be spatially resolved. At some particular separation the two images may be just resolved; that is, one may determine that two objects are present. The most commonly used criterion for spatial resolution is the Rayleigh criterion, sometimes called the least resolvable separation (LRS). This quantity is given by $\text{LRS} = 0.61\lambda/\text{NA}$. Note that LRS becomes larger as λ (the wavelength of the light) becomes larger. Thus the LRS of the visible radiation will be lower than that of the infrared radiation, and two objects that are spatially resolved by the eye may not be resolved by the infrared detector.

A second diffraction effect observed is the formation of an axial dimension of the image. This is commonly referred to as the depth of field or depth of focus of a lens.

An additional problem arises with Cassegrain-type lens. The central obscuration brought about by the secondary mirror not only reduces the energy throughput but has a diffraction effect. The effect produced is that of decreasing the energy in the central disk of the airy pattern by shifting it to the outer rings. Thus diffraction effects are magnified. Since the infrared detector is much more sensitive than the eye, this also increases the effects of diffraction.

From the point of view of the spectroscopist, sample diffraction leads to exposing the detector to radiation originating from a physical area larger than that defined by the aperture. This will clearly affect the observed spectrum and may be noticed in two different ways depending on the nature of the sample. In the case of a freestanding sample, the excess radiation received by the detector can be considered as radiation that has passed around the sample and not through it. Typically this is called stray light by spectroscopists, and the observed effects on the spectrum are well-known—an offset of the 0% transmission line and a resulting photometric inaccuracy and nonlinearity throughout the spectrum. Figure 7.3 illustrates the effect with spectra of two samples of cellulose acetate film [5]. The bottom spectrum is taken from a film that has a large cross-sectional area in comparison to the aperture dimensions so that the edges of the film are far away from the aperture. The upper spectrum is a portion of the same film whose dimensions are exactly those of the aperture ($15 \times 15\,\mu m$). Since diffraction is wavelength dependent and wavelength varies from about 3 to about 20 μm, the effect will vary over the spectrum, becoming greater at the long-wavelength (low-frequency) end of the spectrum. Therefore the lower-frequency end of the spectrum suffers much more from poor definition and loss of resolution. The higher-frequency end is not without distortion, however. The carbonyl stretch at about $1750\,cm^{-1}$ shows 28% T at its peak in the bottom spectrum but an apparent 36% T in the upper spectrum. At lower frequencies the effects become more severe. Note the changes in the relative intensities of the methyl bending mode at $1375\,cm^{-1}$ and the C–O stretching mode at $1225\,cm^{-1}$. At still lower frequencies the absorption bands tend to fade into the background and may not be resolved.

If the sample is embedded in a matrix, the diffraction-produced radiation has passed through the matrix, and therefore one has the added problem of spectral contamination. This may be severe if the matrix has very strong absorption bands relative to those of the sample.

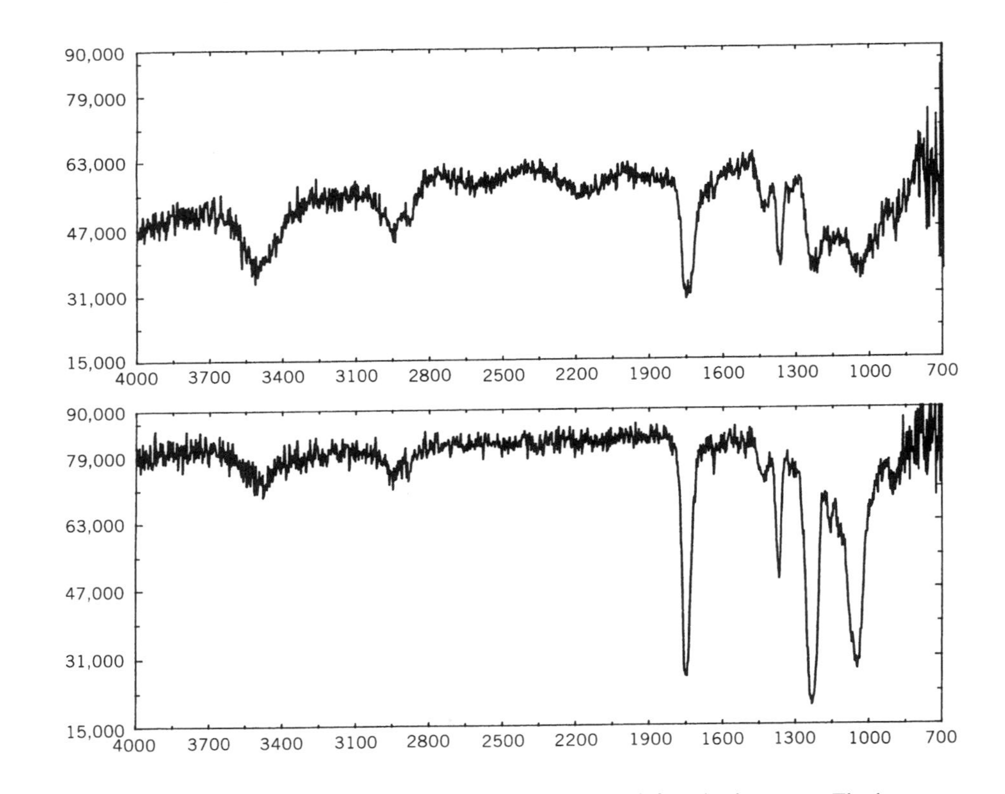

Fig. 7.3. Infrared spectra of cellulose acetate films recorded with an infrared microscope. The lower spectrum is taken from a film whose area is large compared to the aperture size. The upper spectrum is taken from a film whose dimensions are those of the aperture. Reprinted from [5, p. 196] by courtesy of Marcel Dekker, Inc.

Although the effects of diffraction cannot be completely eliminated, they can be greatly reduced. This is done by placing a second field stop (aperture), which is identical in size of opening to the other aperture, at the primary image plane of the objective. This is generally called Redundant Aperturing in the infrared microspectroscopic community. This is a copyrighted term of Spectra-Tech, Inc. In optical microscopy it is termed matched field illumination. As an aside it can be noted that one can find varying descriptions of apertures, condensers, and objectives in the literature. In this article the term "objective" always refers to the lens immediately prior to the detector in the light path, and "condenser" refers to the lens immediately prior to the sample in the light path. Some microscopes reverse the light path when observing the sample visually from that used for infrared detection. The function of the lenses then change. In addition some microscopes use one lens as both the objective and the condenser by utilizing reflection (see Section 7.4). A single aperture can be at the primary image plane of either lens for visual observation, but for infrared work it has been shown that its optimal position to reduce stray light is at the condenser image plane, that is, prior to the sample [7]. In effect this conclusion implies that stray light increases as the size of the focal spot increases. Since off-axis elements (parabolic sections) generally have larger focal spots at the sample than do the on-axis lenses, it is expected that systems utilizing off-axis elements as condensers will have poorer performance with respect to stray light than those using lens condensers. Figure 7.4 is a diagram of the optical path of a typical infrared microscope possessing dual apertures.

The refractive index of the sample arrangement affects the results in two ways. Most infrared spectroscopists are familiar with the problem that occurs when the infrared spectrum of a group of small, irregularly shaped particles is desired. Reflection and refraction of the infrared beam by these particles leads to excessive scattering. This manifests itself in a high, sloping background and, sometimes, other spectral anomalies such as the Christiansen effect [see Chapter 2]. The usual way of treating this situation is to suspend the particles in some medium with a refractive index intermediate between air and that of the sample, such as Nujol or KBr. The most widely used procedure is that of grinding the sample until it is a relatively fine powder, mixing it thoroughly with about 100 times its weight of KBr, and then pressing it in a die, at high pressure (about 60,000 psi). The resulting pellets will usually give a much improved infrared spectrum of the sample. In infrared microspectroscopy one often observes similar effects due to sample refractive index. The problem can be treated in a very similar way. Three or four crystals of KBr are placed on a smooth surface alongside the sample crystal or crystals. These crystals are then pressed, with a rolling motion, by a fine pointed probe. With a little practice, one can make a suitable micropellet that gives a much improved infrared spectrum. A

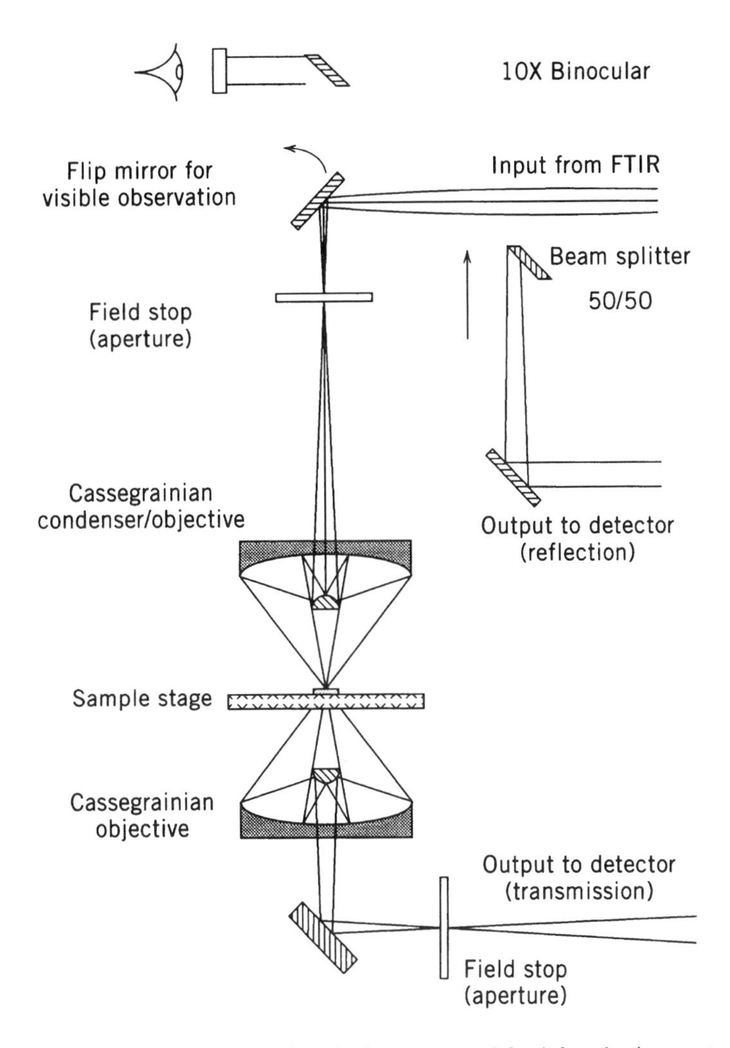

Fig. 7.4. Optical diagram of an infrared microscope used for infrared microspectroscopy. Reprinted from [5, p. 182] by courtesy of Marcel Dekker, Inc.

relatively small force on the probe handle is converted into a suitably high pressure at the probe tip to cause the KBr to flow suitably around the sample. Other soft alkali halides, such as KI, CsI, and CsBr, can also be used, sometimes with better effect because they are softer than KBr.

The effect of the refractive index on the focal plane of the microscope is a second potential problem. If the microscope is out of focus, results will always be poor. The problem does not normally arise with the sample, but with the support used. The sample thickness is usually about the same as the depth of field/focus of the infrared lens typically used (10–20 μm), so changes in the refractive index of the sample do not greatly affect the focus of the microscope except in extreme cases. For transmission studies, however, the sample is almost always placed on some infrared transparent support, usually KCl. The foci of the two lens are then adjusted to be 5–10 μm above the upper surface of the KCl. If the substrate is changed, however, to a material with a different refractive index, the microscope's focus may be changed drastically. For instance, if the substrate is changed from a 2-mm KCl support to a 2-mm ZnS support, the focus is shifted 0.4 mm [5], a very large shift in relation to the typical sample thickness. It is therefore important to be sure the microscope is in focus for any collection of spectral data. Changing the thickness of the substrate, even though it is identical in composition, will also change the focal plane.

There are two factors of interest concerning the shape of the sample: cross-sectional geometry and thickness. The latter is a much more formidable problem than the former. Since the apertures are ideally of the same geometry as the sample so that space surrounding the sample is masked from the detector's view, their geometry is made variable. One can obtain either circular apertures with variable diameter or rectangular apertures with variable lengths of sides. These are placed in the microscope such that they can be readily adjusted and/or interchanged. Many samples have approximately circular cross sections as received and circular apertures are often used because they require only one adjustment (diameter of the circle). On the other hand, rectangular apertures are much better for certain types of samples such as fibers or cross-sectional slices of a multilayer film. It is desirable therefore to have both types of aperture available.

Sample thickness is of more concern, since too thick a sample can give deceptive results, particularly to the inexperienced microspectroscopist. As mentioned earlier, there is always stray light present because of diffraction. If the sample is thick, the amount of stray light accepted by the lens will be larger. In addition the sample may be thicker than the depth of field/focus of the lens, giving further distortion. The instrument operator who is unaware of these problems may use the computer to expand the ordinate to make a more presentable spectrum. This can lead to false conclusions. Figure 7.5 presents a rather extreme example of the effect described. The spectra were taken with a single circular aperture system so that the thickness problem is shown more clearly. The upper spectrum is that of a 30-μm-diameter single fiber as received. The bottom spectrum is that of the same fiber which has been flattened while it was on the KCl substrate by rolling a probe over it with finger pressure only. The instrumental variables are the same for each spectrum. The sample is totally absorbing (0%T) over

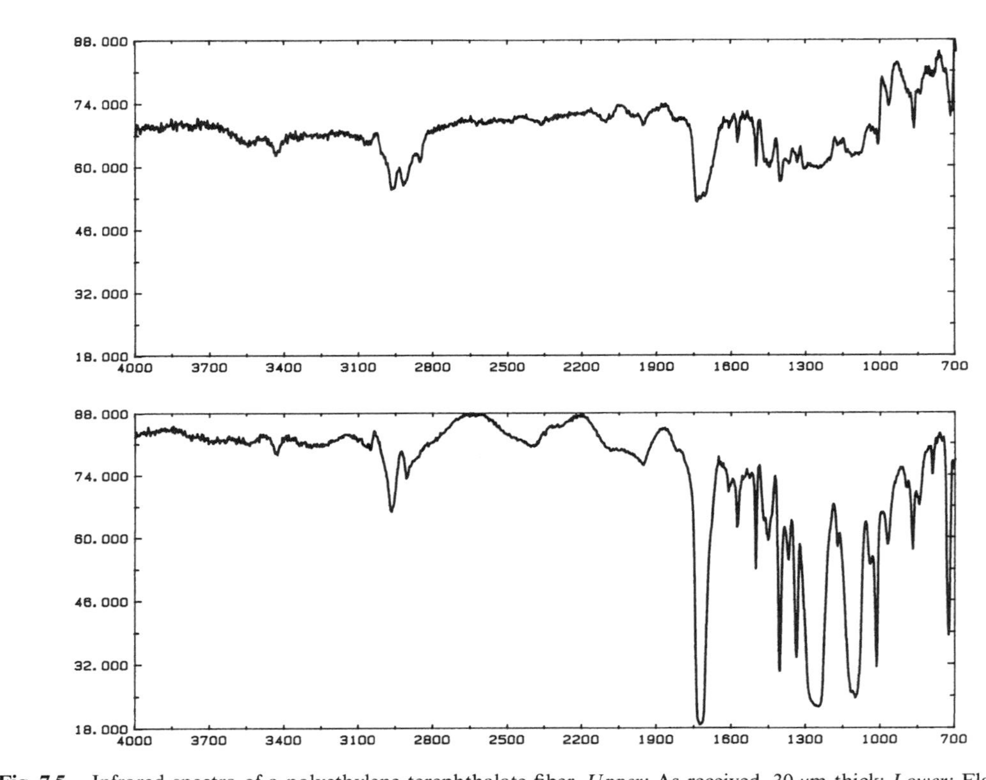

Fig. 7.5. Infrared spectra of a polyethylene terephthalate fiber. *Upper*: As received, 30 μm thick; *Lower*: Flattened by rolling a probe across the fiber. Thickness seems to be about 10 μm. Reprinted from [5, p. 182] by courtesy of Marcel Dekker, Inc.

a significant frequency range in the upper spectrum, but this is certainly not apparent on superficial examination.

The ideal sample thickness varies rather widely depending on the strength of its absorption bands. A good approximate guess is 10 μm. The microspectroscopist needs to be aware of the thickness problem in all samples, however. If the sample is relatively soft, it can normally be suitably thinned by pressing it with a probe or roller type accessory. Nearly all organic samples can be treated this way. If, however, the sample is elastomeric, a method to maintain a suitable pressure on the sample must be utilized. This requires a pressure cell using infrared transparent windows. The most convenient cell of this type utilizes type II diamond windows and is commercially available. The diamonds have only one strong absorption band, centered at about 2000 cm^{-1}, a frequency that does not interfere with the spectrum of most materials. Two types of cell are available, one for low to moderate pressure (suitable for nearly all elastomeric materials) and one for high pressure (suitable for very hard samples, e.g., inorganic materials). The strength of the diamond absorption of the low-pressure cell is not sufficiently great to totally absorb the energy at 2000 cm^{-1}, so the band can be subtracted from that of the sample spectrum, if desired. The spectrum of a 0.8-mm-thick elastomeric material (refrigerator door gasket) which has been strongly compressed in such a low-pressure diamond cell is reproduced in Fig. 7.6 [8]. The diamond absorption has been used as a background, and the sample spectrum is then ratioed to the background. The higher noise around 2000 cm^{-1} reflects the fact that less energy is available in this region due to the diamond absorption.

Most infrared microspectroscopic investigations are carried out on solids. If the solid is a mixture of fine crystals, such as is often observed with residues from solutions whose solvent has been evaporated, one must record the data from a number of crystals by isolating them using the apertures. Often one can differentiate materials visually with the microscope, and this is an aid in selecting crystals to isolate for spectral study. Care must be taken to record data from several crystals to avoid misleading conclusions, however. Small amounts of impurities do not generally hinder macro infrared spectroscopic identification of materials, since the impurity absorptions are dominated by those of the major component. With microspectroscopy, however, one can study both major and minor components individually. This can be a disadvantage if the identification of only the major component is desired, but it is of great utility if the identification of impurities is also important. If the mixture is so intimate that individual components cannot be isolated by the apertures, then separation techniques must be considered. These are discussed in Section 7.5.

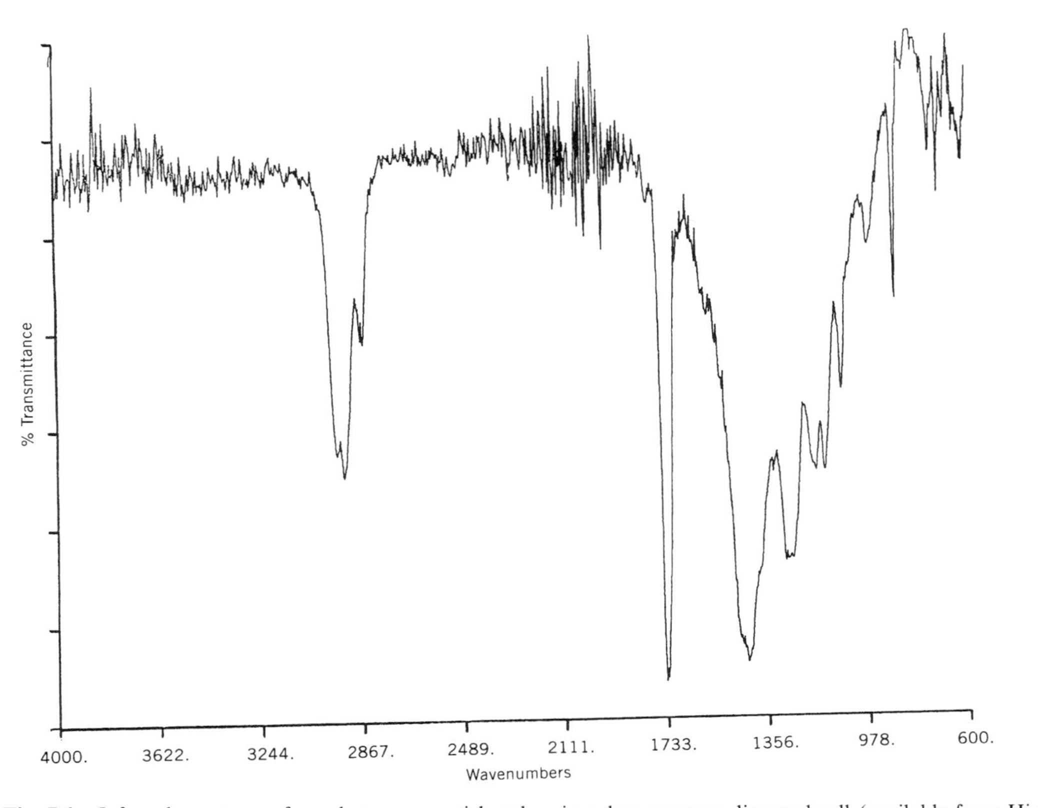

Fig. 7.6. Infrared spectrum of an elastomer particle taken in a low-pressure diamond cell (available from High Pressure Diamond Optics, Tucson, AZ).

7.4. METHODS OF OBTAINING IR SPECTRAL DATA

In conventional infrared spectroscopy the sample is studied by the transmission technique if it is feasible. There are two basic reasons for this. First, this technique gives a true absorption spectrum (assuming proper care is taken to prepare the sample and record the spectrum). Second, a transmission spectrum can usually be obtained faster than by any other means. The most widely used other methods of obtaining spectra utilize one of the reflection techniques, and they invariably require some accessory instrumentation that must be aligned in the instrument.

A schematic for a simple addition to infrared microspectroscopy is given in Fig. 7.4. The most important is the beam splitter which is inserted into the beam at some suitable point between the infrared source and the sample. The beam splitter will normally transmit 50% of the radiation and reflect the remaining 50% in some innocuous direction. The transmitted radiation will then proceed to the sample, reflected energy will be returned along the optical path to the beam splitter, and the 50% reflected energy will now be directed to the detector by some suitable arrangement of one or more mirrors. With many samples this procedure removes the necessity of sample preparation, including thinning a very thick sample, since this optical procedure reduces the beam intensity by a factor of four as half of the available energy is rejected (twice) by the beam splitter. This will of course reduce the S/N of the spectrum. There is, however, a potentially more serious problem. Reflection techniques do not give true absorption spectra, and this must always be kept in mind by the spectroscopist. The addition of the beam splitter to the microscope allows the operator to readily obtain reflection-absorption spectra (thin films of samples on a highly reflecting substrate), specular reflection spectra (dielectric materials with a smooth surface), and diffuse reflection spectra (powders or fine crystals). In all cases the microscope's condenser lens also serves as the objective lens.

Mathematical manipulation of the data by the FTIR computer system can convert specular reflection spectra (Kramers-Kronig transformation) and diffuse reflection spectra (Kubelka-Munk transformation) to absorption spectra, but the results are only accurate if that type of reflectance is the only one present. In practice, many samples give experimental spectra that contain both kinds of reflectance data. Such spectra cannot be converted to true absorption spectra, although they may be closely approximated in many cases.

Two other types of reflection infrared spectroscopy are feasible with the microscope by the utilization of specialized accessories. Grazing angle reflection-absorption spectroscopy is a useful technique for studying very thin films of material on a reflective substrate. Whereas the normal procedure for reflection-absorption spectroscopy involves an incidence and reflectance angle of the infrared beam of about 10–30 degrees, the grazing angle procedure utilizes an incidence and reflectance angle of about 75°. The infrared beam therefore traverses a considerably longer path through the sample than it does with small angles and produces a stronger spectrum. With infrared microspectroscopy it does lead to poorer spatial resolution than that obtained with low angles of incidence and reflection. The minimum film thickness required to obtain a reasonably good infrared spectrum varies with the material but is usually in the tenths of a nanometer range. The sample area required is about $1 \times 10^4\ \mu m^2$ [9]. A grazing angle objective is commercially available and simply replaces the normal Cassegrain lens in the microscope for such studies.

Attenuated total reflection (ATR) infrared spectroscopy has proved to be a very valuable technique for recording spectra of highly absorbing, relatively intractable materials. The infrared beam passes through a dense (high-refractive index) but infrared transparent material and strikes a second less dense (lower-index) material at the interface of the two. The beam is controlled such that it strikes the interface at some angle greater than the critical angle. At such angles the radiation will be totally reflected but will be attenuated by the second material at its absorption frequencies. The result is an infrared spectrum that usually closely resembles an absorption spectrum. The method requires intimate contact of the materials at the point of reflection, and therefore the best results are obtained with relatively soft, easily deformable samples or samples with very smooth surfaces.

An ATR objective that utilizes various high-density hemispheres (ZnS, KRS-5, germanium, diamond, etc.) is also commercially available. The Cassergrain lens is again simply replaced with this objective to make ATR spectral data obtainable. Due to size limitations the unit is limited to a single reflection from the interface. Good spectra can be obtained with a contact area as little as $400\ \mu m^2$ or less [9].

7.5 QUALITATIVE AND QUANTITATIVE ANALYSIS OF MIXTURES

The analysis of mixtures by infrared microspectroscopy can present some new problems when compared to the macro case. If the sample consists of a mixture of relatively large ($10\ \mu m$ or larger) pure crystals, their identification is relatively straightforward by the simple procedure of aperturing around each crystal, in turn, and recording their infrared spectra. If the mixture is more intimate, however, one must devise a method of separating the components, either physically or spectroscopically. By far the easiest method for qualitative analysis is spectral subtraction. This method is often used in ordinary infrared spectroscopy and is relatively well-understood.

The usual procedure in ordinary infrared spectroscopy is to dissolve the sample first in some infrared transparent, chemically inert solvent at low concentration. This procedure minimizes intermolecular interactions that can give rise to shifts in absorption frequencies, changes in absorption intensities, and so on. By using relatively large samples initially, one also minimizes sample inhomogeneity if the original mixture is a solid. If the sample consists of one or two very small particles inhomogeneity is essentially meaningless, however, and, since solids do not ordinarily exhibit measurable intermolecular interactions, spectral changes due to the mixing are not a problem. The method works best when the components of interest are all present at relatively high levels. In such a case one can usually obtain a spectrum in which few or none of the bands possess an absorbance much greater than one unit. Absorbances higher than this cannot be completely subtracted, and some residual will always be left. More important, any overlap of other component bands with such a strong band will not be detectable. This means that some portion of the spectrum will not be available for spectral information. Identification of other components must then be made from what is, in effect, an incomplete spectrum. The magnitude of this problem is highly variable; it depends on the number of components present, their individual spectra, and the availability of other data pertaining to the system. The method has been successfully used to identify components present at the 1.0% level in certain cases.

The coupling of various chromatographic techniques with infrared microspectroscopy for separation and identification of mixtures has been considered and applied in a few cases but has not been widely studied. Enough has been accomplished, however, to indicate that there is a potential for coupling chromatographic methods with infrared microspectroscopy for the identification and, in some cases, quantitation of components in mixtures. Two quantities are of importance in this work: limit of detection (LOD) and minimum identifiable quantity (MIQ). In some cases the chromatographic detector is much more sensitive than is the infrared detector and so the LOD may be that of the chromatographic detector. In all cases, however, the chromatographic method is poor for identification and so infrared microspectroscopy is necessary for MIQ determination.

Capillary gas (GC) and supercritical fluid (SCF) chromatographic methods have been used to separate components of mixtures that have then been identified by infrared microspectroscopy, but only when used in the trapping mode. Although GC/FT-IR analyses of mixtures “on-the-fly” are well-known, they utilize a light pipe so that the pathlength of the infrared beam through the sample is quite long. The trapping of components is accomplished with a variety of “traps.” Clearly, to get good MIQs, the sample must be concentrated over as small an area as possible. One of the advantages of GC and SCF methods is that the carrier is a gas and is easily removed from the desired sample with minimal spreading effects. The MIQs obtained are of course dependent on the strength of the sample's spectrum. Thus polar compounds will, in general, have lower MIQs than nonpolar compounds. If the collector is a mirrored surface on which a thin film of powdered KCl has been placed, the eluent spots will be smaller than those obtained with other "traps." It has been indicated [10] that if such a plate is cooled to 90 K, one can obtain a MIQ of less than 100 pg for polar compounds. The purpose of the KCl is to provide an adsorbent that allows diffuse reflectance spectra to be collected. Such spectra are considerably more intense than reflection-absorption or specular reflection spectra.

High-performance liquid chromatographic (HPLC) separation followed by infrared microspectroscopic identification of components has been applied in both the trapping and on-the-fly modes. The results are much poorer, primarily because the typical carriers (water and aqueous mixtures) are more difficult to remove and are highly absorbing in the infrared. The former affects the trapping mode because of the solubility of KCl and similar salts and because the removal of the carrier leads to spreading of the sample over a larger area. Peak broadening and high background absorption of the solvent limits the application of the on-the-fly methods to relatively large quantities of components (1–2 ppt).

Thin-layer chromatographic (TLC) separation with infrared spectroscopic identification of components has received some attention. Although the procedure is sometimes useful with direct infrared spectroscopic investigation (by diffuse reflection) of the TLC plate, this approach has not been generally very satisfactory because of the high infrared absorbance of the usual substrates, alumina or silica. As a result the problem has more generally been approached by dissolving the separated components from the plate and evaporating the solvent prior to infrared spectroscopic analysis (by transmission). Such a procedure is not very attractive when dealing with small amounts of materials. More recently it has been shown that zirconia can serve as a TLC substrate and has much less infrared absorption than either silica or alumina. By using small channels filled with zirconia, the sample spots may be kept reasonably small so that infrared microspectroscopy may be applied. Preliminary work [11] has indicated a MIQ in the tens of nanogram range with LODs in the 1000–100 pg range. It is probable that these numbers can be improved with further work. Of particular interest is the fact that the method has potential for automation in that the microscope stage can be computer driven and the channel containing the adsorbent and the separated components can be scanned automatically with significant spectra recorded as the scan proceeds along the channel.

The use of infrared spectroscopy for quantitative analysis utilizing Beer's law has been widespread for many years. The typical problems encountered are well-known and widely discussed, so they do not need to be repeated here. There are, however, two basic assumptions of Beer's law that are satisfactorily met with typical infrared spectroscopy but are quite different in infrared microspectroscopy. One is the requirement that the infrared beam be collimated. The path length of the infrared beam in the sample is then identical to the sample thickness provided that the sample's cross-sectional area is entirely perpendicular to the beam. The sample thickness is usually referred to as the cell thickness, since the sample is normally in solution in some suitable solvent in an infrared transparent cell. The infrared beam in the microscope is, however, rather highly focused at the sample. This implies that the path length of individual rays varies with the angle at which the rays strike the sample. Therefore one can only measure an average path length of the beam through the sample, and it will always be greater than the actual sample thickness. Furthermore the average path length will depend on the NA of both the objective and condenser lens. The sample is normally a solid, its thickness is very low (on the order of 10 μm), and it cannot, in general, be accurately measured. An exception to this statement is the occasional highly polished sample with parallel faces that may allow the thickness to be determined by interference fringes. However, even if the sample thickness is known, the average path length must be determined. Under typical conditions the average (effective) pathlength is 10–20% longer than the sample thickness.

The second problem arises from the absorbing entities themselves. In Beer's law the assumption made is that they do not interact with the medium (solvent) or with each other. They must also be homogeneously distributed. This means that the sample is measured as a dilute solution in a solvent that is as nearly chemically inert as possible. The concentration of the absorbing entities can then be determined if one has a calibration curve made by measuring the absorbance of a series of solutions of known concentration under the same set of experimental conditions. Strictly speaking, it is only necessary to measure the absorbance at one known concentration and from this to calculate the sample's molar absorptivity, which is a characteristic constant for the absorbing substance under the conditions of the experiment. With those items in mind, it is easily seen that infrared microspectroscopic quantitation of samples (normally solids) is likely to present some severe problems. Even if one assumes that there is a constant interaction between molecules in a solid, which can be combined into the molar absorptivity, the concentration of absorbing entities per unit volume is unlikely to be known. If the sample is a pure single crystal (not the usual case), the number of molecules per unit volume depends on the material's density and molecular weight. In other cases one must also consider that the degree of packing, namely the practical "density," is not the true compound density.

Although these factors can be expected to lead to great difficulties for quantitation, there are some procedures that offer hope. The simplest of these, if it can be applied, is the use of an internal standard. The technique is well-known in analytical chemistry and involves the measurement of some property of the analyte (in this case its infrared band intensity) in relation to that of a second substance whose concentration is either constant or known. Clearly this means that the internal standard must be present in a uniform amount throughout the sample. The addition of an internal standard to a microscopic sample is generally not practical, but if a second substance is present in an essentially constant amount, it is often used. An example is the quantitation of pigments in single polypropylene fibers [12]. The absorbance of a selected band of each pigment is ratioed to the absorbance of a polypropylene band. A graph of this absorbance ratio versus pigment concentration is then constructed from fibers in which the concentration of dye is known from the amount added to the molten polypropylene prior to fiber formation. The method gave good results (relative standard deviation less than 2%) for three different pigments at concentration of 1–10%. The procedure gives best results if there is no band overlap of the two substances. If such overlap occurs, computer curve resolution techniques allow analysis of the substance with some loss of accuracy. Chemometric techniques also offer some promise for certain quantitations but they have not yet been applied to microspectroscopy.

A somewhat related technique that has stimulated the interest of many chemists is the "mapping" or "imaging" of samples by means of their infrared absorption. The method can be thought of as a type of spatial qualitative analysis, although it might conceivably also lead to semiquantitative analysis. The procedure is very similar to that used in scanning electron microscopy (SEM) to determine elemental composition of certain portions of a sample. In SEM, an electron beam is rastered across a small sample, and detector output is measured as a function of beam position. In infrared microspectroscopic mapping, the beam is in a constant position, and the sample is moved by the microscope stage in steps whose dimensions are controlled. The instrument is set to measure response at some given frequency corresponding to an absorption frequency of a substance sought. The results define positions in the sample where the analyte's concentration is high. The method has so far been mostly applied to plastics to map the positions of insoluble impurities.

An example is given in Fig. 7.7. The sample is a photographic film that contains a defect. Comparison of the spectrum in the defect region with that in a pure film region shows that the defect region is characterized by an infrared absorption at 1733 cm^{-1}. The absorbance at this frequency is plotted against sample position to give an image of the defeat. It can be seen that it is a three-dimensional image in that absorbance is dependent on

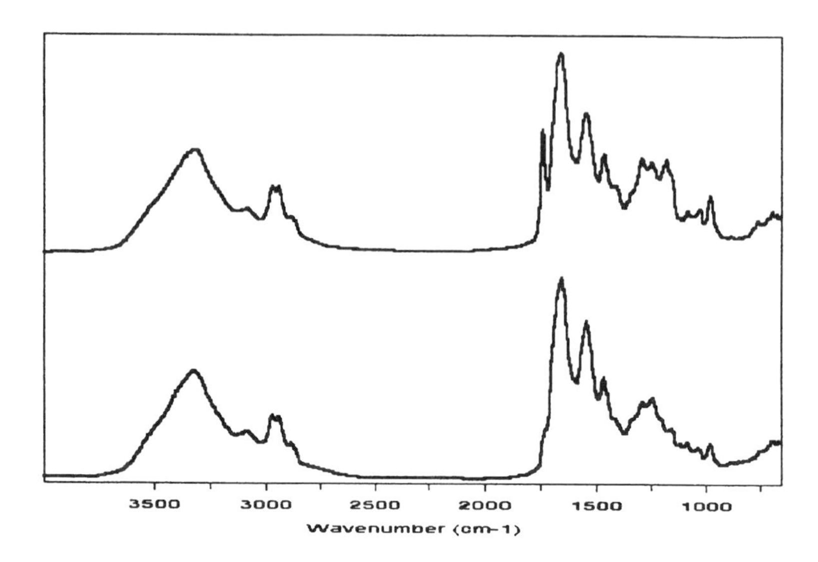

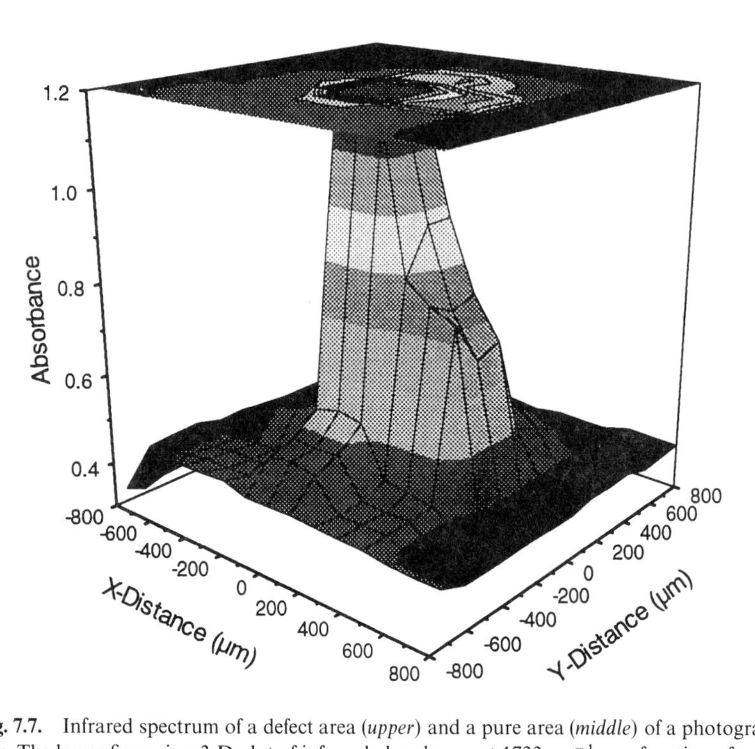

Fig. 7.7. Infrared spectrum of a defect area (*upper*) and a pure area (*middle*) of a photographic film. The lower figure is a 3-D plot of infrared absorbance at 1733 cm^{-1} as a function of sample position in the defect region.

sample thickness assuming the defect density is constant. Thus the defect thickness decreases more slowly in the *X*-direction than in the *Y*-direction.

ACKNOWLEDGMENT

The author expresses his gratitude to Dr. Pamela A. Martoglio and Spectra Tech, Inc. for Fig. 7.7.

REFERENCES

1. R. Barer, A. R. H. Cole, and H. W. Thompson, *Nature*, **163**, 198 (1949).
2. V. J. Coates, A. Offner, and E. H. Siegler, Jr., *J. Opt. Soc. Am.*, **43**, 984 (1953).
3. R. G. Messerschmidt, In *Infrared Microspectroscopy. Theory and Applications*, R. G. Messerschmidt and M. A. Harthcock, eds, Dekker, New York, 1988, pp. 1–20.
4. R. G. Messerschmidt, In *Practical Guide to Infrared Microspectroscopy*, H. J. Humecki, ed., Dekker, New York, 1995, pp. 1–39.
5. J. E. Katon, A. J. Sommer, and P. L. Lang, *Appl. Spectrosc. Rev.*, **25**, 173 (1989–90).
6. R. W. Duerst, W. L. Stebbings, G. J. Lillquist, J. W. Westberg, W. E. Breneman, C. K. Spicer, R. M. Dittmar, M. D. Duerst, and J. A. Reffner, In *Practical Guide to Infrared Microspectroscopy*, H. Humecki, ed. Dekker, New York, 1995, pp. 137–162.
7. A. J. Sommer and J. E. Katon, *Appl. Spectrosc.*, **45**, 1633 (1991).
8. J. E. Katon, P. L. Lang, D. W. Schiering, and J. F. O'Keefe, In *The Design, Sample Handling and Applications of Infrared Microscopes*, P. B. Roush, ed. American Society for Testing and Materials, Philadelphia, 1987, pp. 49–63.
9. J. E. Katon and A. J. Sommer, *Anal. Chem.*, **64**, 931A–940A (1992).
10. A. M. Haefner, K. L. Norton, P. R. Griffiths, S. Bourne, and R. Curbelo, *Anal. Chem.*, **60**, 2441 (1988).
11. S. P. Bouffard, J. E. Katon, A. J. Sommer, and N. D. Danielson, *Anal. Chem.*, **66**, 1937 (1994).
12. S. P. Bouffard, A. J. Sommer, J. E. Katon, and S. Godber, *Appl. Spectrosc.*, **48**, 1387 (1994).

8 Raman Microspectroscopy

Andre J. Sommer

8.1. INTRODUCTION

The field of Raman microspectroscopy is relatively new, but in the 20 years since its development the technique has been employed in a wide variety of scientific disciplines. To date there are more than 700 citations in the scientific literature regarding the use of this technique. Concurrently, developed by groups at the National Bureau of Standards and the University of Lille in France, the technique has found early use in the detection of contaminants associated with air pollution and electronic device fabrication [1, 2]. Although these early instruments and later commercial versions were relegated to strictly research laboratories, recent technological advances have made it possible for the instruments to be employed in the industrial quality control laboratory. The greatest benefit of the technique is that it provides molecular information on particles whose size is comparable to the wavelength of the light being employed for the analysis (i.e., typically one to two micrometers in diameter). In addition, since the technique is based on

Modern Techniques in Applied Molecular Spectroscopy, Edited by Francis M. Mirabella.
Techniques in Analytical Chemistry Series.
ISBN 0-471-12359-5

scattering, little sample preparation is required. Raman microspectroscopy is a hybrid of optical microscopy and Raman spectroscopy. As a result the method has all the concomitant advantages of both techniques.

The purpose of this chapter is to give to the reader a foundation of the underlying principles of Raman microspectroscopy upon which to build. As such, the Raman effect will be discussed in addition to the instrumentation required for an analysis. Following these introductory sections a discussion of experimental considerations in addition to the technique's capabilities and limitations will be presented. For a more comprehensive review of the technique and its applications, the reader is referred to several excellent publications [3–5].

8.2. RAMAN EFFECT

The Raman effect arises when monochromatic (single wavelength) light interacts with a molecule. During this event the molecule can add energy to the incident light by emission of a photon, remove energy from the incident light by absorption of a photon, or simply not interact with the light, in which case its energy remains the same. If the energy of the incident light is expressed in terms of its frequency, ν_{exc}, we can write the following expression:

$$E_{exc} = h\nu_{exc} \tag{8.1}$$

where h is Planck's constant.

The energy of the resultant light after interaction with the sample can then be expressed as:

$$E_{resultant} = h\nu_{exc} - h\nu_{vib} \quad \text{(Absorption)}, \tag{8.2}$$

$$E_{resultant} = h\nu_{exc} + h\nu_{vib} \quad \text{(Emission)}, \tag{8.3}$$

$$E_{resultant} = h\nu_{exc} - 0 \quad \text{(No interaction)}, \tag{8.4}$$

where ν_{vib} is the frequency associated with a molecular vibration (nuclear displacement) of the molecule. If one were to analyze the frequencies of light present after the interaction, one would expect to see light shifted toward lower frequency (higher wavelengths) in the case of absorption, and higher frequency (lower wavelengths) in the case of emission. In addition there will be light present at the excitation frequency corresponding to no interaction.

Whether or not one would observe these frequencies depends on how the atoms in the molecule are displaced during the interaction. When light interacts with the molecule it produces a dipole moment P within the molecule. The dipole radiates light according to the following equation which has been derived from classical mechanics [6]:

$$P = \alpha E_0 \cos 2\pi\nu_{exc} t + \frac{1}{2}\left(\frac{\delta\alpha}{\delta q}\right)_0 q_0 E_0[\cos\{2\pi(\nu_{exc} + \nu_{vib})t\} + \cos\{2\pi(\nu_{exc} - \nu_{vib})t\}]. \quad (8.5)$$

This equation relates the time dependent electric field amplitude E_0 of the excitation frequency ν_{exc} with the polarizability of the molecule. The polarizability α is a measure of how readily the electrons within the molecule are displaced by the electric field of the incident light. The first term on the right-hand side of the equality $\alpha E_0 \cos 2\pi\nu_{exc} t$ expresses light which is elastically scattered from the molecule. That is, it has the same frequency as the excitation light. The second term on the right-hand side of the equality describes Raman scattered light whose frequencies correspond to $\nu_{exc} + \nu_{vib}$ for the case of emission of a photon and $\nu_{exc} - \nu_{vib}$ for the case of absorption. The important feature to note is the term $(\delta\alpha/\delta q)_0$. This term is the change in polarizability associated with a nuclear displacement within the molecule. If the polarizability does not change during a nuclear displacement, this term is zero and no Raman scattering is observed for this particular vibration.

Figure 8.1 illustrates the Raman spectrum of carbon tetrachloride in the 600 cm^{-1} shift region. A wavenumber (cm^{-1}) is proportional to frequency and is the unit commonly used by molecular spectroscopists. Wavenumber shift is typically plotted, since all Raman frequencies are observed relative to some excitation frequency. Raman scattered light shifted to lower frequencies, $\nu_{exc} - \nu_{vib}$, relative to the exciting frequency are known as Stokes shifted frequencies. Those shifted to higher frequencies, $\nu_{exc} + \nu_{vib}$, are known as anti-Stokes frequencies. The Stokes shifted frequencies correspond to an absorption process and are therefore more intense than the anti-Stokes frequencies, which correspond to emission. This intensity variation arises from the fact that a large majority of the molecules exist in a ground state and undergo absorption rather than emission. Since the Stokes transitions are much more intense, their utility for analytical purposes is preferred over the anti-Stokes transitions. Although Stokes shifted frequencies are lower in energy than the excitation frequency, they are assigned a positive value for ease of use. Figure 8.1 also illustrates the specific molecular vibrations that give rise to each Raman transition. It is beyond the scope of this treatise to provide a full description of Raman scattering from a molecular symmetry standpoint. The point to be made, however, is that the fundamental vibrational frequencies are very sensitive to changes within the molecule, or the molecule's local environment. For excellent discussions regarding Raman scattering and molecular symmetry considerations, the reader is referred to Nakamoto, Herzburg, and Guillory [6–8].

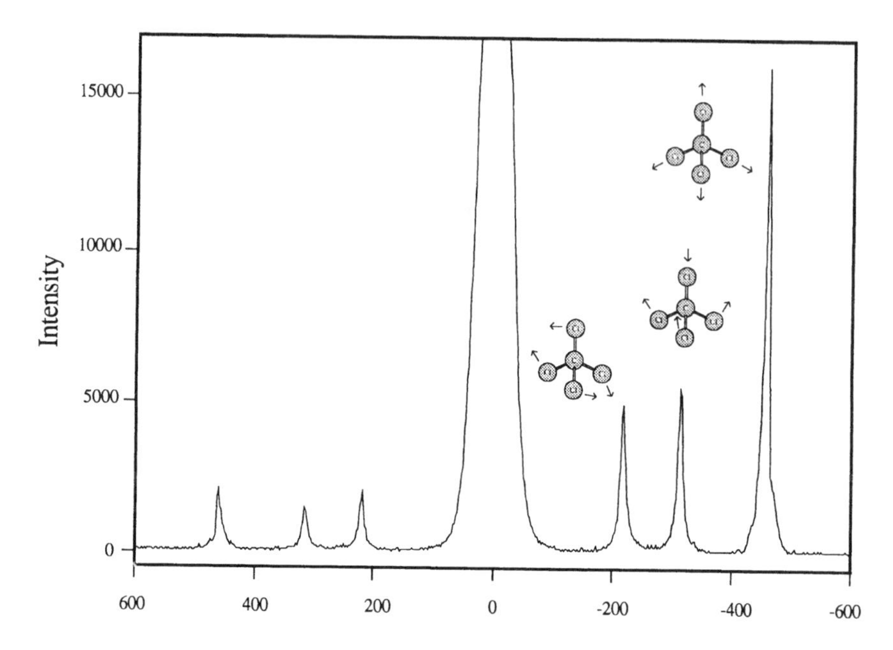

Fig. 8.1. Raman spectrum of carbon tetrachloride.

8.3. SIGNAL PRODUCTION AND COLLECTION

Two fundamental processes involved in the observation of Raman spectra are the excitation of the Raman scattered radiation followed by detection of this radiation by some spectrally sensitive and discriminating device. The Raman signal, in watts sr^{-1}, expected from a uniformly illuminated particle is given by [3]

$$S = NVI_0\sigma, \tag{8.6}$$

where N is the number of particles per unit volume (density), V is the volume of uniformly irradiated sample, I_0 is the excitation laser irradiance (watts cm^{-2}), and σ is the differential Raman scattering cross section ($cm^2\ sr^{-1}$ molecule^{-1}). This latter term is similar to the absorption coefficient in infrared spectroscopy and, as such, is a measure of the strength of Raman scattering produced by a given sample.

When trying to collect this signal, several instrumental parameters enter into the equation:

$$S = NVI_0\sigma\Delta\nu t^{1/2}\Omega\xi(\text{n.e.p.})^{-1}\lambda_{\text{exc}}^{-4} \tag{8.7}$$

These parameters include the spectral resolution employed for the measurement $\Delta\nu$, the measurement time of each spectral element t, the limiting aperture of the optical system Ω, the optical systems efficiency ξ, the noise equivalent power of the detector (n.e.p.), and the excitation wavelength employed in the measurement λ_{exc}. Several of these parameters can be controlled by the analyst (e.g., $\Delta\nu, t$). However, sample limitations, for the most part, will be the limiting factor in achieving as much signal as possible. One parameter that is limiting in the case of a microprobe configuration is the throughput Θ. This parameter is the product of the sampling area and the limiting aperture of the optical system. Since the sampling area can be as small as one micrometer in diameter, it is this parameter that usually limits the throughput of the optical system. The throughput will be discussed in the context of collection lenses in the following section.

8.4. INSTRUMENTATION

Instrumentation in the field of Raman microspectroscopy has, over the past few years, developed into many specialized instruments. These instruments vary with respect to one another in the wavelength of light employed for excitation, the manner in which the sample is illuminated (e.g., punctual, line, or global illumination), and the manner in which the signal of interest is collected and presented. Today's Raman microprobe user is presented with the ability to collect a Raman spectrum in a normal scanning mode (i.e., serial collection), in a spectrograph mode (parallel or multiplex collection), or some hybrid form of each. In addition, due to technological innovations over the past several years in filter and detector technology, the user can also collect two- and three-dimensional molecular maps of a given surface in an imaging mode. The image provides a detailed map of the position of the Raman scattering species of interest.

A detailed discussion of each of these types of systems and their relative merits is beyond the scope of this treatise. However, inherent in each of these systems is a basic design common to all Raman microprobes. The construction of a system involves coupling a research grade optical microscope to a Raman spectrometer. Individual components of a typical Raman microprobe include a laser excitation source, some laser conditioning optics, the microscope, some wavelength selective filter (e.g., spectrometer, interferometer), and a means for detecting the Raman scattered radiation after spectral selection. A general description of their function(s) is as follows: The laser excitation source is focused onto the sample with the aid of the microscope. Raman scattering is then collected with the microscope, which provides the wavelength selective filter with a magnified image of the excited sample. The filter selects the wavelengths of interest, which are then transformed by the detector into some measurable quantity. All modern systems employ some type of computer for instrument control, spectral acquisition, and postprocessing of the data. To understand how each of the components function as an integrated system, it may be beneficial to review some basic characteristics of each component and some basic geometrical optics.

Due to the fact that the signal associated with Raman scattering is inherently weak, it is desirable to excite the sample with a high-intensity source. The laser has served as this source since the early 1960s. Several other characteristics that make a laser ideal for a Raman experiment include the fact that the laser is monochromatic, polarized, and well collimated. Of course these characteristics are relative compared to a conventional extended light source and are not truly ideal. In reality the output of a laser is polychromatic due to the existence of plasma lines, and the polarization may have to be adjusted due to the type of sample being analyzed and/or the beam splitter employed in the microscope. Laser beams may also acquire intensity variations from scattering by defects in optical components or dust particles in the air. This scattering superimposes noise on the laser's intensity profile, which must be removed if diffraction limited performance of the microscope is desired. The purpose of the laser conditioning optics is to correct for and/or remove these defects from the laser source. A general-purpose system consisting of a Pellin-broca prism pair, a spatial filter and a polarization rotator is shown in Fig. 8.2.

The Pellin-broca prisms are designed to divert the laser through 90° and disperse (spatially separate) the plasma lines of the laser. Their use is preferred to a bandpass filter since, bandpass filters are designed for a specific wavelength. Although bandpass filters have an additional drawback of reducing the useful laser power by approximately 50%, newly developed holographic notch filters reduce the useful laser power by at most 10% [9]. This latter characteristic makes them extremely attractive for fixed excitation wavelength Raman microprobes.

Upon exiting the Pellin-broca prism pair the laser beam is spatially filtered. Spatial filtering of a laser beam is conceptually simple if one considers the thin-lens equation:

$$\frac{1}{f} = \frac{1}{s'} + \frac{1}{s}. \tag{8.8}$$

This concept is depicted in Fig. 8.3. Focusing a laser through a lens presents a special case of the thin-lens equation. The laser appears to be positioned at infinity (i.e., $s = \infty$). As such, the image of the focused laser s' is formed at the focal length f of the lens. On the other hand, scattering from dust particles or defects in optical components arise at finite distances within the

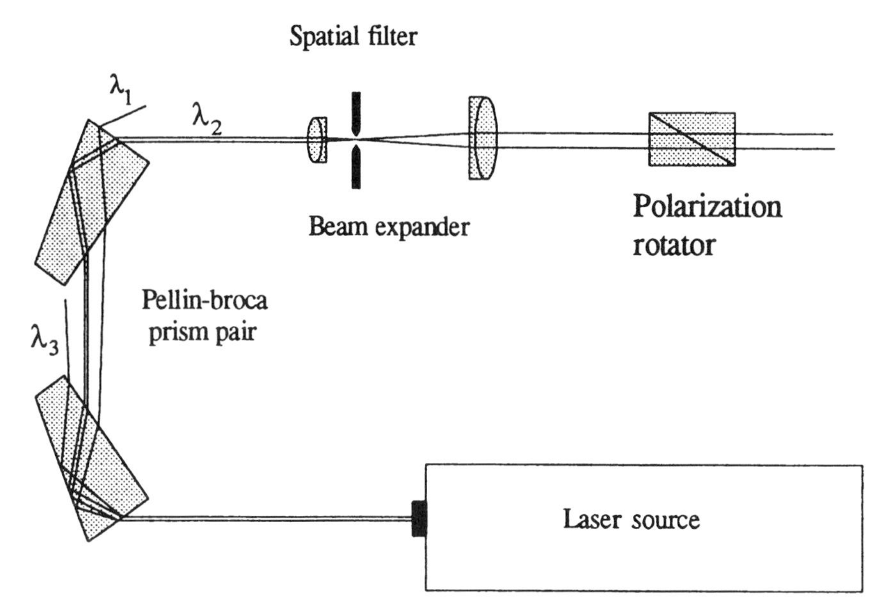

Fig. 8.2. Laser conditioning optics.

optical train (i.e., $s \neq$ infinity). As a result the images of these defects, or scattering particles are formed at distances different from the lens focal length. In addition, if the defect in the optical component or dust particle is off-axis (lower part of Fig. 8.3), the image will also be formed off-axis. The net result is that the image of the laser beam will be formed at the lens' focal length f on-axis, while the images of the defects and scattering dust particles will be formed on-axis, either before or after the lens' focal length, or at the lens' focal length but off-axis. By placing a pinhole at the lens focal length, one allows the laser to pass through, and the light associated with defects or scattering dust particles is blocked by the pinhole or spatially filtered.

The size of the pinhole can be determined from an understanding of the laser beam's intensity distribution as a function of distance from the optical axis. The following description has been adopted from a *Gaussian Beam Optics Tutorial* published by the Newport Corporation [10]. A more detailed analysis has been given elsewhere [11, 12]. Figure 8.4 shows the typical output of a laser beam, operating in a TEM_{00} mode, plotted as a function of radial distance from the beam axis. The shape of the curve is Gaussian. The parameter w is usually called the Gaussian beam radius or waist, and it is the radial distance at which the intensity of the beam has decreased to $1/e^2$, or approximately 13% of its value on-axis. The important feature to note is that the beam intensity decreases rapidly as the distances

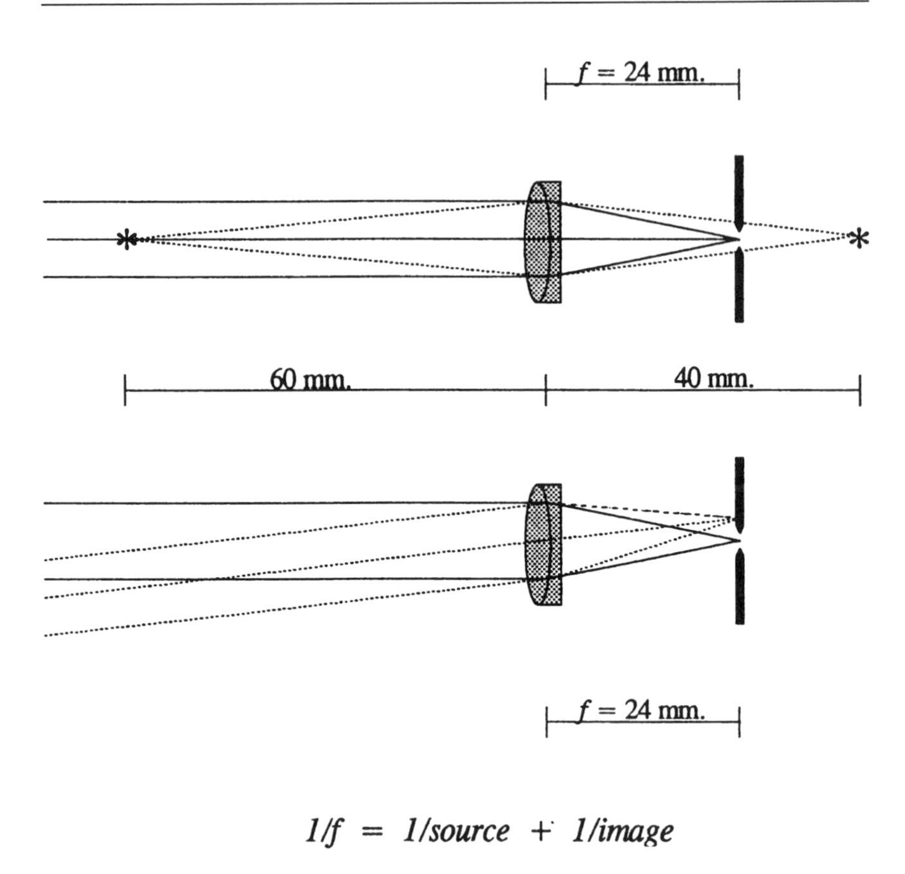

Fig. 8.3. Concept of spatial filtering.

from the axis increases. Integration of the area beneath the curve yields the total power of the beam contained in a specified radius. If the power distribution is normalized to the total power, the distribution is the same curve as the intensity distribution but with the y-axis inverted (see Fig. 8.4, right-hand ordinate). One can observe from the curve that 50% of the laser power is contained within a radius of $r = 0.59w$ and that a radius of twice the Gaussian beam waist $2w$ is necessary to contain 100% of the beam's power.

Focusing the laser beam to a point with a plano-convex lens produces a beam waist at the focus given by

$$2w = \left(\frac{4\lambda}{\pi}\right)\left(\frac{f}{D}\right), \tag{8.9}$$

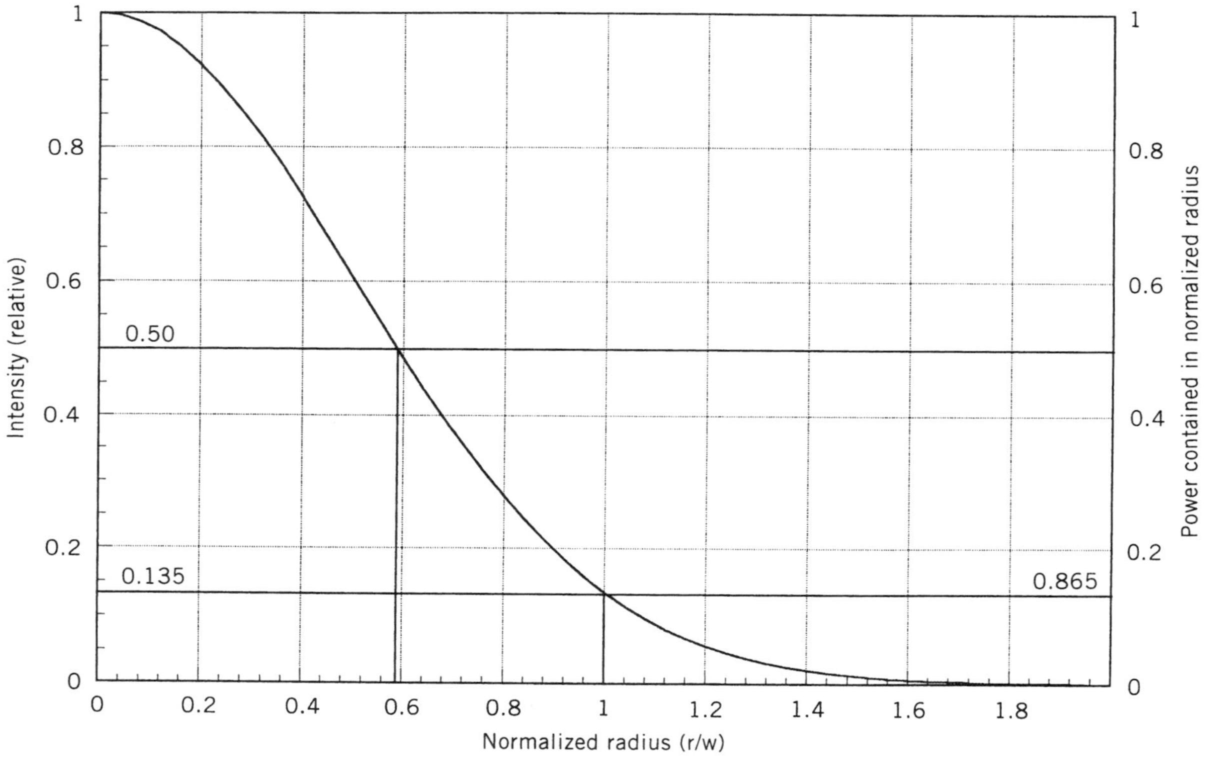

Fig. 8.4. Gaussian intensity distribution for a typical gas laser.

where f is the focal length of the lens, D is the diameter of the input laser beam, and λ is the wavelength of the laser. The ratio of the lens focal length to the input laser beam diameter (f/D) is referred to as the $f\#$ of the system and is the convention employed when using low-power lenses. For effective spatial filtering to take place, the pinhole diameter should be at least one and one-half to two times larger than the Gaussian beam waist $2w$, or

$$\text{Pinhole diameter} = (2\lambda)\left(\frac{f}{D}\right). \tag{8.10}$$

Referring to Fig. 8.4, this diameter corresponds to greater than 95% of the laser beam's power. Care must be taken not to make the aperture too small, since, below this diameter, diffraction effects will take place.

After filtering the plasma lines and cleaning up the intensity profile of the laser, the laser is recollimated by a second plano-convex lens. The focal lengths of the lenses making up the spatial filter are usually chosen so that the laser's diameter is expanded to match the pupil diameter of the microscope objective. Following expansion and recollimation, the laser is sent through a polarization rotator that adjusts the polarization of the laser beam for introduction into the microscope.

An optical diagram of a conventional Raman microscope is shown in Fig. 8.5*a*. The laser beam is reflected into the microscope's objective with the aid of a beam splitter. The objective focuses the laser to a diffraction limited spot on the sample and excites Raman scattering. Raman scattered radiation is collected 180° to the incident excitation by the same objective and transmitted back through the beam splitter. A tele lens forms a magnified image of the illuminated sample. This image can be transferred or reimaged, by several optical components, to match the entrance aperture of the wavelength selective filter (monochromator, interferometer) or viewing optics. The main purposes of the microscope are to excite and collect Raman radiation from the sample and to provide a means for sample positioning and viewing at high magnification.

Several salient features of the design should be discussed in order to fully understand the operation of the microscope. The beam-splitting material is composed of a thin-film dielectric and has a reflection/transmission ratio of typically 20/80. The choice of this ratio is a compromise between excitation power and the amount of Raman scattered radiation that is collected. Since the amount of Raman scattered radiation produced is small, every effort should be made to collect as much of this signal as possible. Compensation for the 80% loss in laser power associated with the beam-splitter configuration can be made by using a high-power laser possessing one to two watts total power.

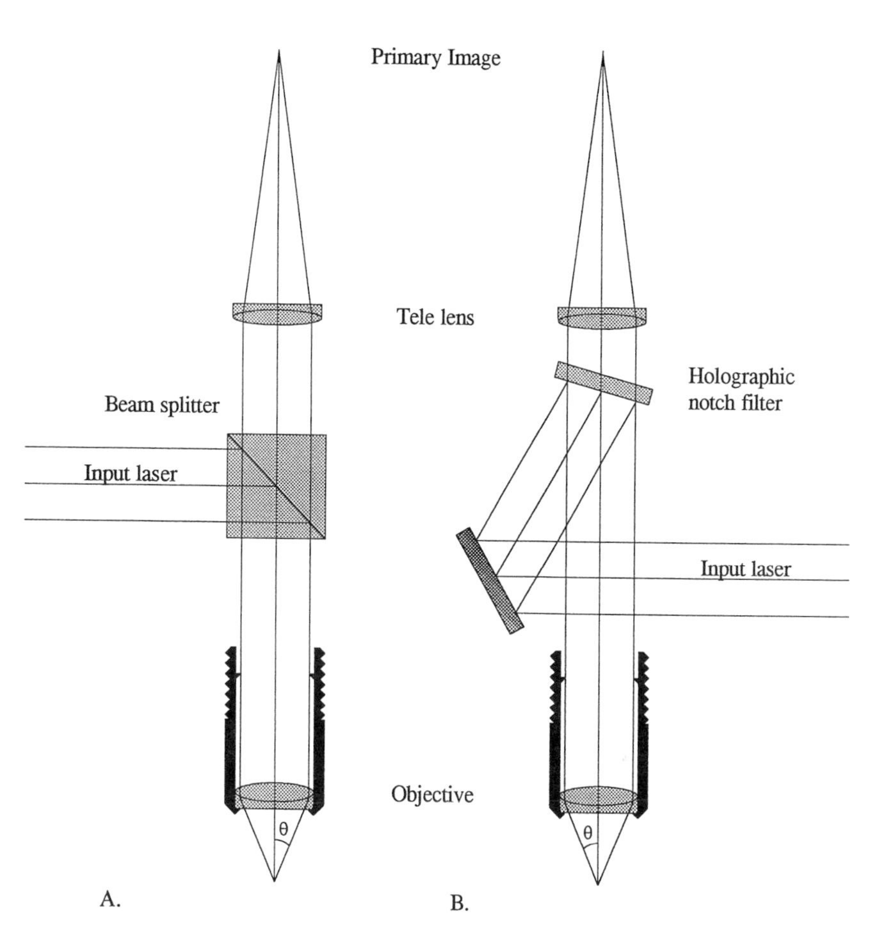

Fig. 8.5. Raman microprobes based on conventional beam splitter (*a*) and holographic notch filter beam splitter (*b*).

Another feature of most Raman microscopes built today is that they employ infinity-corrected objectives. Infinity-corrected objectives image the illuminated sample to infinity, and as such, light rays emerging from the back of the objective are collimated (parallel). The benefit of employing these objectives is that they couple with laser sources very well, and additional optical components such as polarizers and filters can be placed in the exiting beam without changing the image distances of the entire system.

After the scattered light is collected, the Raman scattered frequencies that contain the molecular information must be separated from each other, and

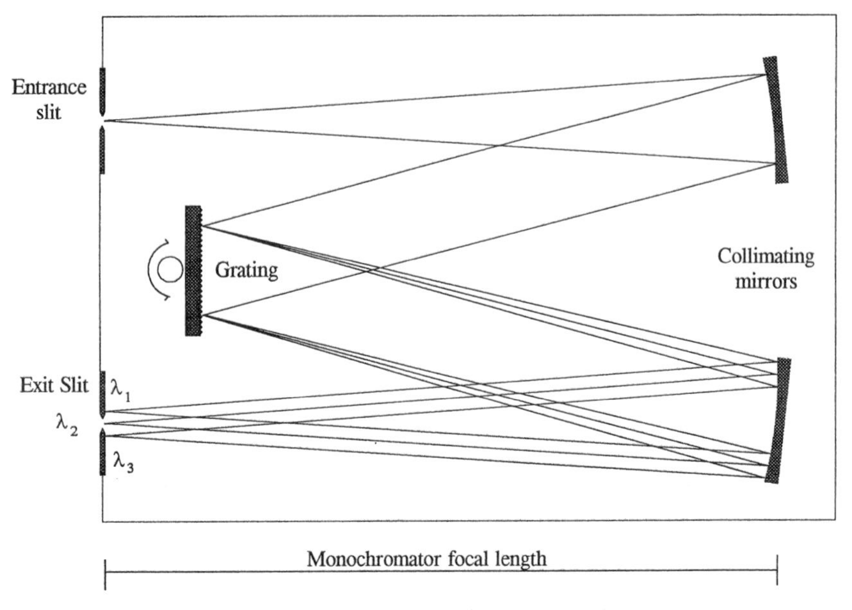

Fig. 8.6. Optical diagram of a single-stage monochromator.

the intense inelastic scattering frequency (i.e., the Rayleigh frequency). This task is comparable to viewing light from a distant star during daylight hours. In this regard the most important component of any Raman microprobe is the device that performs this function. Demands on this component are even more severe due to the configuration of the microscope. For example, if one is examining a highly reflective sample, the Rayleigh scattered light is coupled directly into the microscope. In early Raman microprobes, both functions (i.e., wavelength separation and Rayleigh line rejection) were served by a double monochromator whose focal length was typically one meter in length.

Figure 8.6 shows an optical layout of a single-stage monochromator. Light entering through the entrance slit is collimated by a concave mirror. Upon reflection by this mirror, the light impinges on the diffraction grating, which diffracts the light through various angles as a function of wavelength ($\lambda = c/v$). A second mirror focuses the dispersed light onto the exit slit of the monochromator, which spatially filters the desired frequency. From Fig. 8.6 one can see that as the focal length of the monochromator increases the separation between adjacent frequencies' increases, thereby increasing the spectral resolution of the instrument. Of course there are other factors that play a role in the spectral resolution of the measurement, and these include parameters associated with the grating (e.g., grating groove density and spectral order of use).

A measure of Rayleigh rejection is obtained by scanning the monochromator from the Rayleigh line toward greater frequency shifts (longer wavelengths). The intensity of the light at some distance from the Rayleigh line is ratioed to the intensity of the Rayleigh line. Results obtained by Landon et al. [13] demonstrated that two monochromators in series provided sufficiently enough rejection (1×10^{-11}) to observe Raman transitions in the 25 cm^{-1} shift region.

Until recently most research grade Raman microprobes were constructed with double- and triple-stage monochromators whose focal lengths ranged from 0.75 to 1 meter in length. In 1984 developments in the area of FT-Raman spectroscopy required that efficient long pass or notch filters be developed if the technique was to succeed [14]. In response to this need, holographic filters were developed [9]. These filters not only solved many problems for FT-Raman spectroscopy but revolutionized Raman spectroscopy in general and the manner in which Raman microprobes were designed [9, 15]. In most Raman microprobes being built today holographic notch filters are being employed as the microscope's beam splitter, and the means by which Rayleigh rejection is achieved. The manufacturers of these filters have pushed the technology to the point where a single filter can achieve the Rayleigh rejection of a single one-meter focal length monochromator. By incorporating two filters in the optical path of the microscope, sufficient Rayleigh rejection is achieved, and the sole function of the monochromator becomes that of spectral selection at the desired frequency. This achievement allows the use of a single short focal length monochromator (0.25 m). Figure 8.5*b* shows the optical layout of a Raman microprobe incorporating a holographic notch filter as the beam splitter. Additional gains have been achieved by employing the notch filter as a beam splitter. The filter reflects the excitation wavelength with an efficiency of better than 90% and transmits the Raman scattered radiation with a similar efficiency. No compromise is made between excitation and collection of the Raman scattered light, which allows the use of a low-power laser for excitation. The overall benefit is that the size and cost of Raman microprobes have been reduced, and their sensitivity increased. A case in point is shown in Fig. 8.7, which illustrates spectra of bismuth trioxide obtained with a conventional Raman microprobe having a double monochromator (1-m focal length) and a Raman microprobe employing holographic notch filters and a single short focal length (0.25-m) monochromator. The spectra show that the latter instrument gives comparable results when information in the low wavenumber shift region (i.e., 100-cm^{-1} shift) is desired. However, for extremely low wavenumber shift information (i.e., 2- to 50-cm^{-1} shift), a double one-meter focal length monochromator is required.

Figure 8.7 also demonstrates that even with the microscope configuration, information can be obtained very close to the excitation line. This spectral region provides information similar to that of the very far infrared.

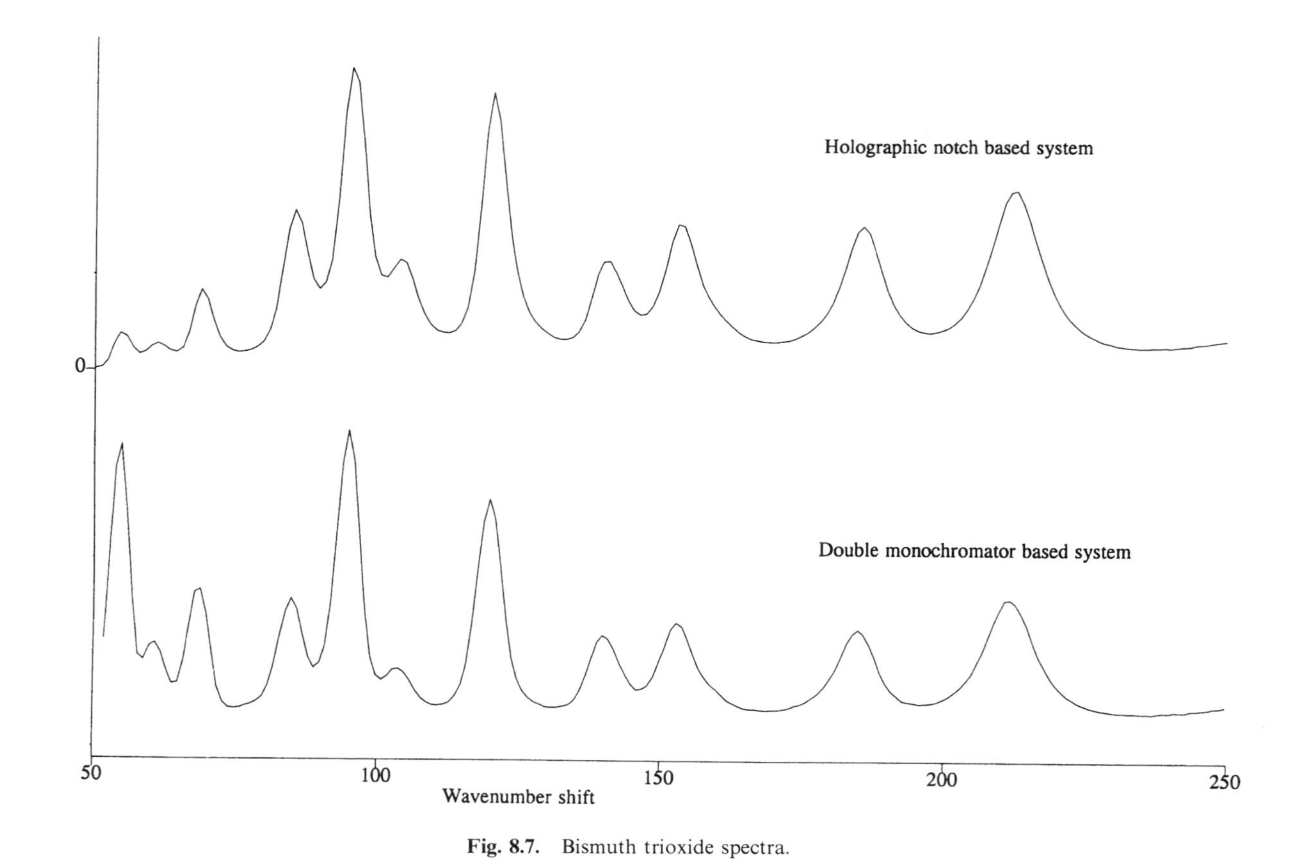

Fig. 8.7. Bismuth trioxide spectra.

8.5. DETECTION AND SPECTRAL COLLECTION MODE

Early Raman spectrometers employed photon counting methods of detection with a GaAs photomultiplier. Since this is a single element detector, spectra were collected in a serial fashion. That is, the instrument was scanned to the desired frequency shift, and the signal was then measured for a given period of time. Several years later linear multielement detectors (photodiode arrays, PDA and intensified photodiode arrays, IPDA) became available, which enabled Raman spectrometers to be employed as Raman spectrographs. In these spectrometers the exit slit is opened wide to allow a range of dispersed wavelengths to fall on the array detector followed by detection for some given time. Since many wavelengths are being detected at once, the method is usually referred to as parallel or multiplex collection. Presently charge-coupled detectors (CCD) are replacing both photomultiplier and photodiode arrays [16, 17]. These detectors are far more sensitive than the latter two and can be obtained as a linear array or two dimensional arrays. In using a two-dimensional array detector, direct two-dimensional Raman images can be obtained [18].

8.6. EXCITATION AND COLLECTION OF THE RAMAN SCATTERED RADIATION

The goal of Raman microspectroscopy is to make the sample appear as a self-luminous point source in the spectral region of interest. This goal is usually achieved by focusing the excitation laser to a very small spot with the aid of a high numerical aperture (NA) objective. The numerical aperture is defined as the product of the refractive index of the measurement environment and the sine of the angle formed by the optical axis and the most extreme ray entering the objective,

$$\mathrm{NA} = n_1 \sin \theta. \tag{8.11}$$

When focusing the laser with an objective (a case similar to focusing the laser with a plano-convex lens discussed earlier), meeting the Gaussian requirement that the pupil diameter of the objective be twice the beam waist, $2w$, is difficult to achieve [3, 12]. As a result the microscope objective truncates the Gaussian profile, and the size of the beam waist at the focus is given by

$$2w = \frac{1.22\lambda}{\mathrm{NA}}. \tag{8.12}$$

The implication of this equation is that a laser, or any light source for that matter, cannot be focused to an infinitely small point. This finite size is known as the diffraction limit which is described by an Airy function. Further the diameter of the laser at its focus is proportional to its wavelength and inversely proportional to the numerical aperture of the objective. The beam waist $2w$ is the diameter of the spot impinging the sample in the x-y plane of the microscope stage. In addition to limits placed on the diameter of the focused laser beam, there is a limitation placed on the depth L to which this diameter is maintained. This depth is commonly referred to as depth of field or depth of focus and is related to the wavelength and numerical aperture through

$$L = \frac{4\lambda}{(\mathrm{NA})^2}. \tag{8.13}$$

The depth of field is best explained through an example using photography and the common 35-mm camera. When taking a photograph during the daytime, relatively large f-stops (low NAs) can be used due to the abundance of light. In this situation the camera has a large depth of field and both the subject of interest and objects situated before and after the subject are in focus and appear as a clear image. If, however, the same camera is employed under low light level conditions, a small $f\#$ (high NA) must be used and only the subject of interest is in focus. The objects before and after the subject appear fuzzy and are out of focus. In the first case the camera has a large depth of field (depth of focus), and in the second case the depth of field is much smaller. Equations (8.12) and (8.13) show that if one wants to achieve the smallest possible beam diameter and depth of focus, the laser must be focused with a high numerical aperture objective.

The above example also demonstrates that under low light level conditions, for example, Raman microspectroscopy, the use of a high numerical aperture objective will enable more light to be collected. Equation (8.14) relates the total flux F of light collected as a function of the refractive index of the medium in which the measurement is conducted and the acceptance angle θ of the objective [19]:

$$F = n^2 \sin^2(\theta). \tag{8.14}$$

Table 8.1 summarizes values calculated for the beam waist, depth of field, illuminated sample volume, irradiance, flux collected, and illuminated cross-sectional area as a function of numerical aperture (i.e., $\mathrm{NA} = n_1 \sin\theta$).

From the data in Table 8.1, one can see that objectives possessing high numerical apertures yield the smallest beam waist and depth of focus and

TABLE 8.1. Diffraction-Limited Parameters as a Function of Numerical Aperture (i.e., sin θ) and Input Laser Diameters

Input Diameter (mm)	Acceptance Angle θ	Numerical Aperture $n_1 \sin\theta$	Beam Waist (μ)	Depth of Field (μ)	Illuminated Volume (μ^3)	I_0 (watts cm^{-2})	Flux Collected	Area Iluminated (μ^2)
1.00	14.5	0.25	3.09	40.5	304	13	0.06	7.50
2.00	27.3	0.46	1.68	12.0	26.8	45	0.21	2.22
3.00	37.8	0.61	1.26	6.75	8.43	80	0.37	1.25
4.00	45.9	0.72	1.07	4.91	4.45	110	0.52	0.91
5.00	52.2	0.79	0.98	4.05	3.04	133	0.62	0.75
6.00	57.2	0.84	0.92	3.59	2.38	151	0.71	0.66
7.00	61.0	0.87	0.88	3.31	2.02	164	0.77	0.61
8.00	64.2	0.9	0.86	3.12	1.81	173	0.81	0.58
8.00	64.2	1.00	0.77	2.53	1.18	214	1.00	0.47
8.00	64.2	1.10	0.70	2.09	0.81	258	1.21	0.39
8.00	64.2	1.20	0.64	1.76	0.57	308	1.44	0.33
8.00	64.2	1.30	0.59	1.50	0.41	361	1.69	0.28
8.00	64.2	1.40	0.55	1.29	0.31	419	1.96	0.24

[a] Objective NA = 0.90 with a pupil diameter of 8 mm.

[b] Calculations based on an excitation wavelength of 632.58 nm and an input power of 100 mwatts.

thus the highest spatial resolution in the *x*, *y*, and *z* planes. As a result these objectives would best be used for the analysis of thin films or small surface contaminants on a larger matrix. In contrast, a larger particle embedded in a transparent matrix, or a solution contained in a capillary, would best be analyzed with an objective possessing a smaller numerical aperture due to the larger depth of field. The choice of which objective, to employ, is usually a compromise based on the sample's size and the sample's thermal and optical properties. A detailed analysis of these considerations will be discussed in the sampling sections that follow.

8.7. CHOICE OF MICROSCOPE OBJECTIVE

One of the strengths, of Raman microspectroscopy, is the fact that little or no sample preparation is required for analysis. If the sample in question can be brought into focus beneath the microscope, ideally an analysis can be affected. The success of the analysis in reality will depend on the sample's size and optical and thermal properties. Having some prior knowledge about the sample will therefore greatly increase the success of the analysis. In this regard selection of the proper microscope objective becomes tantamount to obtaining the best spectral results possible.

Table 8.2 lists the calculated and measured diffraction limited focal dimensions for various objectives. The data in Table 8.2 demonstrate two important concepts. First, the magnification of the objective has no bearing on the actual beam waist or depth of focus that can be achieved. These parameters are determined solely by the numerical aperture of the objective. Second, the diameter of the beam waist should always be experimentally determined, since many factors determine the size of the waist at the sample position.

TABLE 8.2 Calculated and Measured Diffraction Limited Focal Dimensions for Various Objectives

Objective	Numerical Aperture	Pupil	Depth of Field (Calculated)	Beam Waist (Calculated)	Beam Waist (Measured)
10X Olympus	0.25	10 mm	32.9 μ	2.5 μ	7.5 μ
40X Nachet	0.75	8 mm	3.7 μ	0.9 μ	2.2 μ
80X Nachet	0.90	8 mm	2.5 μ	0.7 μ	1.1 μ

[a]Focused spot size measured with a reticle.
[b]Input laser beam diameter of 6 mm.
[c]Calculations conducted with 0.5145 μ wavelength light.

For example, the data differ when comparing the experimental values of the beam waist to those that have actually been measured. The differences reside in the fact that the equations cited for beam waist and depth of field are only valid if the entire pupil of the objective is filled (i.e., the input laser beam diameter matches that of the objective's pupil). In those cases where the input beam diameter is smaller than the pupil diameter, one would expect a larger beam waist and depth of field. This result is also illustrated by data in Table 8.1, which presents values of theoretical calculations for beam waist and depth of field. These calculations were based on an objective pupil diameter of 8 mm and yield values, of the beam waist and depth of field, for a filled and underfilled pupil.

Table 8.3 summarizes results of experiments conducted in which different microscope objectives were employed for the analysis of different types of samples. The basis for the comparison was the amount of Raman signal that could be collected for equal laser power and measurement time, as a function of the objective's numerical aperture. The Raman signal was determined by measuring the area under a given Raman transition (integrated band intensity). The objective that yielded the smallest signal was employed to normalize the signal observed within a given data set. Only the 10X Olympus, 40X Nachet, and 80X Nachet were employed in the comparison due to the fact that their pupil diameters are similar. The materials employed in the study were single crystals of barium fluoride, sapphire, powdered zinc oxide, pelletized zinc oxide, and single-crystal silicon. These samples represent a range of different physical properties that will ultimately affect the amount of scattered radiation that can be observed from each sample.

The data presented in Table 8.3 illustrate some interesting trends that may aid in the choice of an objective for a given sample. For transparent samples that are relatively thick, such as barium fluoride and sapphire, the

TABLE 8.3 Microscope Objective Comparison

		Normalized Integrated Band Intensities				
Objective	NA	BaF_2[a]	Al_2O_3[a]	ZnO Powder	ZnO Pellet	Silicon[b] (100)
10X Olympus	0.25	2.2	2.1	1.2	1.0	1.0
40X Nachet	0.75	2.1	2.0	1.0	3.3	7.2
80X Nachet	0.90	1.0	1.0	—	1.6	9.8
n_d		1.476	1.771	2.008	2.008	3.498

[a]Single crystals of barium fluoride and alumina, each 2 mm in thickness.
[b]A single crystal of silicon 0.5 mm in thickness.

beam penetrates the sample much deeper than for zinc oxide and silicon. As a result of this penetration, the objectives with a larger depth of field yield a better signal. Conversely, these objectives are a poor choice if the sample characteristics do not allow the beam to penetrate the surface. For example, single-crystal silicon is transparent to 514.532-nm light to a depth of approximately 0.7 μ [20]. Thus the entire Raman signal is generated from a sample thickness of only 0.7 μ. The results in Table 8.3 demonstrate that under these conditions the high numerical aperture objectives are more suitable for these types of samples. Of course, if a high spatial resolution is required, these objectives must be employed. The data also suggest that an objective possessing a moderate depth of focus and spot size (e.g., 40X Nachet) can be employed without much loss in signal.

A physical constraint that must be considered in the analysis of a given sample is the working distance of the objective. The working distance is the distance between the objective and the sample when the sample is in focus. If a particle is embedded deep within in a transparent matrix, an objective with a large working distance must be used in order to avoid running the objective into the upper surface of the matrix. Large numerical apertures and working distances are mutually exclusive.

8.8. SAMPLING

Analysis of solid samples is usually conducted by placing an isolated crystal on the microscope slide and bringing the crystal into focus beneath the microscope. If the solid is a powder, significant improvement in signal can be obtained by compressing the powder into a micro pellet or by recrystallizing the sample under the pressure of a pointed sampling probe. The effect of these procedures increases the number of scattering particles per unit volume which, as we know from equation (8.6), will increase the observed signal. Figure 8.8 demonstrates the result of compacting powdered zinc oxide in comparison to direct analysis. The integrated band intensity of the transverse optic lattice vibration for zinc oxide at 437 cm^{-1} shifts is approximately $4\times$ greater for the compacted sample compared to that obtained for the powdered sample. A direct comparison of this improvement, with that expected from equation (8.6) is difficult to undertake, since multiple scattering events usually take place in powdered materials. This situation arises from the close proximity of the particles and the fact that Rayleigh light (elastically scattered light) scattered from one particle can go on to produce Raman scattered light from neighboring particles. For this reason powdered samples should not be employed as a standard for the determination of instrument performance.

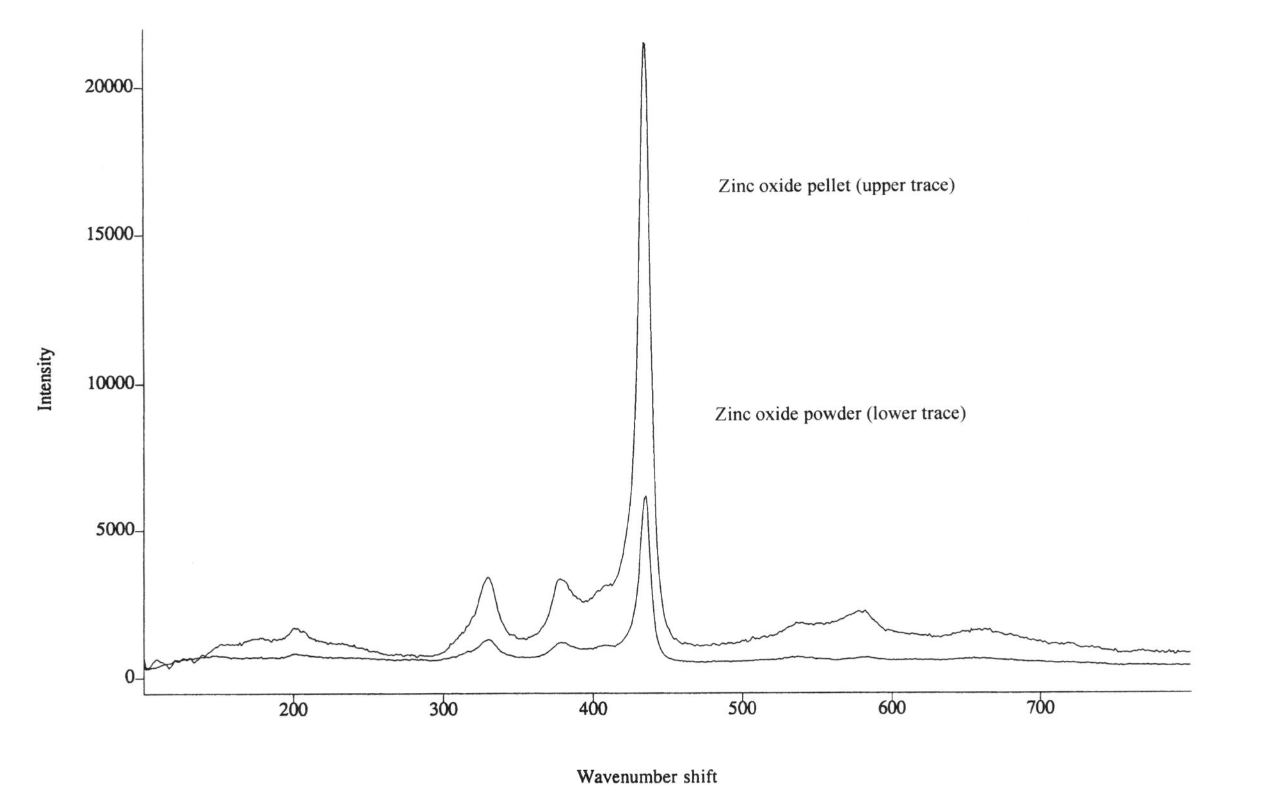

FIG. 8.8. Zinc oxide spectra.

Solution samples are best analyzed in a sealed capillary tube or on a microscope slide with a coverslip placed on top. The analysis of solutions in a capillary will usually require the use of an objective with a relatively long working distance and thus a larger depth of field. Each of these sample holders can be placed on a reflective substrate to increase the occurrence of multiple scattering events and the associated signal. After passing through the sample, the mirror reflects the laser into the sample a second time. Due to the energy contained in the incident beam, and the relatively high vapor pressure of most solutions, it is best to confine the solution in a sealed vessel. Localized heating of the solution will make it flow, thereby, changing the laser focal dimensions and the signal coming from the sample. In the worst-case situation the solution can evaporate at the beam focus and/or be displaced to other parts of the confinement vessel.

The analysis of gases under ambient conditions has not been conducted on a micro scale, due to signal considerations. However, analyses of gas inclusions trapped in matrices where the gas pressure is very high have been reported [21–23]. In these investigations long working distance objectives were employed to probe the inclusions embedded beneath the surface of the matrix. Alternatively, one can polish, or grind away the upper layers of the matrix, thereby negating the requirement of an objective with a long working distance.

8.9. SAMPLING PROBLEMS

Three main problems that confront the Raman microscopist during the course of analysis concern spatial resolution, or the lack thereof, sample decomposition, and fluorescence. A fourth problem that does not affect the ability to acquire data, but mainly its interpretation, is polarization effects. These effects arise when analyzing crystalline and ordered materials.

8.10 SPATIAL RESOLUTION

Raman microspectroscopy yields the best spatial resolution obtainable when compared to other molecular microprobes. From a theoretical standpoint the spatial resolution of Raman microspectroscopy is an order of magnitude better than that of infrared microspectroscopy. Although one can focus the excitation laser source too below one micrometer in diameter, this ability does not control the spatial resolution of the measurement. Like infrared microspectroscopy the sample becomes an optical component of the system, and as such, the sample actually dictates the spatial resolution that can be achieved. The exception to this rule is in the analysis of optically thin and smooth films. Actually these types of samples are employed for the determination of the spatial resolution of most instruments [24]. In 98% of most sampling situations, the sample will scatter, diffuse, and generally spread the beam out to a much larger area than dictated by the general equation for beam waists discussed earlier. In these cases other steps must be taken to increase the spatial resolution.

Two approaches employed to increase spatial resolution involve spatial filtering (i.e., confocal configurations) and immersion microspectroscopy. The first approach is generally more applicable, although the later provides certain benefits as well. In confocal Raman microspectroscopy a limiting aperture (pinhole) is placed at an image plane of the illuminated sample. This mechanism is depicted in Fig. 8.9. The objective of the microscope in conjunction with the tele lens forms a magnified image of the sample. If sample *B*, with near neighbors *A* and *C* are imaged with the microscope, magnified images of all three will be produced at the primary image plane of the microscope. By placing an aperture at this image plane, one can spatially filter the images *A* and *C* such that only scattered radiation from sample *B* is passed on to the detector. In addition to spatial discrimination in the *x-y* plane of the microscope, there is also discrimination in the *z*-axis corresponding to depth within the sample. The right-hand side of Fig. 8.9 shows a laminate sample having layers *A*, *B*, and *C*. These layers are again imaged to the primary image of the microscope. By placing an aperture at this image plane, scattered radiation from only layer *B* will be allowed to reach the detector. Although confocal microspectroscopy will improve spatial resolution, the signal in this measurement will also decrease. This decrease is associated with the decreased illuminated sample volume which is observed by the instrument. (Refer to the data in Table 8.1 for a comparison of sample illumination volumes).

One important concept to note is that the confocal aperture can be placed at any image plane of the illuminated sample. Batchelder et al. [15] have cleverly used this concept and employ the entrance slit of the monochromator and the detector elements of an array detector, both of which occur at image planes of the sample, as confocal apertures. The simplicity of this approach and the use of existing optical components for dual purposes makes the system very efficient in terms of signal management. Williams et al. [25] have demonstrated depth resolution capabilities on a number of polymer laminate systems.

Another approach to increasing spatial resolution is to immerse the sample in a medium whose refractive index is greater than air [26]. From equation (8.12) we know that the size of the beam waist at the laser focus is inversely proportional to the NA of the objective. The NA is the product of the refractive index of the measurement environment and the acceptance angle of the objective. The last six rows of Table 8.1 present data for

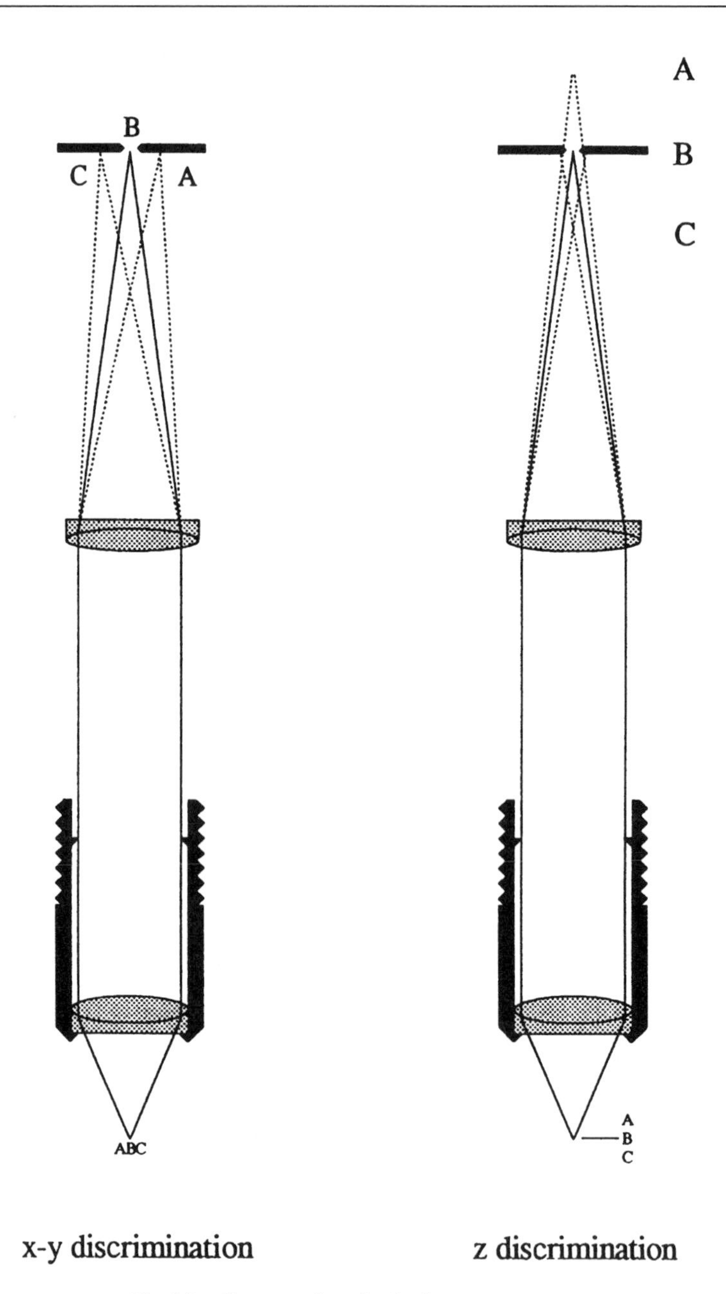

Fig. 8.9. Concept of confocal microspectroscopy.

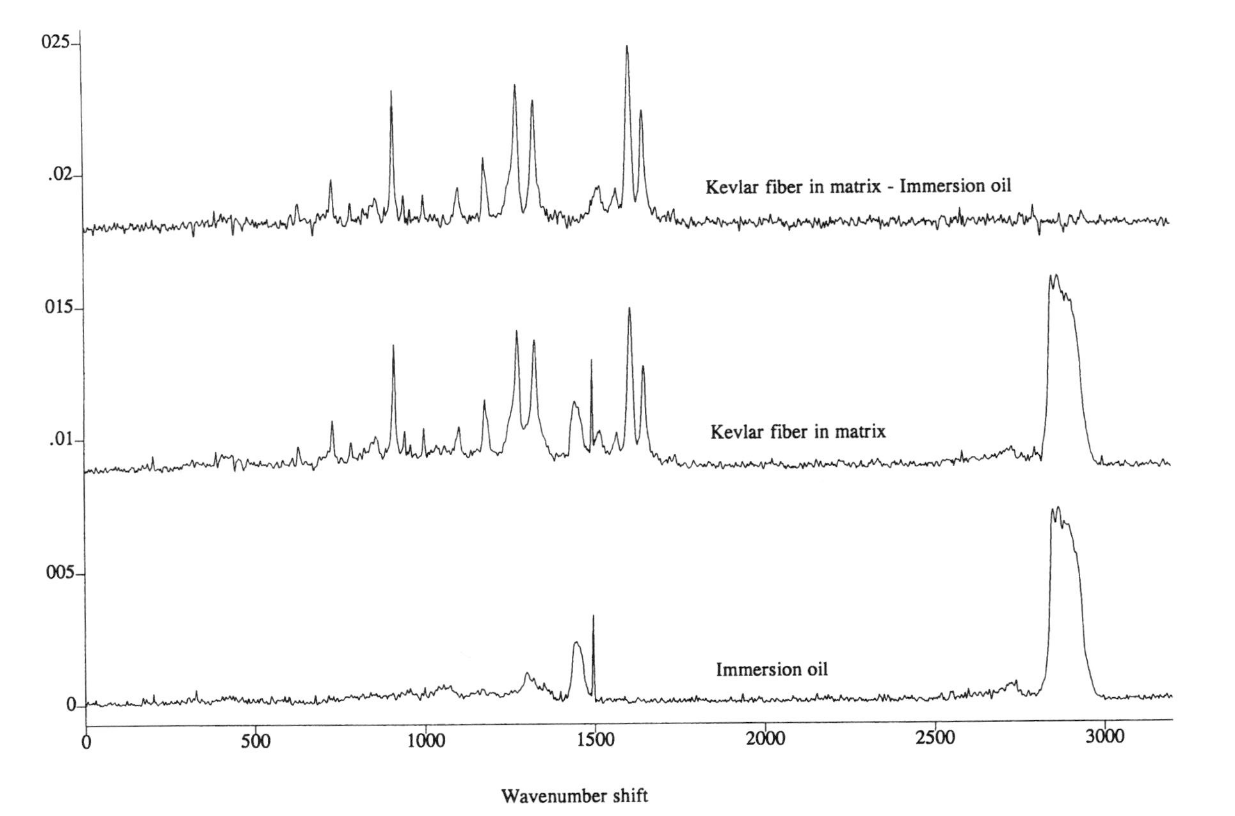

Fig. 8.10. Kevlar fiber immersion experiment.

numerical apertures above 0.90. This value of 0.90 NA is close to the maximum numerical aperture achievable in air and corresponds to an acceptance angle of 64°. If the sample is immersed in paraffin oil (e.g., $n_d = 1.480$), the corresponding numerical aperture is 1.33. This increase results in a moderate gain in the *x-y* spatial resolution. However, a twofold gain is realized in the *z*-axis resolution, the amount of light collected and the incident power per unit area. From a signal standpoint, these gains are completely offset since the scattering volume has decreased by a factor of six. Overall, immersion Raman microspectroscopy allows one to improve the spatial resolution of the measurement without a concomitant loss in signal.

Several factors must be considered in the selection of the immersion medium for grating-based systems. First, the medium should be nonfluorescent and have a spectral window that permits the analysis of the analyte in question. If Raman transitions of the immersion medium and analyte do overlap, subtractions can be employed. In this particular case the wave number precision of the instrument becomes critical. Grating-based measurements should therefore be conducted in a spectrograph mode. Alternatively, for fluorescent immersion media, Fourier transform Raman immersion microspectroscopy can be employed. This method has the inherent wavenumber precision to do the spectral subtractions necessary in the case of overlapping Raman transitions.

Figure 8.10 illustrates FT-Raman spectra of a Kevlar fiber embedded in a polyester matrix. The sample was cross-sectioned, and the fiber was viewed down its axis. An immersion oil was employed to increase the spatial resolution of the measurement. Without the immersion oil, the absolute spatial resolution was determined to be 18 μ [27]. The center spectrum in the figure shows Raman transitions from the Kevlar and immersion oil. No transitions are observed from the polyester surrounding the fiber. A spectrum of the immersion oil (bottom) was then subtracted from the center spectrum, and the resultant spectrum is shown in the top of the figure.

8.11. SAMPLE HEATING

A limitation of Raman microspectroscopy in its early development was related to a general lack of sensitivity. The only manner in which the signal could be increased was to build instruments with better efficiencies or to increase the power of the excitation laser. The former was not technologically possible at the time (circa 1983), and as a result the latter method was employed. The limit to which the laser power can be increased is dependent on the thermal lability and photochemical stability of the sample under study and the ingenuity of the analyst for the following reason.

A 10-mwatt laser beam with a cross-sectional area of 50 mm^2 (diameter 8 mm) has a power per unit area of 2.0 mwatt/mm^2. Assuming a power loss of 80% at the microscope's beam splitter and objective characteristics that yield a 1-μm diameter beam waist at the sample focus, the power density at the focus is then on the order of 2.0 mwatts/μ^2. This power confinement represents a millionfold increase in the power per unit area. Under these circumstances it is entirely possible for the sample to undergo thermal degradation or polymorphic transitions. Laser-induced sample damage is probably the second largest cause of unsuccessful sample analysis in Raman microspectroscopy, especially for colored or opaque samples. The extent of these problems can range from subtle color, or crystalline changes in the material, to the complete ablation of the sample beneath the microscope objective. Corresponding spectral changes include subtle shifts in Raman transitions, the appearance of new transitions, or complete loss of signal. In order to prevent these possibilities from occurring, the following plan of action is suggested.

If the sample is an isolated particle, for which there are no replicate samples, the analysis should be initiated by using the lowest laser power available. Prior to collection of a spectrum, a photograph or mental note of the sample's appearance should be made. This photograph or note can then be compared to the appearance of the sample, after a spectrum has been collected. If any changes have occurred, the sample will appear different. On some Raman microprobes it is possible to view the sample while it is being illuminated with the laser. For these systems one can view the laser focus and determine whether it changes as a function of time. Should material be ablated from the sample, the size of the focused laser will appear larger, since the remaining sample surface is below the microscope's focus. In the case of severe degradation, the microscope objective should be inspected for material that has condensed on the front surface of the lens system. Due to the fact that the objective is directly above the sample and in close proximity to the sample (typically about 1 mm), any sample that is vaporized will condense on the microscope objective. This material must be removed, since it will affect the focal characteristics of the laser and could yield a parasitic background in the Raman spectrum.

In the event that the sample undergoes degradation even at the lowest power levels, several solutions are still available. One of the more obvious solutions is to reduce the laser power still further. This solution can be implemented by defocusing the laser beam. Defocusing the laser is usually accomplished by adjusting the beam expander in the prefiltering optics of the system such that the objective's aperture is under filled. As discussed earlier, by under filling this aperture, a larger beam waist is obtained at the objective's focus. Of course lowering the power of the laser will decrease the signal; however, standard signal averaging practices can then be implemented [12].

Another alternative is to increase the sample's ability to dissipate the thermal energy by heat sinking the sample with a suitable substrate. Substrates such as sapphire and magnesium oxide have been employed where the sample is placed or pressed onto the substrate [28]. The low cost of commercial diamond windows makes these substrates suitable for these purposes. The sample can also be affixed to a glass or steel surface by adding a few drops of water to the sample. When the water has evaporated, the material adheres to the surface much better with improved heat conduction away from the sample. Employing these methods with a thermoelectrically cooled stage has also been successful [29]. Total immersion of the sample in water, or sodium silicate (water glass) in conjunction with immersion objectives, or objectives with long working distances is also very useful [28, 30–31]. Highly colored materials may be diluted and immersed in a KCl matrix. Alternatively, the excitation laser wavelength can be changed to a region where appreciable absorption does not occur [14].

More subtle heating effects are best detected by collecting several spectra at different scan rates and/or power levels. If these spectra are not reproducible, the sample may be changing as a result of irradiation. Many metal oxides, carbon samples, and drug polymorphs can be interconverted as a result of irradiation during sample analysis. If care is not taken during the analysis, the results can be easily misinterpreted. Although sample heating was once a problem in Raman microspectroscopy, it is becoming much less frequent. Technological advances in detectors and instrument design have increased the sensitivity of many Raman microprobes. As a result spectra being collected today are done so with power levels that are an order of magnitude smaller than those of the past.

8.12. FLUORESCENCE ELIMINATION

The major problem associated with conventional Raman spectroscopy has been the elimination of parasitic fluorescence. This problem arises from the fact that conventional Raman spectroscopy is conducted with a visible excitation source. As such, there is sufficient energy to excite electronic transitions which give rise to fluorescence, as well as the vibrational transitions which give rise to the Raman effect. One benefit that Raman microspectroscopy has over conventional macro Raman spectroscopy is an added dimension in fluorescence elimination and rejection. Adar [12] has noted that a majority of materials fluoresce due to the presence of impurities which are inhomogeneously distributed throughout the sample. Employing the spatial discrimination of the Raman microprobe, one can search for areas in the sample with reduced or nonexistent fluorescence. She also notes that fluorescence quenching usually takes less time due to the high-irradiance levels at the sample's focus. Fluorescence quenching is a technique in which the sample is irradiated for a long period of time. After that, the fluorescence signal has been eliminated or reduced to a tolerable level. The use of an aperture at a remote image plane (confocal microspectroscopy) has also been employed in the elimination of fluorescence from near-neighbor locations. Anderson [28] successfully employed this technique to eliminate fluorescence arising from neighboring layers in a multilayer polymer laminate. His results were the first successful demonstration of what is now called confocal Raman microspectroscopy.

If the fluorescence is inherent to the material, there are few alternatives employing conventional means other than bleaching. The excitation wavelength may be shifted to the near-infrared (Nd/YAG excitation, 1064 nm); however, this requires the use of a totally different Raman microprobe. Although fluorescence rejection is quite successful in his region, the spatial resolution of the system is degraded by a factor of 2, and the sensitivity is reduced by a factor of 18, or so.

8.13. POLARIZATION EFFECTS

Polarization effects arise in Raman microspectroscopy from sample and instrumental sources. Since the technique allows the examination of individual crystals, their orientation on the microscope stage can lead to polarization effects in the spectrum. Second, laser sources employed in Raman microprobes are linearly polarized. In addition, optical components employed in the construction of the microprobe exhibit their own polarization properties. As a result the recorded spectrum will be a convolution of polarization dependencies of the laser, the sample and the optical components employed in the Raman microprobe. When analyzing samples that are suspected to be ordered (e.g., single crystals, polymer films, fibers), several spectra should be collected with the sample oriented in different positions. The spectra can then be added to produce a composite spectrum which is relatively free of any polarization effects. If time does not permit this procedure to be conducted, at the very least the spectra collected on these different samples should be done with the sample positioned in the same orientation on the microscope stage.

Quantitative polarization studies must account for all the effects throughout the entire optical train of the Raman microprobe. As an example, Fig. 8.11 presents the optical layout of an Instruments SA U1000 Raman microprobe, which employs a dielectric beam splitter and a grating-based monochromator. Remember that the last component of the prefiltering optics was a polarization rotator. This device enables the laser polarization to be rotated to the preferred orientation of the beam splitter. A dielectric

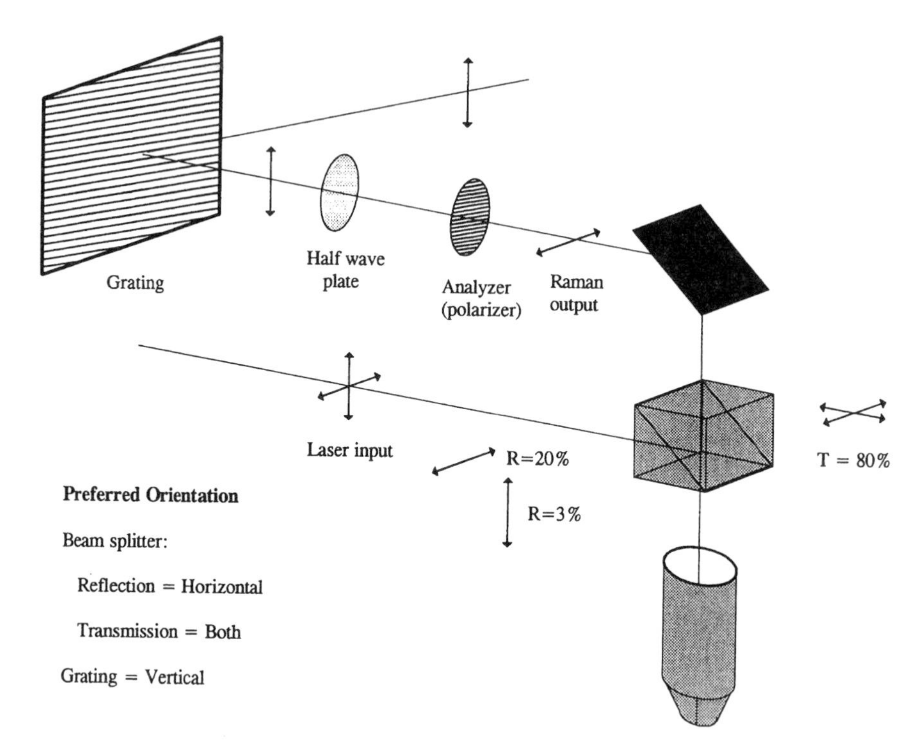

Fig. 8.11. Polarization characteristics of the U1000 microscope.

beam splitter will have preferential reflection characteristics based on the input laser beam polarization. The polarization direction, which is horizontal, is preferred for maximum reflection of the laser at the beam splitter. Because of the preferential reflection by the beam splitter, polarization studies conducted on the microscope should be done by rotating the sample and not the polarization of the laser beam. The beam splitter has been engineered to transmit both polarization components of the Raman scattered radiation equally. It has been observed, however, that the horizontal component of the Raman signal is more intense due to the horizontal polarization of the input beam. Transmitting this orientation into the monochromator would result in a signal loss, due to the fact that the grating is more efficient for the vertical polarization. In order to overcome this signal loss, a half-wave plate can be inserted between the microscope and the monochromator. The half-wave plate rotates the polarization through 90° to that which is preferred by the monochromator.

Analysis of polarized Raman scattering can be accomplished with the use of the half-wave plate and an analyzer (polarizer). The horizontal components of the Raman scattered light can be studied with the half-wave plate and the analyzer set to the horizontal position. The vertical components of Raman scattering can be studied with only the analyzer in place and set to the vertical position. The objective of the microscopes may also introduce polarization effects into the measurement. The high numerical aperture of the objectives employed collect the Raman scattered light over a very wide angle. As a result specific selection rules may break down, and the spectra may exhibit bands associated with both polarizations [28]. Several accounts have been published that detail the analysis of polarization effects arising from the objective, beam splitter, sample, and their combined effects [32–34]. When considered carefully, these effects can be accounted for permitting quantitative polarization measurements to be made.

This discussion about polarization effects is not meant to be comprehensive but rather instructive concerning the measures that should be considered when comparing spectra obtained from different samples. If polarization measurements are to be conducted on a specific instrument, the polarization properties of that instrument must be determined experimentally. This determination can be conducted with a standard sample (e.g., CCl_4) whose depolarization ratios are well-known.

REFERENCES

1. G. J. Rosasco, E. S. Etz, and W. A. Cassatt, *Appl. Spectrosc.*, **29**, 396–404 (1975).
2. M. Delhaye and P. Dhamelincourt, *J. Raman Spectrosc.*, **3**, 33–43 (1975).
3. R. J. H. Clarke and R. E. Hester, eds., *Advances in Infrared and Raman Spectroscopy*. Heydon, London, 1980, ch. 4.
4. D. J. Gardiner and P. R. Graves, eds., *Practical Raman Spectroscopy*. Springer-Verlag, Berlin, 1989, ch. 6.
5. W. T. Mason, ed., *Fluorescent and Luminescent Probes for Biological Activity: A Practical Guide to Technology for Quantitative Real-Time Analysis*. Academic Press, San Diego, 1993, ch. 18.
6. K. Nakamoto, *Infrared and Raman Spectra of Inorganic and Coordination Compounds*, 4th ed. Wiley, New York, 1986.
7. G. Herzberg, F. R. S. C., *Molecular Spectra and Molecular Structure, II. Infrared and Raman Spectra of Polyatomic Molecules*. Van Nostrand Reinhold, New York, 1945.
8. W. A. Guillory, *Introduction to Molecular Structure and Spectroscopy*. Allyn and Bacon, Boston, 1977.
9. J. M. Tedesco, H. Owen, D. M. Pallister, and M. D. Morris, *Anal. Chem.*, **65**, 441A–449A (1993).
10. Gaussian Beam Optics Tutorial, *The Newport Catalog 94/95, Scientific & Laboratory Products*. Newport Corporation, Irvine, CA, 2.66–2.67.
11. J. J. Barrett and N. I. Adams, III, *J. Opt. Soc. Amer.*, **58**, 311–319 (1968).

12. R. H. Geiss, ed., *Microbeam Analysis—1981*, San Francisco Press, San Francisco, 1981, 67–72.
13. D. Landon and S. P. S. Porto, *Appl. Opt.*, **4**, 762–763 (1965).
14. B. Chase, *Anal. Chem.*, **59**, 881A–889A, 1987.
15. K. P. J. Williams, G. D. Pitt, B. J. E. Smith, A. Whitley, D. N. Batchelder, and I. P. Hayward, *J. Raman Spectrosc.*, **25**, 131–138 (1994).
16. R. B. Bilhorn, J. V. Sweedler, P. M. Epperson, and M. B. Denton, *Appl. Spectrosc.*, **41**, 1114–1125 (1987).
17. R. B. Bilhorn, P. M. Epperson, J. V. Sweedler, and M. B. Denton, *Appl. Spectrosc.*, **41**, 1125–1136 (1987).
18. P. J. Treado and M. D. Morris, *Appl. Spectrosc. Rev.*, **29**, 71–108 (1994).
19. M. Pluta, *Advanced Light Microscopy, Vol. 1: Principles and Basic Properties*. Elsevier Science, New York, 1988, p. 221.
20. Private communication with Bob Bennett, Renishaw Transducer Systems Division, Old Town, Wotton-under-Edge, Gloucestershire, UK, GL12 7DH.
21. B. Wopenka and J. Dill Pasteris, *Appl. Spectrosc.*, **40**, 144–151 (1986).
22. A. Masui, M. Noshiro, and M. Kurata, *Anal. Sci.*, **1**, 313–319 (1985).
23. J. Dubessy, C. Beny, N. Guilhaumou, P. Dhamelincourt, and B. Poty, *J. De Physique*, **C2**, 811–814 (1984).
24. D. J. Gardiner, M. Bowden, and P. R. Graves, *Phil. Trans. R. Soc. Lond.*, **A320**, 295–306 (1986).
25. K. P. J. Williams, G. D. Pitt, D. N. Batchelder, and B. J. Kip, *Appl. Spectrosc.*, **48**, 232–235 (1994).
26. Edgar S. Etz, ed., *Microbeam Analysis—1995*. VCH, New York, 1995, pp. 111–112.
27. A. J. Sommer and J. E. Katon, *Spectrochimica Acta*, **49A**, 611–620 (1993).
28. M. E. Andersen and R. Z. Muggli, *Anal. Chem.*, **53**, 1772–1777 (1981).
29. G. R. Loppnow and R. A. Mathies, *Rev. Sci. Instrum.*, **60**, 2628–2630 (1989).
30. R. H. Geiss, ed., Microbeam Analysis–1981. San Francisco Press, San Francisco, 1981, pp. 61–64..
31. A. D. Romig, Jr., and J. I. Goldstein, eds., *Microbeam Analysis—1984*. San Francisco Press, San Francisco, 1984, pp. 125–126.
32. G. Turrell, *J. Raman Spectrosc.*, **15**, 103–108 (1984).
33. C. Bremard, P. Dhamelincourt, J. Laureyns, and G. Turrell, *Appl. Spectrosc.*, **39**, 1036–1039 (1985).
34. C. Bremard, J. Laureyns, J. Merlin, and G. Turrell, *J. Raman Spectrosc.*, **18**, 305–313 (1987).

9 Emission Spectroscopy

S. Zhang, F. S. Franke, and T. M. Niemczyk

9.1. INTRODUCTION

There are numerous examples of manufacturing processes where the ability to perform a rapid, nondestructive determination of product quality is of great value. Infrared spectral data of a sample contain compositional as well as physical property information, hence can often provide the information required to determine product quality. Infrared emission spectroscopy (IRES) is a technique that can often be applied to conveniently collect infrared data from a sample in a process. All that is required to measure an infrared (IR) emission spectrum is to generate a situation where the sample is at a different temperature than that of the infrared detector. In many manufacturing processes the sample temperature is elevated relative to ambient, and IR detectors can be easily cooled, thus generating a situation where an IR emission spectrum of the sample can be readily obtained. All that is required is optical access to the sample.

Modern Techniques in Applied Molecular Spectroscopy, Edited by Francis M. Mirabella.
Techniques in Analytical Chemistry Series.
ISBN 0-471-12359-5

IRES has some significant advantages when compared to the much more common technique of IR absorption spectroscopy, especially when used in a process monitoring application. In many cases the samples are large relative to the area of the sample viewed by the spectrometer, a situation that eliminates the need for precise alignment or, alternatively, allows determination of properties as a function of position on a sample. No sample preparation is required and the sample is the source, alleviating the need of a conventional source. Sample temperatures are hardly ever as hot as the source temperature in a conventional IR spectrometer; hence the amount of radiation that must be detected is relatively low. Because of this, instrument throughput, spectrometer stability, and detector sensitivity all become important parameters. Modern IR instuments are, however, exceptionally good in these areas, making it possible to obtain an IR emission spectrum very rapidly with high signal-to-noise ratio even when the sample temperature is only slightly above the detector temperature.

Even high signal-to-noise ratio IR emission data are complex. The complexity is due to the many parameters that affect the data. Sample composition, thickness, refractive index, and geometry influence the data. The choice of background, the temperature, and any temperature differences or gradients in the background or sample will be manifest in the data. Internal reflection/reabsorption phenomena add to the complexity of IR emission data, generally in a manner that makes some features very nonlinear. Finally, stray radiation will be generated by anything that is not at the temperature of the detector. It is difficult to completely eliminate stray radiation in an IRES experiment. These complexities, and methods that have been presented in the literature to compensate for their effects in IR emission, are discussed in some detail below.

While all these factors contribute to the complexity of IR emission data and make the data difficult to interpret, the complexity is the feature that makes IRES so powerful. Each factor that affects the data can be determined if the experiment is carefully designed and the appropriate data interpretation methods are employed. Some of the examples discussed below demonstrate that multiple sample parameters can be simultaneously determined from an IR emission spectrum if a multivariate calibration approach is applied to the data.

In this chapter we cover the IRES and the instrumental approaches that have been used to collect IR emission data. A variety of cases to which IRES has been applied is discussed. The emphasis of the chapter is on quantitative applications, especially those that are readily adapted to process monitoring. The applications we feel have the most potential in this area are the study of thin organic and glass films. These thin films are critical in a number of areas, such as in the microelectronics industry, and there is interest in the development of tools to monitor product quality in or at the manufacturing process. IRES shows tremendous potential for monitoring thin-film production.

9.2. THEORY

In infrared emission spectroscopy a sample is generally heated to a temperature that is greater than the temperature of the detector. This heated sample emits infrared radiation, due to the thermal excitation of the molecules (and/or atoms) inside the sample, and becomes the infrared source for the spectrometer. The intensity of the emitted radiation depends on the population of the molecules in the higher vibrational energy levels, which follows Boltzmann's distribution and thus depends on the temperature of the sample. The higher the temperature, the greater the intensity of the emitted radiation is. The most efficient infrared emitter for all frequencies is called a blackbody. In the following subsections we will discuss the laws governing blackbody radiation, real-sample radiation, and infrared emission spectroscopy of thin-film samples.

Blackbody Radiation

All materials emit electromagnetic radiation when they are heated. The intensity of the radiation depends on the temperature and the properties of the material. A material that is considered to be a blackbody totally absorbs radiation incident upon it regardless of the wavelength. A blackbody also acts as a perfect emitter for all wavelengths because it must remain in thermal equilibrium with its surroundings. Therefore a blackbody is an ideal source of infrared radiation, and it is used as a reference in infrared emission spectroscopy. In the laboratory a blackbody emitter can generally be approximated with a hollow, insulated enclosure, or an oven that has nonreflecting walls with a pinhole on one wall, or with a thick, optically opaque sample with a very smooth surface.

There are several physical laws governing blackbody radiation, including Planck's, Rayleigh-Jeans, Wien's, Stefan-Boltzmann, and Wien's displacement laws.

Planck's Law

Planck's radiation law states that the spectral energy density of a blackbody, $U_b(\nu, T)$, as a function of frequency ν and its absolute temperature T can be expressed as [1, 2]

$$U_b(\nu, T) = \frac{8\pi h \nu^3}{c^3} \cdot \frac{1}{e^{h\nu/kT} - 1}, \tag{9.1}$$

where h is Planck's constant, k is the Boltzmann's constant, c is the speed of light *in vacuo*, and the subscript b means blackbody. Equation (9.1) can be expressed in terms of the spectral radiance in frequency units, $B_b(\nu, T)$, where $B_b(\nu, T) = U_b(\nu, T) \cdot c/(4\pi)$. It can also be expressed in terms of spectral radiance in wavelength units by $B_b(\lambda, T) = B_b(\nu, T) \cdot c/\lambda^2$. These spectral radiance formulations yield a more useful expression of Planck's law for spectroscopic applications because the spectral radiance is the actual quantity measured by the spectrometers. Planck's law can then be written as follows:

$$B_b(\nu, T) = \frac{2h\nu^3}{c^2} \cdot \frac{1}{e^{h\nu/kT} - 1} \tag{9.2}$$

or

$$B_b(\lambda, T) = \frac{2hc^2}{\lambda^5} \cdot \frac{1}{e^{hc/\lambda kT} - 1} \tag{9.3}$$

Another expression of Planck's law can be derived in terms of wavenumbers, $\bar{\nu}$, where $B_b(\bar{\nu}, T) = B_b(\nu, T) \cdot c$ to yield

$$B_b(\bar{\nu}, T) = 2hc^2\bar{\nu}^3 \cdot \frac{1}{e^{h\bar{\nu}c/kT} - 1}. \tag{9.4}$$

Equations (9.2)–(9.4) show that at a particular frequency/wavelength/wavenumber the spectral radiance depends only on temperature. In Fig. 9.1 blackbody radiation curves derived from Planck's law (9.3) are plotted as a function of wavelength at several different temperatures. Two things can be noted from Fig. 9.1: First, the intensity of the curves at any wavelength increases as the temperature increases, and second, the curve maximum shifts toward shorter wavelengths as the temperature is increased.

Rayleigh-Jeans Law

Rayleigh-Jeans law approximates Planck's law for blackbody radiation at long wavelengths, where $hc/\lambda kT \ll 1$ so that $(e^{hc/\lambda kT} - 1) \approx hc/\lambda kT$ and

$$B_b(\lambda, T) = \frac{2ckT}{\lambda^4}. \tag{9.5}$$

This expression gives results accurate to within 1% when $\lambda T > 7.2 \times 10^8\ \text{nm} \cdot \text{K}$, but it fails at short wavelengths because this law predicts that the spectral radiance continuously increases as the wavelength becomes

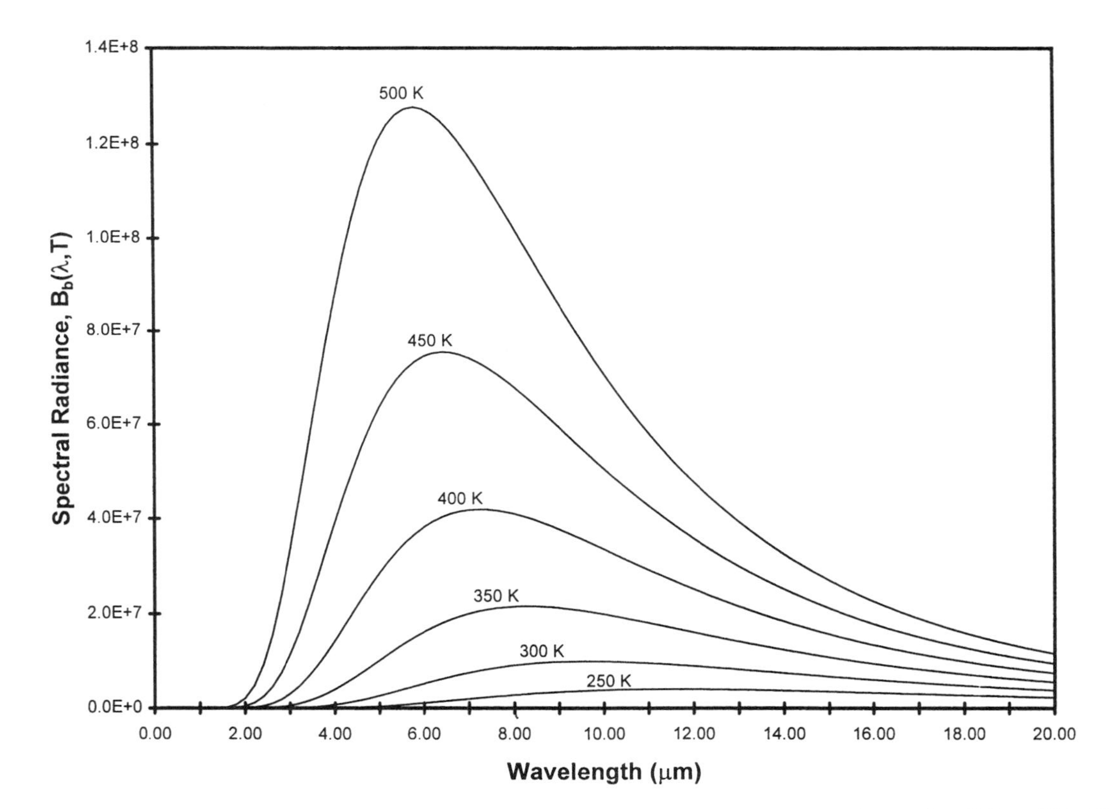

Fig. 9.1. Spectral radiance, $B_b(\lambda, T)$, of a blackbody as a function of wavelength, derived from Planck's law (9.3) at 250, 300, 350, 400, 450, and 500 K.

shorter which is untrue; see (9.5). The failure of the Rayleigh-Jeans law at short wavelengths is known as the "ultraviolet catastrophe" [1, 2].

Wien's Law

Wien's law approximates Planck's law at short wavelengths, where $hc/\lambda kT \gg 1$. Here $(e^{hc/\lambda kT} - 1) \approx e^{hc/\lambda kT}$ and

$$B_b(\lambda, T) = \frac{2hc^2}{\lambda^5} e^{-hc/\lambda kT}. \tag{9.6}$$

This equation gives results accurate to within 1% when $\lambda T < 3.1 \times 10^6\,\text{nm} \cdot \text{K}$.

Stefan-Boltzmann Law

The total radiance of a blackbody as a function of temperature, $B_b(T)$, can be obtained by integrating equation (9.3) over all wavelengths:

$$B_b(T) = \int_0^\infty B_b(\lambda, T) d\lambda = \sigma T^4, \tag{9.7}$$

where σ is called the Stefan-Boltzmann constant, and its value is $(8\pi^5 k^4/15c^3h^3) = 5.6686 \times 10^{-8}\,\text{W} \cdot \text{m}^{-2} \cdot \text{K}^{-4}$. Equation (9.7) is known as the Stefan-Boltzmann law.

Equation (9.7) shows that the total radiance emitted by a blackbody is very sensitive to changes in the temperature of the blackbody, since the total radiance increases as T^4. This sensitivity to temperature change increases the difficulty of performing quantitative analysis using infrared emission spectroscopy. In order to obtain stable spectral data, which is very important for quantitative analysis, the sample temperature must be tightly controlled and must be uniform throughout the sample. The required temperature stability, and especially temperature uniformity, may be difficult to achieve experimentally. Alternatively, as discussed below, the generation of temperature gradients can be used to great advantage in IRES of samples that would be opaque at constant temperature.

Wien's Displacement Law

Figure 9.1 shows that the maxima of the Planck's blackbody radiation curves move toward shorter wavelengths as the temperature is increased. The wavelength maximum, λ_{max}, at a given temperature can be obtained by solving $dB_b(\lambda, T)/d\lambda = 0$, that is, by differentiating equation (9.3), to yield

$$\lambda_{\max} = \frac{2897.9}{T}, \tag{9.8}$$

where λ is in microns. This equation is called Wien's displacement law. In wavenumber units, which is more meaningful for spectroscopic studies, equation (9.8) can be expressed as $\bar{\nu}_{\max} = 1.93T$, obtained by taking the derivative of the spectral radiance (9.4) at a constant interval of $\bar{\nu}(dB_b(\bar{\nu}, T)/d\bar{\nu} = 0)$.

Real-Sample (Graybody) Radiation

When electromagnetic radiation impinges upon a real sample, it will be partially absorbed, partially reflected, and partially transmitted. If the amount of radiation that is absorbed, reflected, and transmitted is measured as absorbance, reflectance, and transmittance, then at any given temperature and frequency the sum of these quantities must equal unity:

$$\text{Absorbance } (A) + \text{Reflectance } (R) + \text{Transmittance } (Tr) = 1. \tag{9.9}$$

According to Kirchoff's law, if the sample is in equilibrium with its surroundings, the amount of radiation emitted by the sample will be the same as that absorbed. Equation (9.9) can then be rewritten as

$$\text{Emittance } (E) + \text{Reflectance } (R) + \text{Transmittance } (Tr) = 1. \tag{9.10}$$

A blackbody does not reflect nor transmit radiation, so $R = 0$, $Tr = 0$, and $E = 1$. For a real sample, the emittance E is usually expressed as the ratio between the intensity of the sample radiation at a particular frequency to that of the blackbody at the same temperature. Since a real sample, which is sometimes called a "graybody" emitter, is not always totally opaque (i.e., partially transmitting), $0 < Tr < 1$, and its surface is not always totally smooth (i.e., partially reflecting), $0 < R < 1$, the intensity of the radiation emitted at a given temperature is always lower than that of a blackbody. Therefore E is always less than unity for a graybody emitter.

There is some confusion in the literature concerning the use of the terms "emittance" and "emissivity." Some papers use these terms interchangeably [3–5], and others make a clear distinction [6–8]. In this chapter we will define emissivity and emittance as follows:

Emittance (E) is defined as the ratio of the emitted radiance per unit area of a sample to that of a blackbody at the same temperature and frequency. Experimentally emittance (E) is obtained from the ratio of the single beam emission spectrum of a sample to that of a blackbody recorded using the same conditions (geometry, temperature, etc.).

Emissivity (ε) is an intrinsic property of a sample, independent of the geometry and temperature. For a blackbody, emittance and emissivity are identical.

Infrared Emission Spectroscopy of Semitransparent Thin Films

The theory of infrared emission spectroscopy of nonopaque (i.e., semitransparent or fully transparent) materials was initially developed by McMahon [9] in 1950 and was extended by Gardon [10] in 1956. McMahon's approach was based on the assumption that the radiation emitted by the sample is unidirectional, normal to the surface of the sample. Gardon extended the theory by considering the radiation emitted by the sample to be hemispherical, which takes into account the diffuse nature of the emitted radiation. In both of the treatments the sample was considered to be a uniformly heated bulk material.

The extension of IRES from bulk samples to thin-film samples or thin films on bulk substrates is of great interest, but the addition of a thin film to the surface of the sample viewed by the spectrometer considerably complicates the IR emission spectrum. However, the theoretical and experimental aspects of infrared emission spectroscopy of semitransparent thin-film systems have been developed and discussed in the literature [4, 5, 7, 11–14].

Generally, in the spectroscopic investigation of thin-film systems the experimental setup of the sample can be classified into one of the following arrangements: (1) unsupported thin films, (2) a thin film on a window, (3) a thin film between two windows, (4) a thin film on a support, and (5) a thin film between a window and a support. These experimental arrangements are shown in Fig. 9.2. The formulas describing each arrangement have been discussed in detail by Rytter [14]. In this chapter we will highlight only the formulas describing a thin-film sample without a support (Fig. 9.2*a*) and a thin film on a solid support (Fig. 9.2*d*). The equations describing the latter situation are most applicable for describing dielectric thin films, such as phosphosilicate glass (PSG) or borophosphosilicate glass (BPSG) thin films, deposited on silicon wafers. These systems have been extensively studied in our laboratory using IRES [15–18], as well as external infrared reflection-absorption [19–21] and conventional infrared transmission [22, 23] spectroscopies. A review of quantitative determination of dielectric thin-film properties using IR spectral data has appeared in the literature [24]. An analogous situation to dielectric thin films on a support is an organic thin film on a support. The theoretical description of IR emission from a thin-

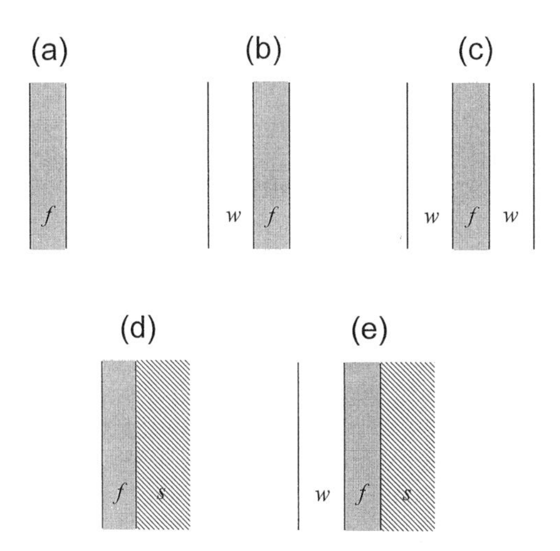

Fig. 9.2. Experimental arrangements for spectroscopic studies of thin films: (*a*) unsupported thin films, (*b*) a thin film on a window, (*c*) a thin film between two windows, (*d*) a thin film on a support, and (*e*) a thin film between a window and a support. f = thin film, w = window, and s = support.

film sample on a support applies equally well to both dielectric and organic thin-film systems.

The emittance of a thin-film sample depends on the surface reflectivity and the internal transmittance of the sample. The internal transmittance, p, which is also called the sample permeability in the literature [14], is defined as $p = e^{-Kd}$ according to Lambert's law, where K is the absorption coefficient and d is the thickness. For an unsupported semitransparent thin-film sample having plane parallel surfaces as shown in Fig. 9.2*a*, with surface reflectance R, and internal transmittance p, the emittance of the sample at any given frequency and temperature follows an expression that is based on McMahon's work [9]

$$E = \frac{(1 - R)(1 - p)}{(1 - R \cdot p)}. \tag{9.11}$$

This expression is derived for reflecting, semitransparent thin films, with the assumption that the sample temperature is uniform throughout the sample. If the sample is totally opaque, $p = 0$, the emittance equation (9.11) is simplified into

$$E = 1 - R, \tag{9.12}$$

which is the same as the more commonly seen form of Kirchoff's law, namely equation (9.10) with $T = 0$. In this case the emittance is equivalent to the emissivity of the sample. Equation (9.12) shows that if a sample is highly reflective, the emittance will approach zero. This fact has a very important role in the study of thin films on reflective supports, such as organic or polymer thin films on metals. Since metals are highly reflective, interferences due to the emitted radiation from the support are negligible in the total emission spectra, and therefore complications in the IR emission data are reduced.

For a thin-film sample on a solid support (Fig. 9.2*d*), the emittance depends on the internal transmittance of the thin-film sample, the reflectivity at the sample-to-air interface r_1, and the reflectivity at sample-to-support interface, r_2. In addition radiation from the solid support needs be considered in the total emittance, since the support does emit some radiation unless it is totally reflective. Although the radiation emitted by the support will be significantly attenuated passing through the thin-film sample, it will contribute some fraction of the observed emittance. Therefore the emitted radiation coming from the solid support must be included in the total emittance of the sample. Several authors have proposed different expressions for the total emittance of such a system by including the emissivity and reflectivity of the support [6, 14, 25, 26]. The details of their derivations are beyond the scope of this chapter. However, the emittance of a thin-film sample on an optically opaque support can be summarized for three different cases of thin-film permeability [14]:

1. For a partially transmitting sample, where $p = e^{-Kd}$,

$$E = \frac{(1 - r_1)(1 - r_2 p^2)}{(1 - r_1 r_2 p^2)}. \tag{9.13}$$

2. For an opaque sample, where $p = 0$,

$$E = (1 - r_1). \tag{9.14}$$

3. For a totally transparent sample, where $p = 1$,

$$E = \frac{(1 - r_1)(1 - r_2)}{(1 - r_1 r_2)}. \tag{9.15}$$

These expressions are derived with the assumption that there are no temperature gradients within the thin-film sample and that the sample and the support are in isothermal equilibrium. To simplify the system, the support is usually chosen to be totally opaque and highly polished, and thus totally self-absorbing and as such does not emit any radiation. In this case the emittance becomes

$$E = \frac{(1 - r_1)(1 - p^2)}{(1 - r_1 p^2)}, \tag{9.16}$$

which is similar to equation (9.13) with $r_2 = 1$.

Stray Light

In addition to the potential interference from the infrared emission of the support material, it is also possible to observe undesired stray light emanating from the walls of the spectrometer or from any of the spectrometer optical components. This stray light can also interfere with the infrared emission measurement, causing a baseline distortion in the emittance spectra similar to that observed in absorbance or reflectance spectroscopy. In order to eliminate the effects of stray light, Chase [27] has suggested a correction using a four-measurement method, in which the spectra of both sample and blackbody background are collected at two different temperatures, T_1 and T_2. The radiant intensity, $S(\bar{v}, T)$, at each temperature follows the expression

$$S(\bar{v}, T) = R_f(\bar{v}) \cdot [E(\bar{v}, T)B_b(\bar{v}, T) + G(\bar{v}, T) + I(\bar{v}) \cdot R(\bar{v})], \tag{9.17}$$

where

$\bar{v}$ = frequency in wavenumbers,
$R_f(\bar{v})$ = instrument response function,
$E(\bar{v}, T)$ = emittance of the material,
$B_b(\bar{v}, T)$ = Planck's law,
$G(\bar{v})$ = background radiation,
$I(\bar{v})$ = background radiation reflected off the sample, and
$R(\bar{v})$ = reflectance of the sample.

For the blackbody reference material, $E(\bar{v}, T) = 1$ and $R(\bar{v}) = 0$. By making the assumption that $E(\bar{v}, T)$ and $R(\bar{v})$ of the sample are independent of the operating temperature, the emittance can be calculated as

$$E = \frac{S_{s2} - S_{s1}}{S_{b2} - S_{b1}}, \tag{9.18}$$

where S_{b2} and S_{b1} are the single-beam spectra of the blackbody reference at temperatures T_2 and T_1, respectively, and S_{s2} and S_{s1} are the single-beam spectra of the sample at temperatures T_2 and T_1, respectively. Compton et al. [28] have suggested an alternative method to minimize the effect of stray light present in the spectrometer system. They suggest measuring the infrared emittance of the spectrometer background in addition to that of the sample and blackbody reference. Correction for stray light is then made by calculating the ratio of the difference of the sample single-beam S_s and background single-beam S_{bg} to the difference of the blackbody single-beam S_{bb} and the background single-beam S_{bg}. All single-beam spectra are measured at a given temperature. The emittance is calculated as

$$E = \frac{S_s - S_{bg}}{S_{bb} - S_{bg}}. \tag{9.19}$$

In practice, the authors of both proposed methods suggest the use of interferograms instead of single-beam spectra in calculating the corrected emittance to avoid phase correction problems [3, 27, 28].

Emission Band Distortions

Ideally the emission spectrum of a sample is similar to its absorption spectrum. At frequencies where absorption bands are intense, the emittance is also high, and so on. However, in many cases distorted bands are observed in infrared emittance spectra of real samples. These band distortions, which are especially prominent at strong absorption bands, include band splitting, band intensity variations, and even total band inversion. These distortions arise from multiple effects, including excessive sample thickness [12, 13, 25, 26], large changes in the absorption coefficient and thus correspondingly complex refractive index changes at strong absorption bands [12, 13], and dispersion of surface reflectivity [6, 14, 26]. These distortions have also been attributed to the reabsorption of the radiation emitted inside the sample by the cooler surface molecules when there are temperature gradients in the sample, such as between the bulk and surface of the sample when the sample is heated from the rear [4, 6, 14, 29]. Figure 9.3 shows the band distortions observed by Kozlowski [25] in emittance spectra of molten KNO_3 at 560°C with different thicknesses. Several authors have suggested an experimental method that may minimize spectral distortions [6, 7, 14]. These authors suggest the use of an opaque, thick sample as a reference instead of a blackbody so that the corrected emittance E^* is then obtained by ratioing the thin-sample single-beam emission spectrum, S_{thin}, to that of a thick, opaque sample, S_{thick}, rather than to that of a blackbody.

$$E^* = \frac{E_{\text{thin}}}{E_{\text{thick}}} = \frac{S_{\text{thin}}}{S_{\text{thick}}}. \tag{9.20}$$

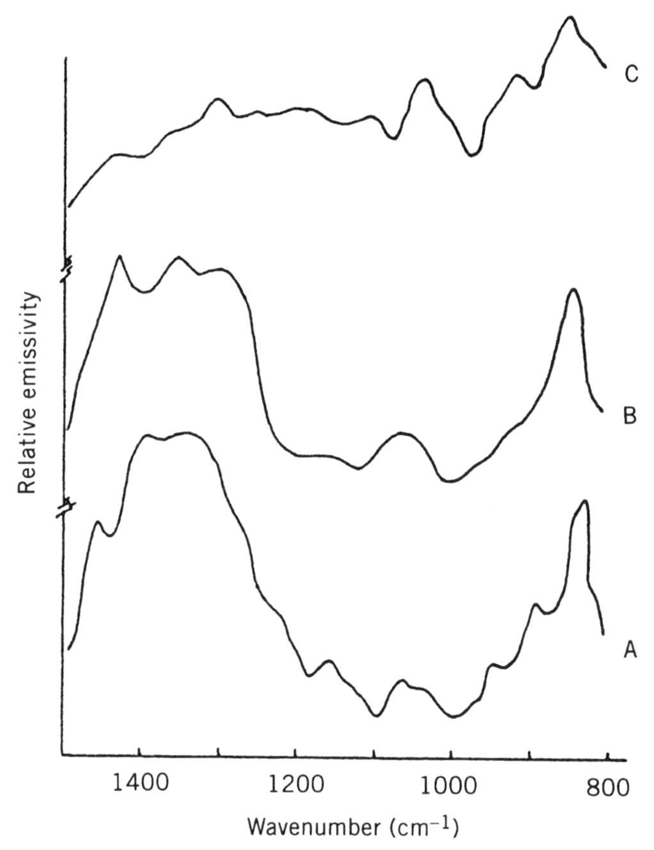

FIG. 9.3. Relative emissivities of molten KNO_3 at 560°C for different film thicknesses, 0.048 cm (*A*), 0.107 cm (*B*), and 0.163 cm (*C*). Reprinted from [25] with permission.

This method has been shown experimentally, the results of which are provided in Fig. 9.4 [14], as well as theoretically using a damped harmonic oscillator model, to successfully eliminate band distortions for both strong and weak emission bands [6, 14]. The success of these approaches can be explained by the following: When the single-beam spectrum of a thick sample is used as a reference, variations in the sample reflectivity at strong absorption bands have been incorporated into the reference spectrum. Therefore, when the single-beam spectrum of the thin sample is ratioed to this reference spectrum, the band distortions will be a part of both the sample and reference, and thus ratio out.

It was mentioned at the beginning of this section that the sample is generally heated above the detector temperature in IRES. However, Low and Coleman [30] in their study of minerals and rocks and Chase [27] in his study of condensed phase materials have demonstrated that it is possible

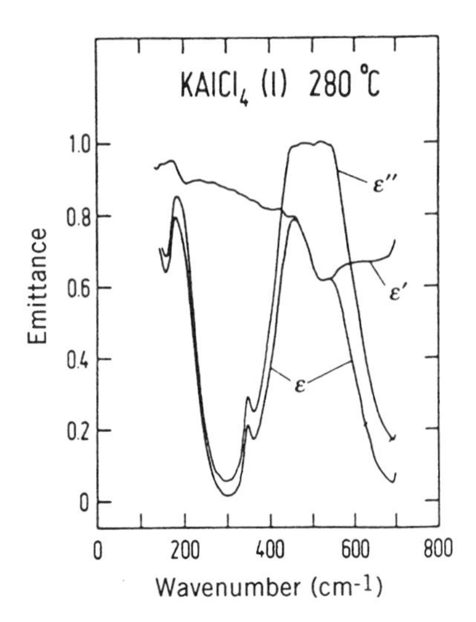

Fig. 9.4. The emittance (ε), emissivity (ε′), and corrected emittance (ε*) of molten $KAlCl_4$. Note that the band splitting appearing in the ε spectrum has been corrected in the ε* spectrum. Reprinted from [14] with permission.

to obtain emittance spectra of a sample by either heating or cooling it with respect to the detector. Furthermore Chase [27] has shown in a study of a polystyrene film using a room temperature triglycine sulfate (TGS) detector that the spectra obtained with the sample temperature 15°C higher and lower than the detector temperature are essentially identical (see Fig. 9.5). In an emission experiment the radiation detected is essentially a relative flux of radiation as a result of the temperature difference between the sample and the detector. A large temperature difference is desirable, since it will increase the flux of radiation and hence the signal-to-noise ratio of the emittance spectra. However, in practice usually there is a limit on the useful temperature range. For example, high temperatures might cause deterioration in many organic materials. In such a case a cooled detector can be used to increase the signal-to-noise ratio of the emittance spectra.

Temperature Determination

There are numerous examples, especially in process monitoring, where measuring a sample temperature is desirable. The single-beam emission spectrum from a sample can be compared to a best-fit computer-generated blackbody curve based on Planck's law to determine the temperature of the sample. Indeed this approach has been suggested in the literature [8], and it is essentially the basis of optical pyrometry. Optical pyrometers are

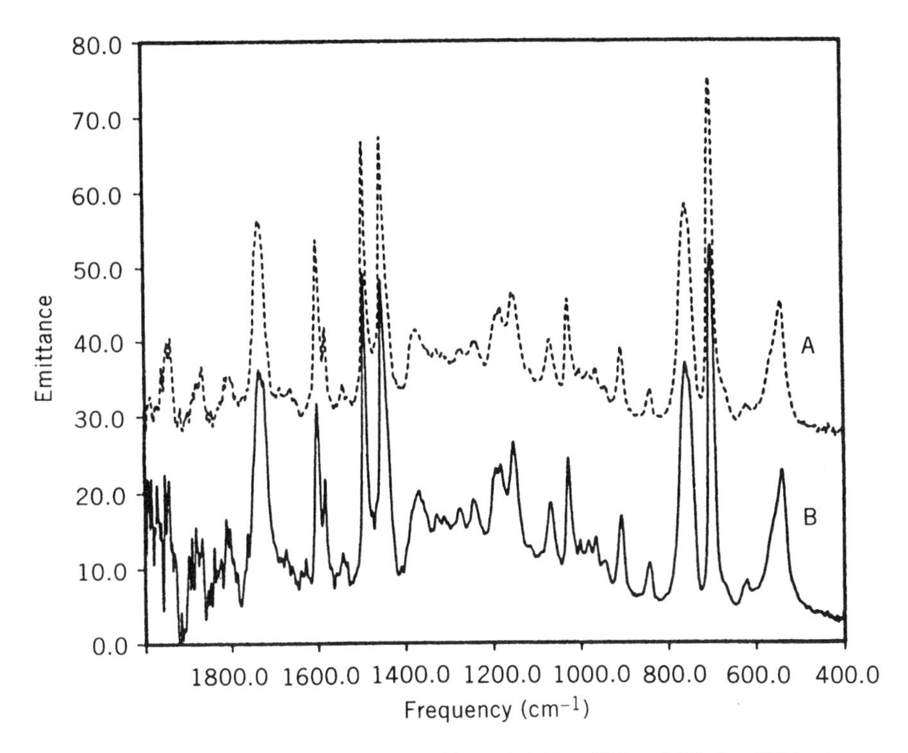

Fig. 9.5. Emittance spectra of a polystyrene film at (*A*) $T = 40°C$ and (*B*) $T = 10°C$ using a room temperature TGS detector. Reprinted from [27] with permission.

generally calibrated for a particular sample type and temperature range but can provide rapid noncontact temperature determinations as long as the sample type is identical to that used in the calibration. In order to achieve accurate results from an IR emission spectrum, the instrument response function and detector temperature must be taken into account [31]. Even this approach can fail when a temperature determination is made based on the emission spectrum of a new sample if the emissivity of the new sample has changed from that for which the corrections were developed.

Any quantitative application of IRES will necessarily be faced with samples of varying composition. Sample emissivity will change with changing sample composition. When the samples are thin films, changes in film thickness affect the emissivity far more dramatically than changes in the film composition. Thus the determination of temperature from IR emission data obtained from thin-film samples, such as the *in situ* monitoring of a thin-film deposition process, is a significant challenge. A theoretical model correlating variations in the IR emission data to changes in film composition and thickness would be very complicated. Temperature determinations can, however, be made from IR emission data obtained from thin-film samples of varying composition and thickness using an empirical calibration, as will be discussed below.

9.3. EXPERIMENTAL CONSIDERATIONS

Spectrometers

The signal level in an infrared spectrometer is dependent on the temperature difference between the source and the detector. The sources employed in conventional infrared spectrometers are generally one of two types, a glowbar or a Nernst glower. A glowbar is a silicon carbide rod that is generally operated at 1200–1400°C. A Nernst glower is a rod of zirconia, yttria, or thoria operated at 1000–1700°C. Even if a room temperature detector is employed, the temperature difference between one of these conventional sources and the detector is large, thus the radiation flux available at the detector is relatively large. Most IRES experiments have been carried out by heating the sample and using it as the source of the spectrometer. Many samples of interest cannot tolerate high temperatures, or the sample becomes blackbodylike with increasing temperature [29, 32]. Thus the spectrometer source in IRES, the sample, is almost never as hot as conventional spectrometer sources. Because the total radiance emitted from a sample is dependent on T^4 (see equation (9.7)), the radiation flux at the detector in IRES is relatively low. As a result the potential for IRES as an analytical technique was not really recognized until commercial Fourier transform infrared spectrometers (FTIRs) became widely available in the late 1960s and early 1970s.

One of the major advantages of FTIRs is high throughput. For this reason, and the fact that FTIRs have a signal-to-noise ratio advantage over dispersive spectrometers, almost all IRES studies have been carried out using FTIRs. Also liquid nitrogen cooled detectors are often employed in IRES studies. These detectors are more sensitive, have lower noise, and increase the temperature difference between the heated sample and the detector. The theory, practice, and instrumentation of Fourier transform infrared spectroscopy have been discussed in detail in a number of publications [33], so they will not be discussed here.

A consideration when using a cooled detector in a room temperature spectrometer is that the spectrometer walls, the optics, etc., will all radiate to the detector. This background radiation can complicate sample spectra and be a source of variation if not constant. Generally, the major sources of this background radiation are components in the optical train of the spectrometer. With the exception of the beam splitter, the optics in most FTIRs are reflective, so the beam splitter is the major source of background radiation. Operation with a room temperature detector eliminates this problem, but the trade-off with signal level is sometimes severe.

Constant Temperature IR Emission Attachments

A number of commercial IR emission attachments are available. In general, these attachments have been designed for qualitative applications. The major limitation of most of these attachments, when quantitative applications are considered, is the lack of precise and accurate temperature control. In 1992 we reported [15] the modification of a commercial attachment [34] with the objective of increasing the precision of the sample temperature. After modification the surface temperature of a sample was found to vary ± 1.8°C over a two-hour period. This apparatus functioned well enough to produce quantitative results as long as the temperature maximum was kept below 225°C [16, 17]. At temperatures above 225°C the heater in the attachment did not have sufficient power to maintain a stable sample temperature. The main goal of the IRES studies in our laboratories was to demonstrate that IRES could be used as an *in situ* monitor for the chemical vapor deposition (CVD) of thin dielectric films. The CVD apparatus used to deposit these films generally operate at temperatures significantly higher than 225°C. In order to more closely mimic the CVD process, we designed and constructed an IRES attachment with a considerably higher power heater. A schematic drawing of this apparatus is shown in Fig. 9.6. The heater, capable of heating a sample to 600°C, was produced by winding an inconel heating element into a coil and mounting it in a hollowed-out copper block. The sample is placed directly on top of the heater. Two thermocouples were employed in this system. One was soldered into a hole near the top of the copper block and connected to the PID (proportional with integral and derivative) temperature controller. The second was used to monitor the surface temperature of the sample. The aluminum housing was made to be vacuum tight so that experiments could be carried out in a vacuum or in a controlled atmosphere. This apparatus was used to collect the IR emission data for the 21 sample sets of BPSG thin films discussed below.

These attachments, like most of the commercially available attachments, were mounted externally to the spectrometer, and a mirror, shown in Fig. 9.6, was used to direct the radiation from the sample into the spectrometer through an auxiliary port. An alternative to the externally mounted apparatus has been described by Rewick and Messerschmidt [35]. This attachment mounts in the sample compartment, which in some situations might be more convenient.

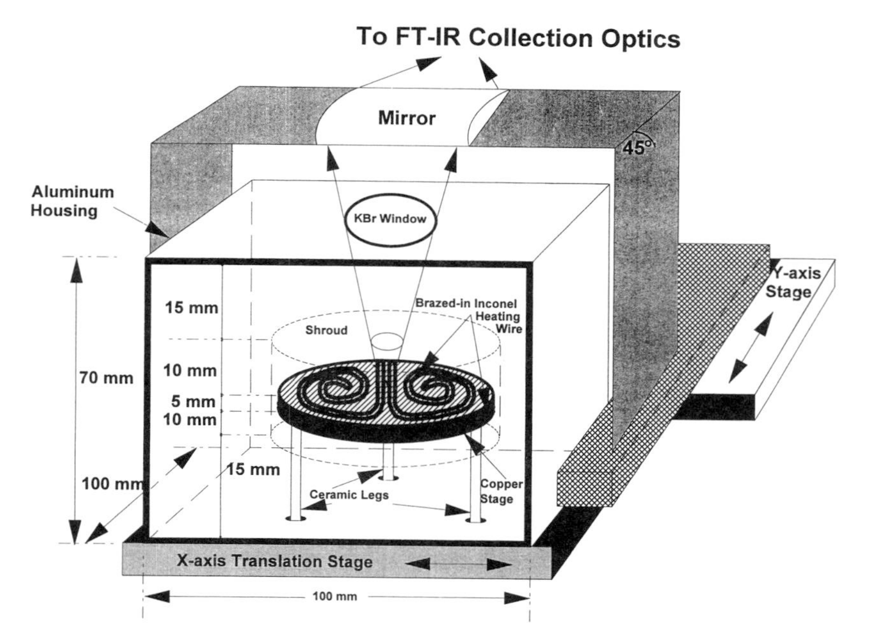

Fig. 9.6. Schematic drawing of the constant temperature/controlled atmosphere IR emission apparatus.

A number of other IRES attachment designs have been discussed in the literature. Of particular interest is the system described by Karpowicz [36] specifically designed for the study of polymers. Also of note, an electrical furnace specifically designed for gas phase studies over the temperature range of 100 to 1000°C has been described by Tilotta and coworkers [37].

Transient Infrared Emission Spectroscopy

In conventional IRES the sample is held at an elevated uniform temperature which limits the application of the technique to optically thin samples. When a material is heated to an elevated temperature, all parts of the sample both emit and reabsorb infrared radiation. This self-absorption phenomenon causes severe truncation of emission bands and eventually leads to a blackbodylike spectrum that contains little information about the sample properties [29, 32, 38]. One method of reducing the effects of self-absorption in IRES is to thin the sample, or limit applications to the study of thin films. In the case of thin films, such as much of the work covered below, the samples are coatings on low-emittance substrates, and self-absorption is not a significant problem. In many cases, however, the samples are bulk materials, and there is no opportunity to thin them.

An alternative to thinning the sample is to heat only a thin layer of the sample to the desired emission temperature. Koga et al. proposed a method of front surface heating to deal with bulk samples [39]. More recently lasers have been employed to rapidly heat a thin layer of material on the sample surface [38, 40, 41]. The IRES measurement is accomplished by recording the emission from the heated thin layer of the sample before thermal diffusion causes the heated layer to thicken and cool. The process is thus transient, and it has been referred to as transient infrared emission spectroscopy (TIRES).

An example of the experimental arrangement used to carry out TIRES using a continuous laser is shown in Fig. 9.7. When a continuous laser is used to heat the sample, the laser must be scanned across the sample, or the sample must be moved through the laser beam. In the experimental arrangement shown in Fig. 9.7, the sample is mounted on a rotation stage that moves the sample relative to the laser beam. The system was set up so that the area of the sample irradiated would be rotated immediately into the field of view of the IR spectrometer. A jet of liquid nitrogen chilled helium was positioned so that the sample would be rotated into the jet immediately following the field of view of the spectrometer. Note that the use of a moving sample closely simulates the conditions that might be encountered when monitoring a product at-line.

An alternative approach to heating the sample is the use of a hot-gas jet in place of the laser. Jones and McClelland describe the use of a hot-gas jet for transient heating of a thin layer of sample in an experimental arrangement nearly identical to that shown in Fig. 9.7, with the exception that the

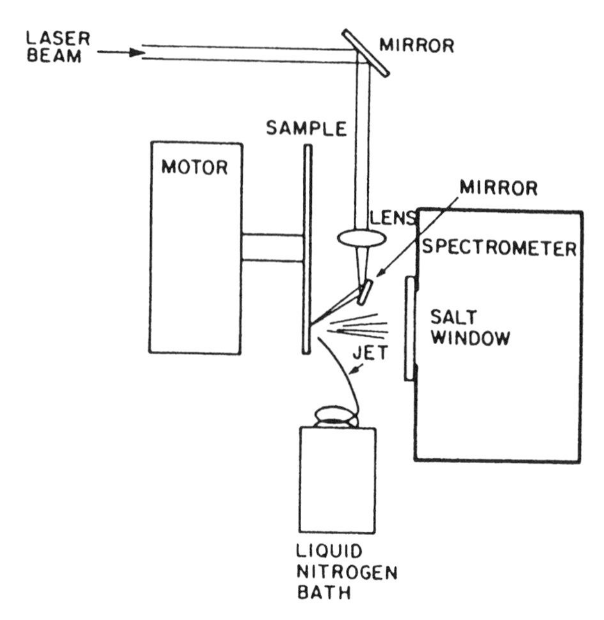

Fig. 9.7. Experimental arrangement for TIRES using a continuous laser. Reprinted from [40] with permission.

laser is replaced with a jet of heated nitrogen [42]. Hot-gas jet heating has the significant advantage, when compared to laser heating, that it is not limited to materials that absorb at the laser wavelength.

A related method of generation of a transient heat gradient in a sample is to cool a thin layer on the surface of a heated sample. Jones and McClelland were the first to show that a ratio of a spectrum obtained after cooling to that obtained before cooling closely resembles a transmission spectrum of the material [43]. In such an arrangement the heated bulk sample becomes the source, and the cooled surface layer becomes the sample. Pell et al. provide additional examples of this approach, and a better theoretical understanding of the experiment [44]. Such an approach might be of significant value in process monitoring applications where the samples are heated relative to ambient. In such situations further heating of the sample might be of little value, while the transient cooling approach might produce useful IR spectral data.

Multivariate Calibration

As stated above, there are many factors that affect IR emission data, making interpretation of the data, especially quantitative interpretation, a difficult process. In the 1970s when FTIRs were becoming commonplace and the potential for analytical IRES was being discovered, both Griffiths [29] and Hirschfeld [45] pointed out that IRES could be a powerful quantitative tool if algorithms were developed to extract quantitative information from the complex data. Multivariate calibration techniques, in particular the partial least squares (PLS) and principal components regression (PCR) algorithms, have been found to be especially powerful for quantitative interpretation of complex spectral data. Almost all quantitative applications of IRES have employed multivariate calibration algorithms to interpret the data. These algorithms have been thoroughly described in the literature [46, 47]. Interested readers are referred to these references for general discussions of multivariate calibration. The specific application of multivariate calibration to IR emission data to extract both qualitative and quantitative information has been presented elsewhere [18].

In order to avoid overfitting the data during a calibration process, and to assess the ability of a calibration model to predict an unknown sample, a process of cross-validation is used [48]. In cross-validation one (or more) of the calibration samples is removed, and a calibration model is produced from the remaining samples. The model produced is then used to predict the value of the removed sample(s). This process is repeated with a different sample(s) removed at each cycle until the values for all samples have been predicted. The cross-validated standard error of prediction (SEP) is then calculated.

$$\mathrm{SEP} = \sqrt{\frac{\Sigma_{i=1}^{n} (y_i - \hat{y}_i)^2}{n}} \tag{9.21}$$

where y_i are the reference values obtained by an independent method of analysis, $\hat{y}_i$ are the values predicted in the cross-validation process, and n ($n - 1$ if the data are centered) is the number of samples.

9.4. APPLICATIONS

IRES has been used in a variety of applications. A number of these applications are discussed below. We have attempted to cover a wide range of applications so that the utility of IRES to different types of studies can be demonstrated. We have also chosen to focus on those applications that demonstrate the potential for IRES as a process-monitoring tool.

Gas Phase Studies

Infrared emission spectroscopy has been applied to physical transition and kinetic studies of small molecules for a relatively long time. Generally, an energy source such as a microwave discharge or a pulsed laser is used to excite the molecules, and the infrared emission spectra are collected with high resolution, and sometimes as a function of time. In 1970 Koopmann and Saunders [49] presented a quantitative method for the determination of relative populations of carbon dioxide vibrational levels using infrared emission as a function of temperature over the temperature range of 1000 to 3200 K. Roux et al. [50] analyzed the rotational structure of bands in $^{14}N_2$ and $^{15}N_2$ emission data and reported the rotational constants for the transitions. Infrared emission of a mixture of N_2 and SiH_4 was studied by Elhanine et al. [51] using radio-frequency discharge activation coupled with a high-resolution Fourier transform interferometer. They identified the transient species NH, SiN, and SiNH and derived molecular structure parameters with high accuracy. Rawlins and coworkers [52] used a time-resolved dispersive infrared spectrometer coupled with a liquid nitrogen cooled 12-element HgCdTe array detector to study shock-heated, dilute NO/Ar mixtures over the temperature range of 500 to 2200°C. They used the spectral intensity distributions to measure the vibrational temperature of NO_v and from these data determined the thermally averaged Einstein coefficient for the optical transition $v = 1$ to $v = 0$. They found that this Einstein coefficient is not dependent on temperatures up to 2500°C. Wang et al. [53] studied vibrational quenching of NO ($v = 1 - 11$) in the gas phase, generated from the reaction of N_2O and $O(^1D)$ using a laser photolysis process. They reported rate constants for quenching which correspond to a relaxation process where the NO energy is transferred to the ν_1 and $\nu_1 + \nu_2$ modes of N_2O. Hartland et al. [54] obtained vibrationally excited NO_2 using collision-induced internal conversion and measured the energy distribution of excited NO_2 during collisional deactivation. Schatz et al. [55] were the first to report some 200 new far-infrared emission lines of NH_3 in a mixture of $^{14}NH_3$, $^{14}NH_2D$, $^{14}NHD_2$, $^{15}NH_3$, and $^{15}ND_3$ heated using a 20-atm CO_2 laser. They also studied the polarization and transient behavior of the mixture.

The photodissociated free radical $CH_3^{\bullet}$ produced from CH_3I was detected by Hermann and Leone [56] using low-resolution infrared emission measurements. They observed three distinct Q-branch features in the ν_2 umbrella mode. Henderson et al. [57] describe an experiment in which they calculated vibrational-rotational energy levels and constructed an overtone emission spectrum of $OH^{\bullet}$ free radicals. They were able to explain some unusual spectral phenomena related to the unpaired electron in the radicals, such as vibroniclike spectral band heads and "spin-orbital" multiplets.

IRES has also been applied to the study of a number of additional source types. Vandooren and Vanpee [58] used H_2/F_2 and CH_4/F_2 premixed flames to study vibrationally excited HF, while Braun and Bernath [59] used the pyrolysis products of CH_3Br in a stainless steel tube furnace at

1000°C to study vibrational-rotational emission spectra and rotational constants for both $H^{79}Br$ and H^{81} Br using a Fourier transform spectrometer. The emission spectrum of gas phase HfO produced in a hollow cathode lamp was investigated by Ram and Bernath [60]. The rotational constants for HfO were reported. Similar studies of PtO [61], NiO [62], and CoO [63], also using hollow cathode discharge lamps, have been reported by the same research group. Later Zhang et al. [64] measured the Dunham coefficients and internuclear potentials of BF and AlF over spectral region of 850 to 1670 cm^{-1} using a Fourier transform spectrometer. The spectra were generated by heating a mixture of boron and CaF_2 powder in a carbon boat positioned in a 1600°C alumina tube. Recently an electrodeless discharge through H_2S was used to generate excited HS molecules, and the infrared emission spectrum recorded over the range of 1850 to 2800 cm^{-1} region was used to calculate vibrational constants for the HS molecules [65].

Small Molecules on Surfaces

Kunimori et al. [66] studied the mechanism and dynamics of the oxidation of CO on Pt and Pd surfaces using time-resolved IRES. They examined the relationship between the surface oxygen coverage and the flow rate of CO introduced with a steady-state beam of O_2, and found no relationship between the production rate of CO_2 and the surface oxygen coverage. Chiang and coworkers [67–70] published a series of papers describing studies of the vibrational modes of CO adsorbed on single-crystal Ni surfaces. IR emission spectra over the region of 330 to 3000 cm^{-1} were collected using a novel apparatus consisting of a liquid helium cooled grating spectrometer coupled to an ultrahigh vacuum system. They showed that the CO infrared emission transitions and spectral linewidths of the transitions are a function of the CO coverage on the Ni surface. The detection of C–Ni and C–O stretching vibrations from monolayer and submonolayer coverage of CO on Ni was demonstrated. Tobin et al. [71] investigated the structure of surface-adsorbed CO bridge bonding and the ordering within CO overlayers. They found that the bridge-bonded CO was due to surface contamination. Later, Tobin [72] reported vibrational studies of C–Pt and O–Pt stretching bands of CO adsorbed on single-crystal Pt. He based his interpretation of the intermolecular interactions to correlations between the C–O frequency shifts and CO exposure.

Characterization of Solid Samples

Fabbri and Baraldi [73] studied the process of thermal decomposition of $CuSO_4 \cdot 5H_2O$ over the temperature range of 100 to 500°C in different environments and found that the nature of the decomposition products to

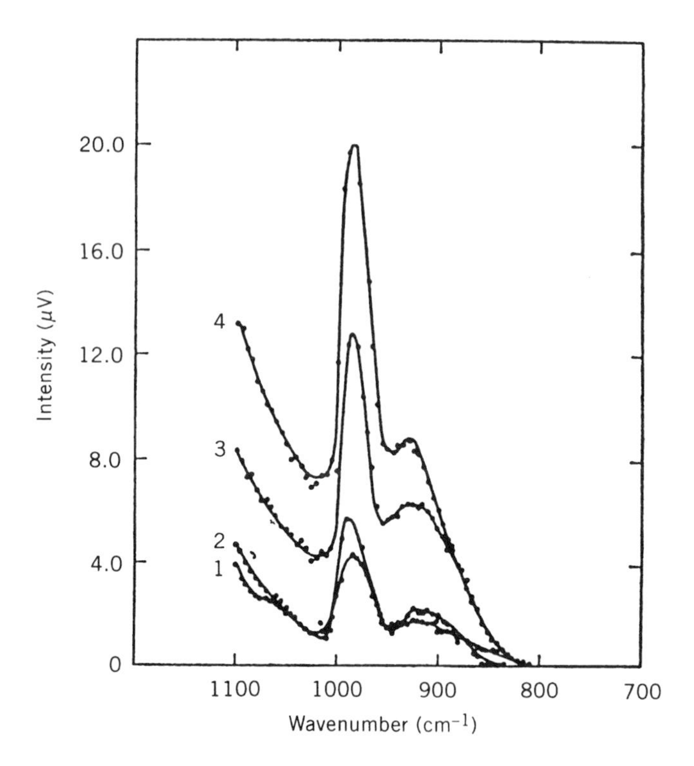

Fig. 9.8. Emission spectra of MoO_3 at 485°C obtained from different mass samples. Sample mass: (1) 0.2 mg, (2) 0.4 mg, (3) 0.8 mg, (4) 1.1 mg. Reprinted from [78] with permission.

be dependent on the gaseous environment. Later [74] they investigated the relationship between the concentrations of components and the intensities of infrared emission bands in K_2SO_4–KNO_3 and K_2SO_4–KBr mixtures. They suggested that the concentration dependence of K_2SO_4 emission is due to self-absorption of the infrared radiation. Aina et al. [75] recorded infrared emission spectra from AlGaAs/GaAs, InP/InGaAs, and InAlAs/InGaAs heterostructures. They found that the 6082.6 cm^{-1} (0.755 eV) emission from AlGaAs/GaAs to be the most prominent. The intensity of this feature correlated very well with both the electrical and optical properties of the samples.

Metal oxidation and metal oxides have been investigated by many research groups using IRES. Fabbri and Baraldi [76] followed the oxidation of Cu in air or inert environments at different temperatures using IR emission. Analogous IRES studies of Mo oxidation were reported by Gratton et al. [77]. Ueno et al. [78, 79] studied both oxidation and reduction of molybdenum (IV, VI) oxides in an IR emission apparatus. They

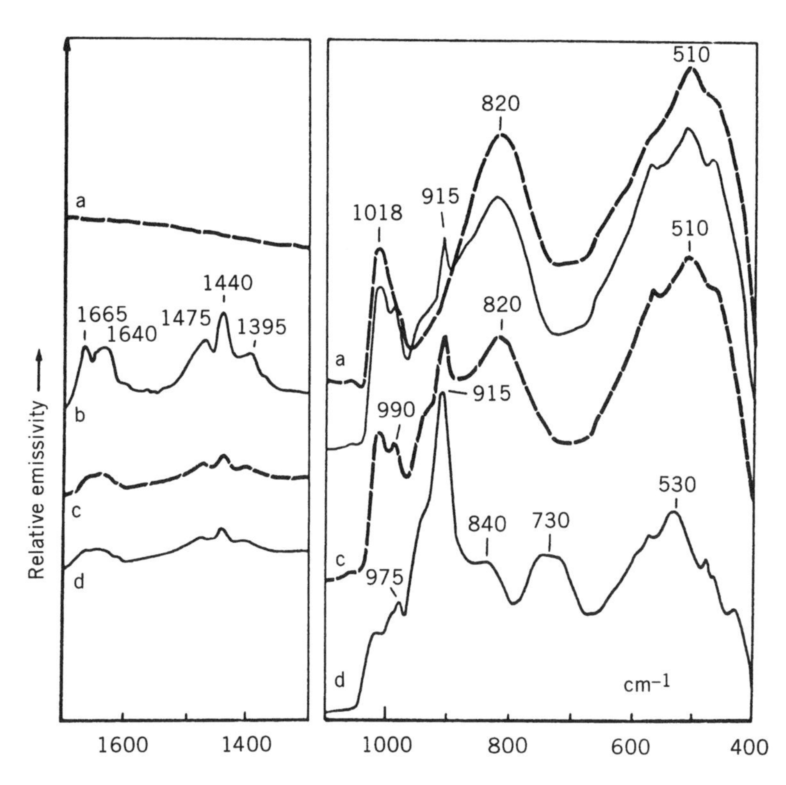

Fig. 9.9. Emission spectra obtained during the interaction of V_2O_5 with propene. All spectra were recorded at 100°C and ratioed against the spectrum of a blackbody at the same temperature. (*a*) V_2O_5 under 20 torr O_2 pressure. (*b*) V_2O_5 after contact with 20 torr of propene at 120°C for 3 hours. (*c*) V_2O_5 after contact with 20 torr of propene at 200°C for 30 minutes. (*d*) V_2O_5 after contact with 20 torr of propene at 250°C for 1 hour. Reprinted from [81] with permission.

found the peak intensities of the emission bands were proportional to the sample weight, as shown in Fig. 9.8. They calculated the activation energies for both reduction and oxidation and studied the behavior of oxygen atoms during the reduction. Bleux et al. [80] studied the thermal properties and solid-liquid transition of glassy oxides over a wide range of temperatures. They found that there was no significant distortion in the lattice structure of the glassy oxides when the material was undergoing fusion. Infrared emission investigations of V_2O_5 [81] and other metal oxide catalysts [82] have also been carried out. IRES was employed as a means to study the reactions on the catalysts' surface and reported to be a very suitable method for studying the interactions of the reactants with the catalysts. For example,

Fig. 9.9 contains IR emission spectra obtained during the interaction of propene with a V_2O_5 catalyst.

Atmospheric Studies

The study of point sources of air pollution is an obvious application for IRES because of the possibility of remote observation. In 1974 Bernes et al. [83] reported the remote determination of sulfur dioxide in power plant plumes with reasonable accuracy using a dispersive infrared emission apparatus. A few years later Herget and coworkers [84, 85] used a Fourier transform spectrometer for the remote optical sensing of emissions (ROSE) system installed in a van to measure gaseous pollutants at distances up to 1000 m from a pollutant source or to obtain emission spectra of gases exiting industrial stacks at elevated temperatures. High-resolution (0.07 cm^{-1}) data were collected and used for identification and quantification of pollutants. This system was used to detect propylene and ethylene near an oil refinery and to measure the efficiency of a flare intended to combust vinyl chloride [85]. The introduction of fluoride-containing compounds to the atmosphere, mainly as HF, are believed to be harmful to the ozone layer. The presence of fluoride-containing compounds being introduced to the atmosphere from gypsum ponds at phosphate fertilizer plants and from jet engine emission at an airport was also confirmed [84].

IRES has also been employed to study the makeup of the stratosphere using balloon-borne or airplane-borne high-resolution Fourier transform infrared spectrometers. Kendall and Clark [86] used far-infrared (30–110 cm^{-1}) emission spectra for detection of HCl, NO_2, OH, H_2O_2, and CO. High-resolution data are required to separate the emission features of these molecules from the much more prominent features of H_2O, O_3, and O_2. Bangram et al. [87] reported the concentration profiles of O_3, H_2O, CF_2C_2, HF, HCl, and HNO_3 over an altitude range of 10 to 40 km using continuously collected IR emission data. Carli and Park [88] and Chance et al. [89] conducted similar studies to determine the same suite of molecules in addition to HDO, HCN, OH, and O_3 heavy isotopes using the far-infrared (40–210 cm^{-1}) spectral region. The high-resolution spectra obtained by these researchers show sharp bands that agree extremely well with theory, hence allowing unambiguous identification. The distribution of $ClONO_2$, HNO_3, and O_3 in the Arctic during winter [90] and determinations of other species such as, HOCl, ClO, N_2O, HBr, and HO_2 in the stratosphere [91] have also been reported in the literature.

Planetary Observations

Much of our knowledge of planetary objects has been obtained using remotely obtained optical, including IR, emission data. IRES measurements of the moon in the mid-infrared spectral region using earth-based spectrometers have been made since the 1960s. Goetz [92] collected IR emission spectra from 22 lunar spots using a dispersive infrared spectrometer coupled with a telescope. He used differences in the spectra as the basis for his determinations. Two years later, Murcray et al. [93] collected IR emission spectra from six selected areas on the lunar surface using a balloon-borne spectrometer. Like Goetz, they based their lunar surface compositional determinations on spectral differences. Celestial objects beyond the moon have also been the source of IR emission studies. Andrillat and Swings [94] described near-infrared (9090.9–12500 cm^{-1}) observations of T Tauri and related stars. Chandler et al. [95] detected differences in CO emission band head profiles of five young stars at 4347.8 cm^{-1} that correspond to their stellar velocities. Christensen and coworkers [96, 97] report the use of infrared emission data for the determination of the composition of the Martian surface.

Minerals and Mineral Deposits

The formation of mineral deposits in coal-fired power plants and coal-fired boilers is of considerable concern to the coal supplier and the plant operator. These deposits stick on the walls of the furnace and provide a site for further ash deposition. Vassallo and coworkers [98] describe an infrared emission cell for *in situ* studies of minerals and deposits over the temperature range of 100 to 1500°C. A schematic diagram of the cell is shown in Fig. 9.10. The cell uses a modified atomic absorption graphite rod furnace as the heater, with a 6-mm-diameter platinum disk on top of the graphite rod as the sample holder. The samples were finely ground to minimize the effects of self-absorption. Using this apparatus as a mimic of a coal combustor, they followed the thermal transformation of quartz to cristobalite, of aragonite to calcite, and of kaolinite to metakaolinite to mullite and finally to cristobalite. Emission spectra illustrating the transformation of aragonite to calcite over the temperature range of 400–600°C are shown in Fig. 9.11. Cole-Clarke and Vassallo [99] used the same apparatus to trace changes in the C–H group and the mineral matter of an Australian black coal over the temperature range of 200 to 700°C. This investigation demonstrates the potential of this technique for monitoring coal carbonization. Another study conducted by the same research group [100] focused on transformations in sulfur containing minerals, such as gypsum ($CaSO_4 \cdot 2H_2O$), anhydrite ($CaSO_4$), thenardite (Na_2SO_4), and glauberite ($Na_2Ca(SO_4)_2$), when subjected to heating over the temperature range of 50 to 1400°C. They detected the metastable forms of the minerals as well as reversible transformations using IRES.

Another application involving the monitoring of deposits in a combustor was reported by Baxter et al. [101]. They used IRES as an *in situ* monitor to follow the build-up of coal-ash deposits containing silica, sulfates, and

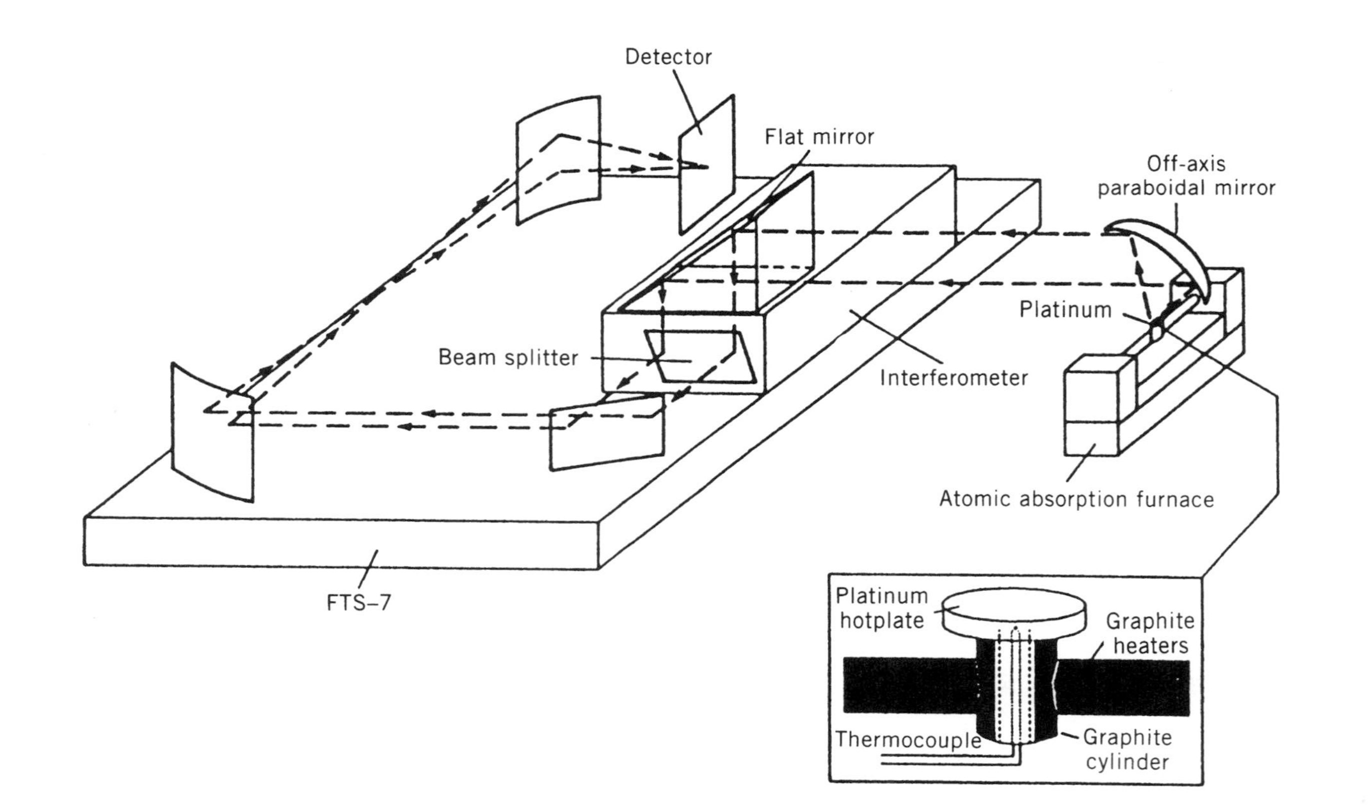

Fig. 9.10. Schematic diagram of the emission cell and the modified FT-IR spectrometer used to mimic the walls of a coal combustor. The details of the platinum hotplate and thermocouple are shown in the inset. Reprinted from [98] with permission.

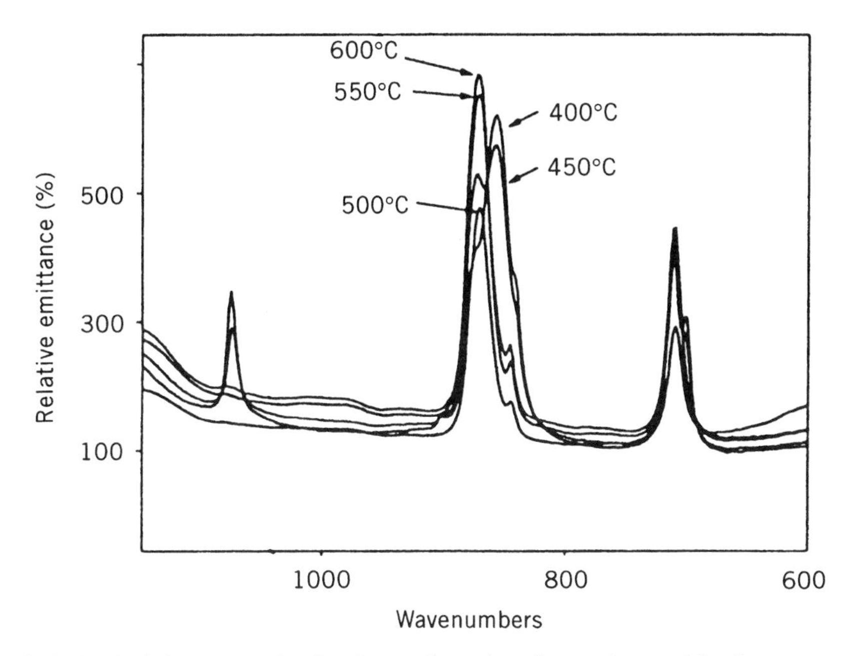

Fig. 9.11. Emission spectra showing the transformation of aragonite to calcite. Spectra were recorded at 400, 450, 500, 550, and 600°C. Reprinted from [98] with permission.

silicates as a function of time in a cylindrical multicombustor. Significant changes in formation of sulfates and silicates in the ash deposits from a western coal were noted to be a function of both deposition time and combustion conditions. Deposit properties such as strength and tenacity were correlated to the IR emission data. This application is a good example of the use of IRES as an *in situ* process monitor.

Organic Thin Films

Infrared emission spectroscopy applied to studies of organic thin films can be traced back to as early as 1964, to the work of Low and Inoue [102] who reported IRES of oleic acid on aluminum plates. They also studied various other materials including silicone lubricant, paint, rubber, and polyethylene. Although it was recognized that the IRES technique offered great utility as an analytical method, it was still less favorable when compared to transmission and reflection spectroscopies because good-quality emission spectra were difficult to obtain at the time [103]. In another early example of the application of IRES to the study of thin films, Griffiths [29] reported on the analysis of samples placed on a heated reflective substrate, such as silicone grease on heated aluminum foil. The results of this work showed how changes in film thicknesses affect the measured infrared emission spectra.

One important application of IRES of organic thin films is in studies of lubricants. Lauer and coworkers have extensively studied the physical properties of bearing lubricants using IRES. In 1975 Lauer and Peterkin [104] discussed the collection of infrared emission spectra of lubricant fluid films in simulated elastohydrodynamic (EHD) contact regions by using a heated, high-pressure diamond anvil cell. They proposed that it is possible to relate changes in spectral bandwidths to changes in the phase, structure, viscosity, and temperature of the fluid. A detailed description of the experimental apparatus used in these studies was presented elsewhere [105]. Lauer and Peterkin used this apparatus to study the infrared emission spectra of polyester and naphthenic hydrocarbon oils, each having an equal amount of polymethylstyrene as a spectral indicator [106]. The temperature of the fluid and metal surfaces were estimated under different loads and speeds to determine changes in the fluid composition and physical state. The results showed some principal differences between these two materials. In 1979 Lauer [107] reported an IRES study of polyvinyl ether fluid contained in an operating, dynamic EHD sliding contact. A detailed description of the experimental apparatus used in the study was presented. Distinct changes in the structure and intensity of a doublet at about 750 and 780 cm^{-1} were observed for different loads, sliding speeds, and bath temperatures. Spectral changes as a function of polarization angle, measured with respect to the plane of the EHD conjunction line, were correlated to operating conditions of the sliding contact, as described in Fig. 9.12. Careful analysis of the thickness and temperature of the lubricant film, which were based on the IR emission spectra, as a function of shear rates was performed. The results suggested that the spectral changes were consistent with a phase transition of the lubricant from a viscous liquid to a visco-elastic liquid or to an elastic solid. Lauer and King [108] further reported on modifications to an FTIR instrument to produce an IR emission microspectrophotometer for surface analysis of very thin polymer layers. An extended theory for thin films on a metal support was discussed, which included cases where known temperature gradients existed within the thin-film sample and/or between the thin-film sample and the surface of the metal support. The microspectrophotometer was also applied to study the properties of lubricants in a bearing. Polarization phenomena were observed in the bearing lubricant spectra, and the extent of the polarization was compared to unpolarized spectra obtained for different bearing loads and sliding speeds.

Several authors have reported different ways to increase the sensitivity of IRES for studies of thin polymer films. Durana [109] proposed the use of a cooled sample chamber and spectrometer to minimize background radiation and spectral noise. Results obtained from 4000-Å-thick polystyrene films on

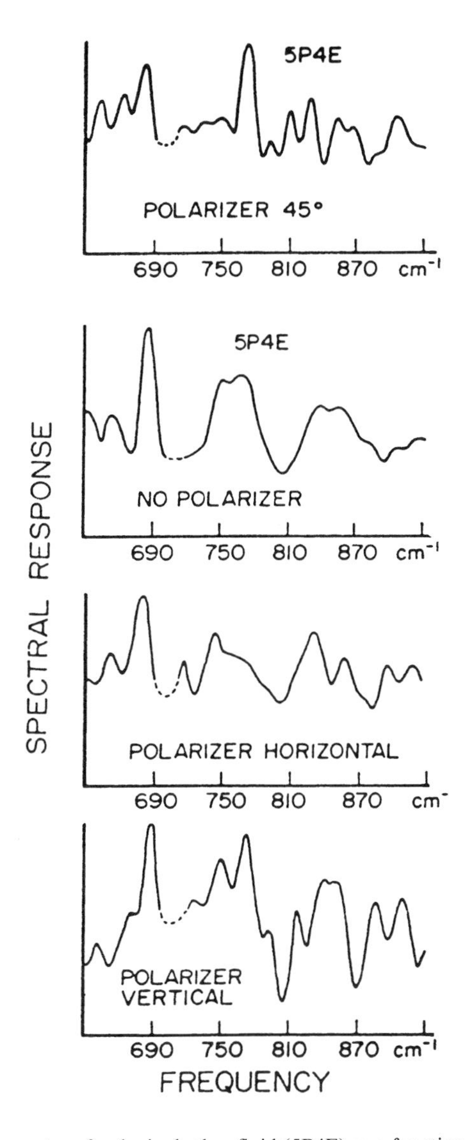

Fig. 9.12. Emission spectra of polyvinyl ether fluid (5P4E) as a function of polarization angle measured relative to the plane of the EHD conjunction line. Operating conditions were 380 rpm (1.2 m/s sliding speed), 11 kg load, and 52°C. Polarizer vertical corresponds to the plane of polarization parallel to the conjunction line. Reprinted from [107] with permission.

an aluminum support before and after subtraction of the background were shown. The use of the cooled spectrometer and sample chamber resulted in the collection of emission data with higher sensitivity. Chase [27] demonstrated that IRES can be obtained from samples at the microgram level without heating (or cooling) the sample significantly above ambient. A method to correct the IR emission spectra from the background emission of the spectrometer was proposed, and the emittance spectra of a polystyrene film at three different temperatures (60, 90, 130°C) were shown to be similar. Chase also demonstrated that the emittance spectra of a polystyrene film obtained by setting the film temperature 15°C above and below the detector temperature were identical. In addition he presented the emittance spectra of poly(methyl methacrylate), carbazole, and polystyrene in microgram quantities to illustrate the sensitivity of the technique. Nagasawa and Ishitani [110] studied poly(acrylonitrile-*co*-styrene) (PAS) thin films on aluminum using IRES and a large viewing angle (70°). The use of this large viewing angle allowed them to measure the IR emission spectra of PAS films as thin as 100 Å. They also discussed the potential of using this technique for the analysis of thin films on nonflat, rough metal surfaces such as lubricants on steel tire cords and polymer films on copper wires. A comparison of IRES and infrared reflection-absorption spectroscopy (IRRAS) was presented. The detection limits of the two experimental approaches were shown to be comparable, but the authors pointed out that IRES has significant advantages over IRRAS for studies involving nonflat samples. In their view, the analysis of such samples using the IRRAS method would be compromised because there is a decrease in the reflected light intensity and no well-defined angle of reflection for a nonflat, rough surfaced sample. The potential of their IRES technique for semiquantitative analysis of such samples was also reported. Tochigi et al. [111] studied organic thin films on metal surfaces using a polarization-modulation technique combined with IRES. The polarization-modulation technique was used to eliminate background radiation. They compared a step-scan and a continuous-scan interferometer, each combined with a photoelastic modulator, and the former was found to be more stable and to give a better signal-to-noise ratio. The step-scan instrument was employed to collect emission spectra of a 12-nm-thick perfluoropolyether film and a 100-nm-thick poly(vinyl acetate) film on a gold mirror at 120°C.

A new infrared emission technique was developed for studying optically thick, solid samples, termed transient IR emission spectroscopy (TIRES) [40]. The experimental details of the TIRES approach are discussed above. The basis of the technique is, however, to heat only a thin surface layer of the sample, thus reducing self-absorption that would normally occur in thick, solid samples. Jones and McClelland [40] used the TIRES method to study a 3.0-mm-thick phenolic plastic disk having a smoothed surface.

Other materials were also studied, including paint (blue-green baked-enamel paint on a 3-mm-thick aluminum substrate) and electrical tape (0.18-mm-thick pigmented, plasticized poly(vinyl chloride) sheet attached by its own adhesive to a 1.6-mm-thick aluminum disk). Samples were measured while being moved through a field of view of the spectrometer, and different translation speeds were investigated. The TIRES spectra were compared with conventionally recorded IR emission spectra and photoacoustic absorption spectra of the same samples. The authors showed the single-beam TIRES spectra were about 2.7 times more intense than the conventionally recorded emission spectra and that the TIRES emittance spectra contained features that were obscured in the conventionally recorded emittance spectra due to self-absorption. Figure 9.13 shows emittance spectra of the electrical tape collected with TIRES and conventional IRES, and also a reference spectrum collected with photoacoustic absorption spectroscopy. Although the TIRES approach has significant advantages when compared to conventional IRES for the study of bulk materials, it has limitations when compared to absorption spectroscopy in two important ways: (1) the signal-to-noise ratio degrades significantly at high wavenumbers, and (2) the TIRES spectra appear to be more saturated. In a following paper [41] the use of pulsed-laser TIRES for studies involving various plastics such as polycarbonate, phenolic-plastic disk (as above), and electric tape (as above) attached to an aluminum disk was reported. The pulsed-laser TIRES spectra showed less saturation than photoacoustic absorbance spectra of identical samples. In an attempt to better understand TIRES, Pell et al. [44] discussed results obtained from modeling transient IR emission data using non-steady-state heat flow calculations, harmonic oscillator modeling, and radiation transfer theory. Simulated spectra derived from the models were shown and compared to experimental results using transient IRES (obtained by cooling a thin surface layer) of a solid polymer sample (ethylene-vinyl acetate copolymer) and two viscous liquid samples (silicone oil and a polyol). A comparison of these results with thin-film transmission spectra showed the transient IRES spectra to have high saturation levels relative to the transmission spectra.

Although a number of IRES studies have involved polymer samples, quantitative determinations of polymer samples based on IRES have only recently been reported. Pell et al. [112] applied a partial least squares calibration to single-beam IRES data from heated ethylene-vinyl acetate copolymer samples of moderate thickness (0.1–0.5 mm) with the goal of predicting the vinyl acetate content, thickness, and temperature of the samples. They reported the SEPs to be 1.1% for the vinyl acetate content over the range of 9 to 28%, 0.02 mm for the sample thickness over the range of 0.1 to 0.5 mm, and 4°C for temperature over the range of 100 to 120°C. Jones and McClelland [42] applied TIRES coupled with a PCR data

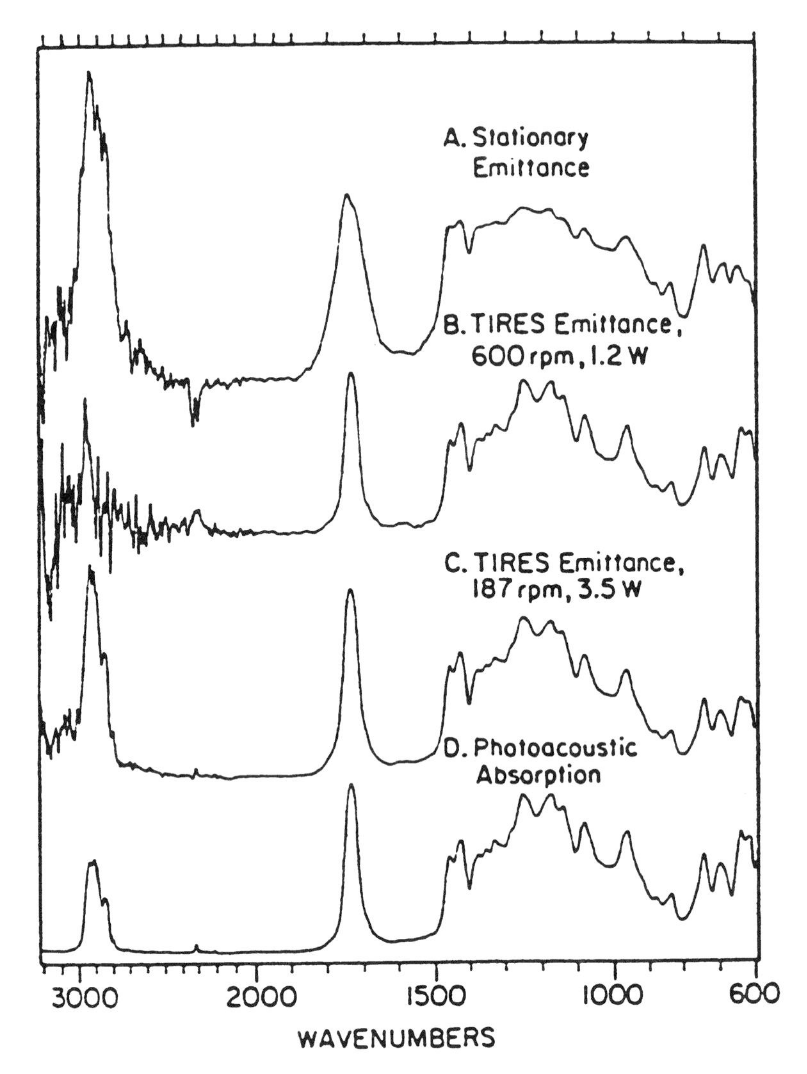

Fig. 9.13. Emittance spectra of electrical tape obtained by (*A*) conventional IRES, (*B*) TIRES at 600 rpm rotation, and (*C*) TIRES at 187 rpm rotation. (*D*) corresponds to the photoacoustic absorption spectrum of the tape. Reprinted from [40] with permission.

analysis to determine the composition of two different types of copolymers. The SEPs for these determinations were reported to be 0.75 mol% for methyl methacrylate content over the range of 0 to 100% in poly[(methyl methacrylate)-*co*-(butyl methacrylate)], and 0.83 wt% for vinyl acetate content over the range of 8.73 to 51.06 wt% in poly[ethylene-*co*-(vinyl acetate)]. The experimental setup for the hot-gas jet TIRES used in these studies is discussed above. The authors also noted that increasing the sample translation speed, which decreases the time between heating and data collection, reduces the amount of self-absorption and thus reduces the extent of spectral saturation.

In 1991 Pell et al. [113] suggested the use of IRES combined with multivariate statistical modeling as a noninvasive process monitoring tool to study the rate of curing of a commercial paint product. The emittance data were found to possess a nonlinear concentration/response relationship, and a linearization procedure was attempted based on a theoretical model. While the linearization step improved the agreement between the absorbance and emittance spectra, the linearized emittance data introduced large errors in the precision of the nonlinear model parameters used to describe the reaction rate of the curing process when compared to those obtained by using raw emittance data.

Dielectric Thin Films

Doped silicon oxide thin films, such as phosphosilicate glass (PSG) and borophosphosilicate glass (BPSG), are widely used as dielectric materials in the manufacture of microelectronic devices. The film thickness and dopant concentration(s) can dramatically affect the physical and chemical properties of the thin films, such as melting point, reflow temperature, reactivity with aluminum interconnects, thermal expansion coefficient, and film stress. These thin films are generally deposited on silicon wafers using chemical vapor deposition (CVD) processes. The thermal decomposition rate and efficiency of the gas phase CVD precursors such as SiH_4, O_2, PH_3, and B_2H_6 are highly temperature dependent. Currently no method has been found to be suitable as an *in situ* monitor for the properties, composition, and thickness of thin films produced in a CVD process. It would also be desirable to monitor the temperature of the film during the deposition process. Noncontact temperature measurements can be made using optical pyrometry, but they generally fail in an application such as monitoring the deposition of a thin film due to the large changes in the emissivity of the sample caused by changes in the film thickness. IRES is an ideal candidate as an *in situ* monitor for the CVD deposition of thin films. IR data contain information corresponding to the film content, film thickness, and film temperature. In the following section we will show that IRES can be used to make rapid, noncontact, and simultaneous determinations of dielectric thin-film content, temperature, and thickness.

PSG Thin Films

Phosphorus-doped silicon dioxide is a binary silicate glass, known as phosphosilicate glass in the microelectronics industry. It has been widely used in the manufacture of integrated circuits because the phosphorus serves the functions of passivation, lowering of reflow temperature, and immobilization of sodium ions. The phosphorus content of the films is generally kept in the region of 2–8 wt%. Films with P content less than 2 wt% do not function effectively, while those with P content greater than 8 wt% tend to form acids in the glass, which lead to corrosion of the aluminum interconnects that underlie the film. The properties, especially reflow temperature, of these films are highly dependent on the P content. Therefore precise monitoring and control of the phosphorus content are required.

The first application of IRES to the quantitative determination of dielectric film content was the determination of phosphorus concentration in a set of PSG thin films [16]. Twelve PSG thin films on silicon wafers with phosphorus content varying from 1.8 to 6.4 wt% and film thickness ranging from 0.90 to 0.94 μm were provided by Harris Semiconductor (Melbourne, FL). These samples were carefully produced so that the P content of the films was evenly spaced across the concentration range, while the film thickness remained relatively constant. Emittance spectra of the 12 films were collected at 225°C using the constant temperature apparatus described earlier [15]. The emittance data were processed using the PLS algorithm, resulting in a phosphorus content calibration with an SEP of $\pm$0.11 wt%. The best SEP that could be obtained using a univariate calibration of these data was $\pm$0.7 wt%. Although this was a very simple experiment, it served to illustrate that precise quantitative results could be produced using IRES if multivariate calibration is applied to the data.

A much more challenging, and useful, experiment is to include temperature as a variable in the experiment. Emittance spectra of the 12 PSG samples were obtained at six temperatures by ratioing the sample single-beam spectra collected at each temperature to a reference (blackbody) single-beam spectrum of a thin graphite plate at the same temperature, resulting in a data set of 72 spectra [17]. Figure 9.14 [18] shows emittance spectra of three different PSG thin films obtained at 140°C. The predominant P=O stretching band at 1329 cm^{-1} is readily apparent and grows with increasing phosphorus concentration. The presence of the Si–Si phonon band at 610 cm^{-1} indicates that the radiation from silicon substrate also contributes to the emittance spectrum. The dominant feature in the absorption spectrum of any one of these films is the broad Si–O stretch near 1100 cm^{-1}. In an IR emission spectrum this band becomes a derivativelike feature, as shown in Fig. 9.14. This phenomenon is also seen in reflectance spectra of these thin films [20], and it has been explained by Wong and Yen [114] as an absorption/reflection process caused by the large refractive index changes at the air/PSG and PSG/silicon substrate interfaces. The intensity of Si–O band near 1100 cm^{-1} also varies with phosphorus content. This is due to the fact that the film is a binary mixture; namely the SiO_2 content decreases as the P_2O_5 content increases.

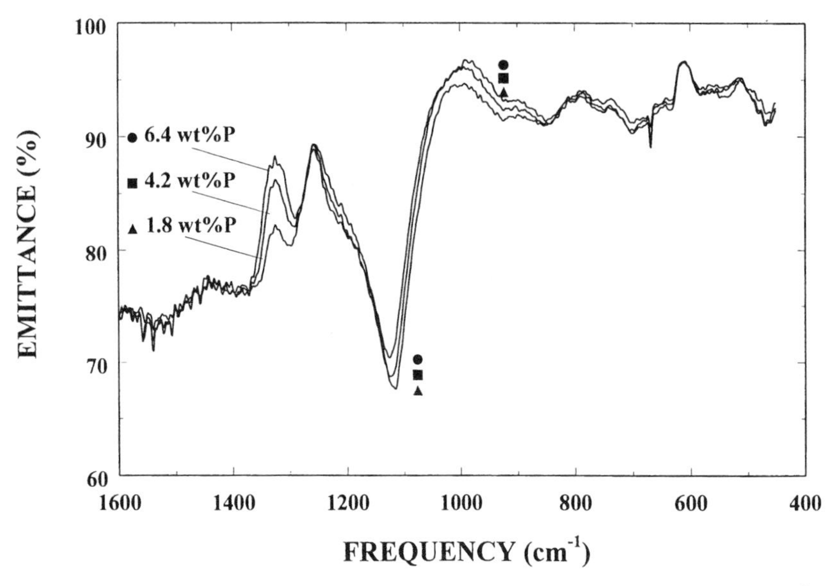

Fig. 9.14. Emittance spectra of three PSG thin-film samples recorded at 140°C. Reprinted from [18] with permission.

Figure 9.15 [18] shows the emittance spectra measured for a film with 2.3 wt% P at four different temperatures. Note that the most significant change is an increase in the intensity of the derivativelike Si–O feature near 1100 cm^{-1} with increasing temperature. There is also a small increase in the emittance across the region of 700 and 1000 cm^{-1} with increasing temperature. Any calibration process applied to these data must be able to separate the effects of changing temperature from the effects of changing P content.

Calibration for both P content and temperature were carried out by using the PLS algorithm. Figure 9.16 [18] shows the prediction results of phosphorus content from the PLS determination. The SEP for phosphorus content is 0.13 wt%, which is comparable to the reported precision of the electron microprobe reference method ($\pm$0.1 wt% phosphorus). The SEP for phosphorus content was determined by removing all six spectra (six temperatures) for each individual film during the cross-validation rotation process. Calibration for temperature was carried out in the same manner as that for phosphorus content, but 12 spectra (all 12 samples at each temperature) were removed during each SEP rotation. The temperature prediction resulted in an SEP of 2.9°C, and the calibration plot is shown in Fig. 9.17 [18]. The errors in the temperature predictions are mainly due to temperature fluctuations on the surface of the thin-film samples and the surface of the blackbody reference during the spectral measurements. The temperature fluctuations at the sample surface were measured with a

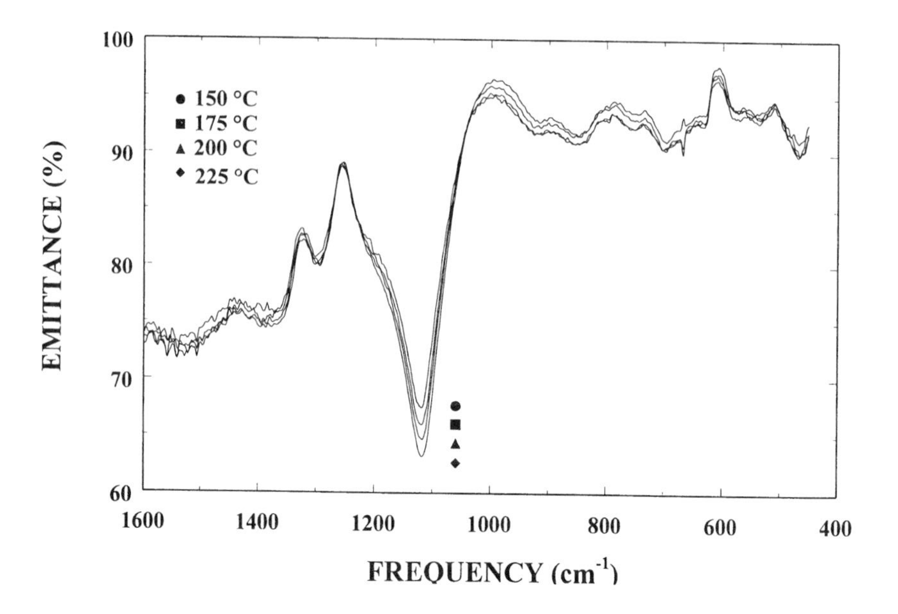

Fig. 9.15. Emittance spectra recorded at four different temperatures for a PSG sample with 2.3 wt% P. Reprinted from [18] with permission.

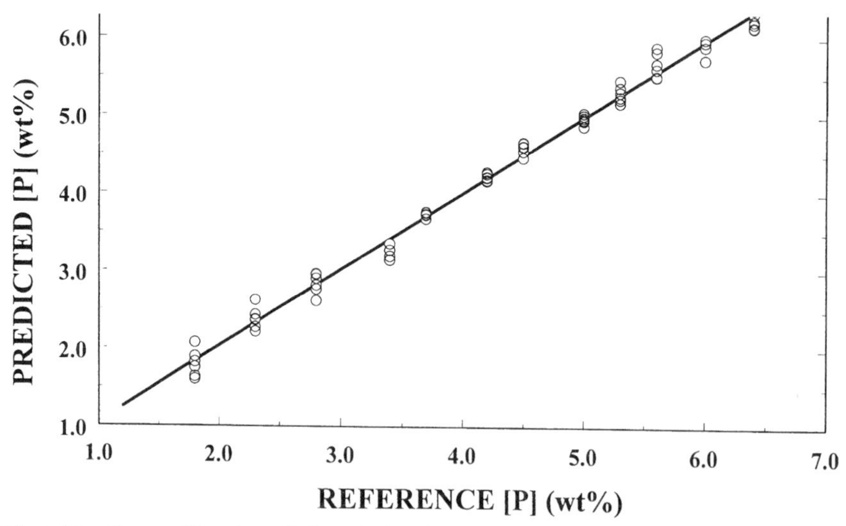

Fig. 9.16. Cross-validated prediction of phosphorus content versus the reference phosphorus value for the PSG sample set. The six different predictions at each reference phosphorus value represent the six different temperatures used in developing the calibration data set. Reprinted from [18] with permission.

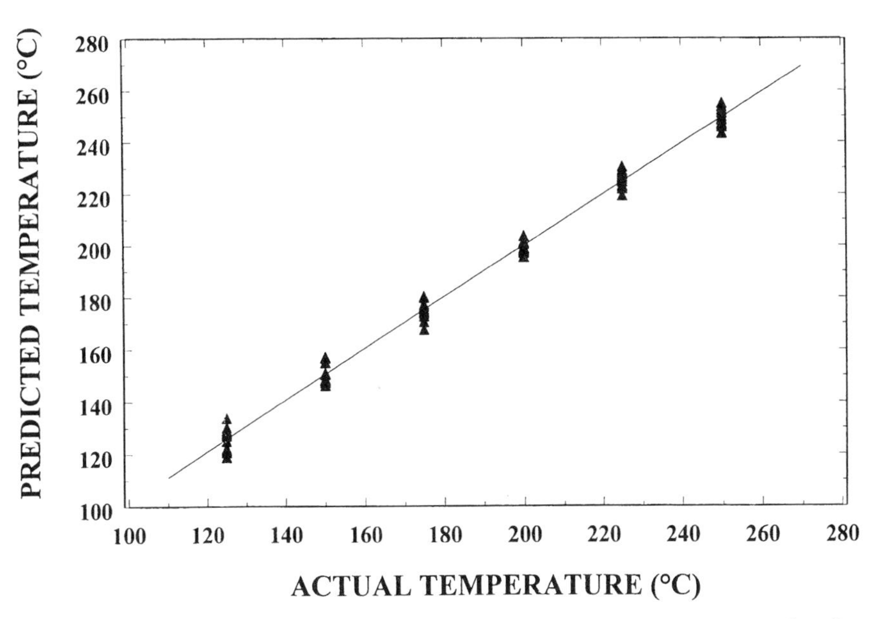

Fig. 9.17. Cross-validated prediction of temperature versus the temperature measured at the sample surface for the PSG sample set. The 12 points at each reference temperature represent the 12 different PSG samples. Reprinted from [18] with permission.

thermocouple and found to be $\pm 2°C$ from the set temperature. Another factor that likely contributes to the SEP for temperature is any nonlinearity in the data that is the result of temperature changes. The high and low temperatures are modeled by extrapolation during the cross-validation process. If the spectral response is nonlinear, the PLS algorithm will have trouble modeling the extreme points. This can be tested by reducing the range over which the calibration is carried out [17]. The highest and the lowest temperatures were removed from the calibration data set and the remaining 48 spectra were used to develop a model for temperature prediction. The resulting SEP was 1.9°C, which is comparable to the temperature fluctuations at the sample surface.

The PLS calibration process develops a set of vectors, called weight-loading vectors, that are used to model the experimental data [46]. If the experiment has been carefully designed (no correlation between the parameters to be predicted), these weight-loading vectors will contain a great deal of qualitative information. The first two weight-loading vectors for the phosphorus and temperature prediction models are shown in Fig. 9.18 [18]. The first weight-loading vector for the phosphorus model, $\mathbf{w}_1\mathbf{(P)}$, indicates the spectral features that have the highest correlation with the changes in phosphorus content. The P–O band at 1329 cm^{-1} (Fig. 9.14) is clearly the

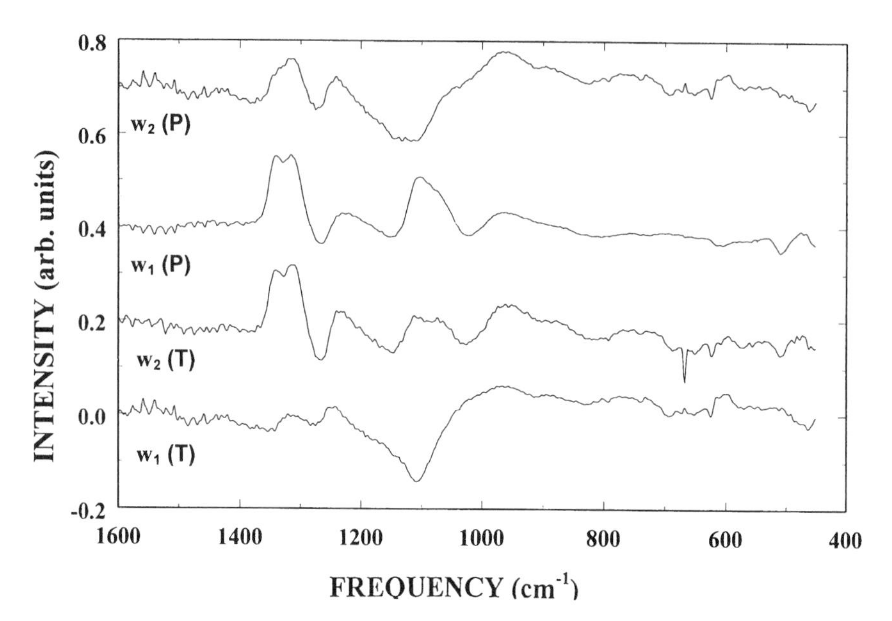

Fig. 9.18. The first and second PLS weight-loading vectors for the prediction of phosphorus content and temperature of the PSG samples. $\mathbf{w}_1$**(T)**, first weight-loading vector for temperature; $\mathbf{w}_2$**(T)**, second weight-loading vector for temperature (offset by +0.2); $\mathbf{w}_1$**(P)**, first weight-loading vector for phosphorus content (offset by +0.4); $\mathbf{w}_2$**(P)**, second weight-loading vector for phosphorus content (offset by +0.7). Reprinted from [18] with permission.

feature that changes the most with phosphorus content. There are smaller changes (all positive correlations with increasing P content) in the regions near 1240, 1110, and 950 cm^{-1}. There are known P–O features at 1110 and 950 cm^{-1}. The $\mathbf{w}_1$**(P)** vector in Fig. 9.18 has positive features at each of these frequencies. An analogous examination of the changes in the spectral data due to temperature, Fig. 9.15, and the first weight-loading vector in the temperature model, $\mathbf{w}_1$**(T)** in Fig. 9.18, show a similar correlation between spectral changes and features in the weight-loading vector.

The first weight-loading vectors in the models correlate with the spectral changes due to the parameter being modeled, as expected. It is, however, interesting to see how the PLS model for one variable incorporates changes in the data due to changes in variables other than that being modeled. This can be seen in the second weight-loading vectors in the phosphorus and temperatures models, $\mathbf{w}_2$**(P)** and $\mathbf{w}_2$**(T)** in Fig. 9.18, respectively. For example, the second weight-loading vector in the temperature model very closely resembles the first weight-loading vector of the phosphorus model. The PLS model for temperature first finds the features that correlate best with changes in temperature and then performs a correction to allow for changes in the phosphorus content. Vectors $\mathbf{w}_1(\mathbf{T})$ and $\mathbf{w}_2(\mathbf{P})$ are also very similar, and an analogous argument for the calibration model correcting the spectral data for temperature changes when predicting P content can be made.

BPSG Thin Films

BPSG thin films serve similar functions as PSG thin films do in the manufacture of integrated circuits. However, the presence of boron in the thin films further lowers the reflow temperature of the glass, which allows better step coverage and greatly reduces the diffusion of dopants into the underlying device structures. Also boron in the thin films lowers the fusion temperature more effectively than phosphorus does. This allows partial substitution of boron for phosphorus in the thin films, which minimizes the problem of potentially forming corrosive phosphorus-containing compounds. The concentrations of boron and phosphorus in the silica matrix can be varied over a wide range, thus producing thin films that can be tailored to almost any specific application. Analogous to the PSG films discussed above, the dopant concentrations must be tightly controlled if the film is to function properly. Infrared absorption spectroscopy has been shown to be capable of precise determination of B content, P content, and thickness of BPSG thin films [23].

The infrared spectra of BPSG thin films are considerably more complicated than the IR spectra of PSG thin films. The IR emission spectrum of a BPSG thin film on a silicon substrate obtained at 225°C is compared to the room temperature IR absorption spectrum of the identical film in Fig. 9.19. Note that the strong B–O feature near 1390 cm^{-1} overlaps the most significant P–O feature at approximately 1320 cm^{-1} in both the absorbance and emission spectra. Also note that the strong Si–O band at 1110 cm^{-1} in the absorption spectrum becomes a derivativelike feature in the emission spectrum, analogous to that seen in the PSG emission data.

The first IR emission experiment on BPSG thin films was performed on a set of 20 BPSG thin films [18]. These films were a subset of a larger set of films produced for an IR absorbance calibration experiment [22]. The B content of the films covered the range of 1–4 wt%, the P content covered the range of 3–6 wt%, and the film thicknesses spanned the range of 600–1000 nm. IR emission data were collected from each film at six different temperatures which resulted in 120 calibration spectra.

The results obtained from a PLS calibration of the 120 spectral files were very encouraging. The SEPs were 0.08 wt% for the boron determination, 0.17 wt% for the phosphorus determination, 22 nm for the thickness determination, and 4.8°C for the temperature determination. The SEP for the B

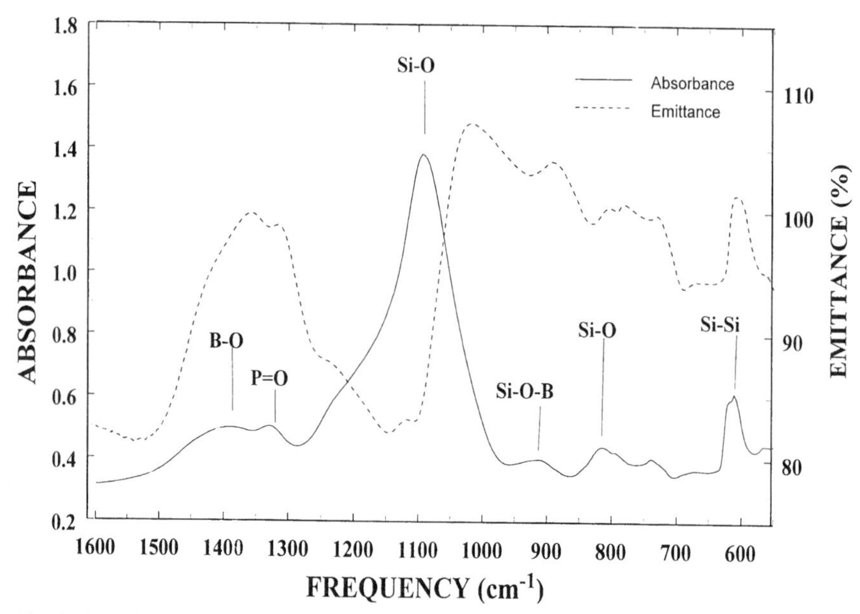

Fig. 9.19. Absorbance and emittance spectra obtained from the same BPSG sample. The measured properties of this sample are B content = 2.0 wt%, P content = 6.7 wt%, and thickness = 1.00 μm. The emittance spectrum was obtained at a sample surface temperature of 225°C.

content determination is probably limited by the reference method used to determine the B content of the film; that is to say, the SEP is as good as possible. The P content and thickness determinations are not as good as those obtained using IR absorption data from the entire sample set (44 samples). These SEPs are probably affected by the limited number of samples used in the calibration. Twenty samples might be a large enough set of samples, but the fact that they were a subset of a larger design almost assures the samples do not appropriately span the individual property spaces. The relatively large temperature SEP simply reflects the ±3.2°C temperature fluctuations at the sample surface measured during these experiments. The results of these experiments show the potential for the simultaneous determination of B content, P content, thickness, and temperature of a BPSG film. The results, however, would be improved through the use of a better designed sample set and better temperature control in the IR emission apparatus.

An experimental design for a 21-sample BPSG calibration set was produced, eliminating any correlation between boron, phosphorus, or film thickness but maximizing the distribution of samples across the range spanned for each component. The calibration ranges are 2.8–5.2 wt% for boron, 3.2–5.3 wt% for phosphorus, and 0.76–0.92 μm for film thickness. Two BPSG calibration sample sets based on this design, one set on monitor wafers and the second set on product wafers, were prepared at National Semiconductor (Santa Clara, CA). The monitor sample set consisted of 21 BPSG thin films over 0.1 μm thermal oxide on undoped silicon wafers. The second wafer set consisted of 21 BPSG films deposited over device structures. The infrared emission apparatus (Fig. 9.6) described above was used to collect the IR emission data. This apparatus has improved temperature stability, ±1°C, when compared to the apparatus used to collect the initial BPSG emission data. As in the other studies, a roughened graphite flat was used as a blackbody reference.

Emittance spectra for the set of BPSG films on monitor wafers were obtained by signal averaging 128 4-cm^{-1} resolution scans (approximately two minutes of sampling time) at each temperature for each individual sample. Spectra were collected at six temperatures across two different temperature ranges, 125–225°C and 300–400°C, yielding 126 spectra for each temperature range. The temperatures at which data were recorded within each range were randomized to avoid any time correlation in the data.

Figure 9.20 contains emittance spectra of two different films, with spectra collected from one of the films at two temperatures. Note that there are large changes in the spectral data with temperature. In addition to the baseline shift, all spectral features appear to be red shifted. The Si phonon band at 610 cm^{-1}, which is readily apparent in the emittance spectrum obtained at 125°C, is absent in the spectra obtained at 400°C. As the temperature of the samples is increased, the number of free electrons in the Si substrate, even in the undoped monitor wafers, increases to the point that the substrate becomes mirror-like. Thus the phonon band disappears, and the spectra appear as if the BPSG films were deposited on metal substrates. A comparison of the spectra of the two films at 400°C gives an indication of the subtle changes that must be used by the calibration model to predict the B content, P content, and film thickness. At the same time the changes in the spectral data due to changing temperature are dramatic; hence, the precise determination of the four film parameters predicted from these data is a significant challenge to any calibration routine.

Despite the complex nature of the data, excellent quantitative results were obtained when the spectral data were subjected to PLS calibration. Figures 9.21 and 9.22 contain the calibration plots obtained over the 300–400°C temperature range for phosphorus content and sample temperature, respectively. As can be seen, the calibrations are very good. The calibration plots for B content and film thickness look exactly like the calibration plot for phosphorus content. The temperature calibration is of particular interest for

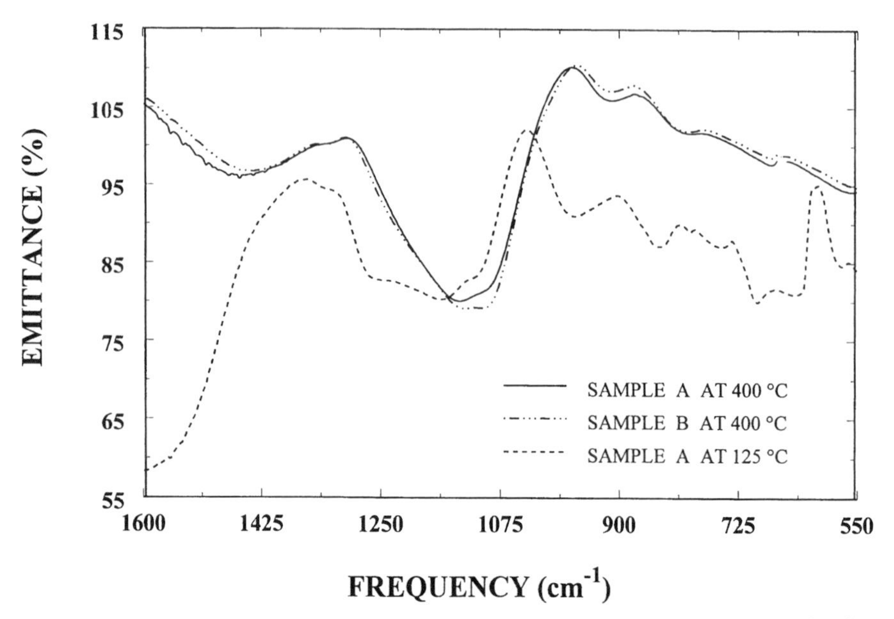

Fig. 9.20. Infrared emittance spectra of BPSG sample *A* (4.77 wt% B, 5.25 wt% P, 0.8862 μm) at 125 and 400°C and BPSG sample *B* (2.99 wt% B, 3.73 wt% P, 0.8252 μm) at 400°C.

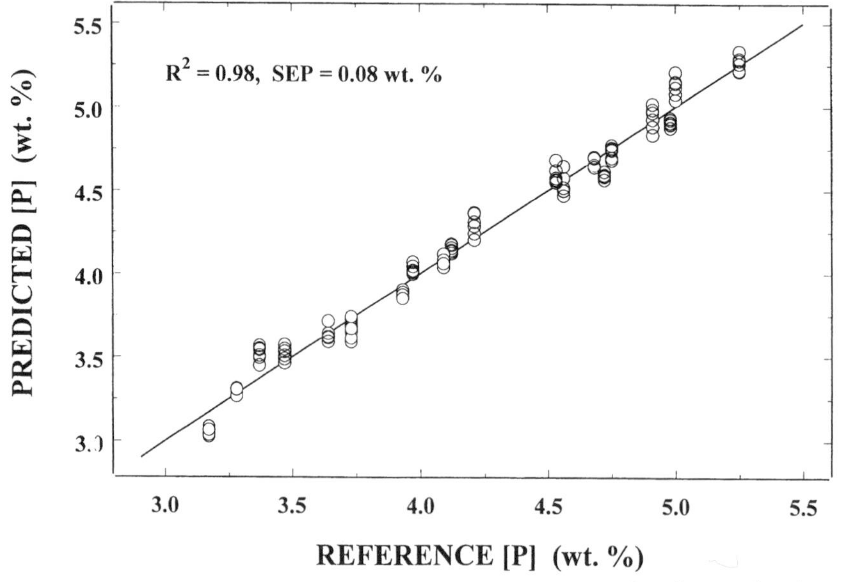

Fig. 9.21. Cross-validated prediction of phosphorus content versus the reference phosphorus value using the emittance spectra obtained from the BPSG monitor wafers collected over the temperature range of 300–400°C.

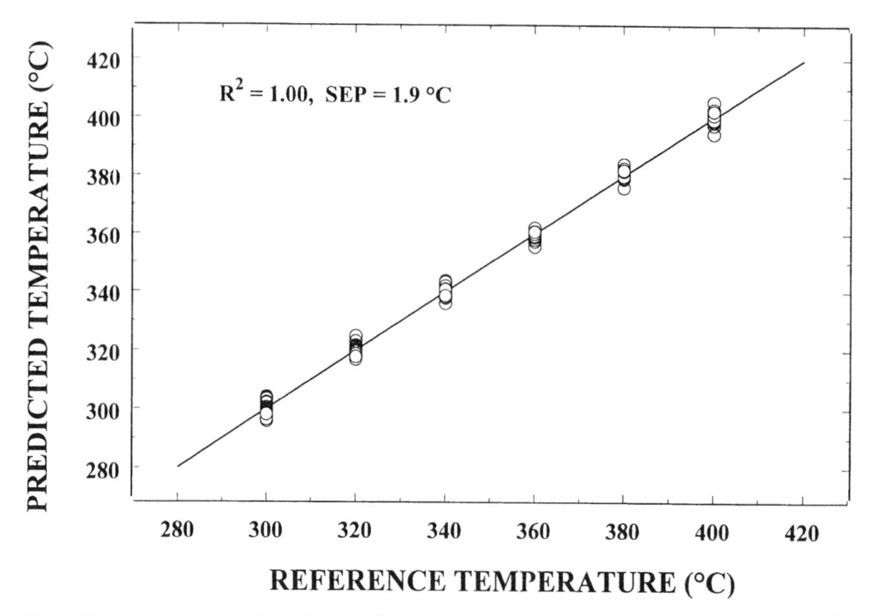

Fig. 9.22. Cross-validated prediction of temperature versus the temperature measured at the sample surface using the emittance spectra obtained from the BPSG monitor wafers collected over the temperature range of 300–400°C.

it is an example of optical pyrometry. Conventional optical pyrometry, even two-wavelength pyrometry, depends on constant sample emissivity. It would not perform well if samples such as the set of BPSG films were being studied because each sample has a different emissivity. Sample emissivity changes somewhat with changes in dopant composition, but it changes dramatically with film thickness. The PLS model for temperature across the temperature range of 300–400°C produced an SEP of 1.9°C. This is almost at the limit expected for the sample temperature stability achieved across the temperature range of 300–400°C used in the measurements. The temperature stability was measured to be ± 1°C, and this degree of uncertainty was present in both the sample and blackbody spectra used to calculate the emittance data.

The SEPs calculated for the various calibrations using the 4-cm^{-1} resolution data obtained across the two temperature ranges are summarized in the first two columns of Table 9.1. The precision of the reference methods used for phosphorus and boron content determinations have been estimated to be ± 0.1 wt%. As can be seen, the SEPs obtained for these parameters are equivalent to the reference method precisions; hence the IRES method is limited by the reference method and not by noise in the spectral data or by

TABLE 9.1 Calibration Results for BPSG Thin Films on Monitor Wafers

	SEP (4 cm^{-1}) (125–225°C)	SEP (4 cm^{-1}) (300–400°C)	SEP (32 cm^{-1}) (300–400°C)
Phosphorus (wt%)	0.10	0.08	0.08
Boron (wt%)	0.08	0.09	0.09
Thickness (Å)	34	36	42
Temperature (°C)	0.8	1.9	1.8

PLS model error. The film thicknesses were measured with an elipsometer. Elipsometers are capable of producing very accurate and precise results, but thickness measurements at multiple points across the wafer indicate that the film thickness varies by ±200 Å across the wafers. The IRES measurements were not necessarily carried out at the same locations where the thickness reference measurements were made; hence the thickness determination SEP is probably limited by variations in the film thickness rather than by imprecision in the reference or IRES measurement.

The sensitivity of the determinations to the spectral resolution at which the data were collected was examined by processing the interferograms at decreasing resolution, and then performing the PLS calibration on the resulting spectral data. The results of the lowest resolution tested, 32 cm^{-1}, are included in Table 9.1. As can be seen, the results do not suffer when the resolution of the spectral data is degraded. Consequently data can be collected much faster due to the fact that the scanning mirror in the FTIR does not have to be moved as far, or if a dispersive instrument is employed it can be a relatively low-resolution system.

The structure underlying the BPSG films on the product wafers very likely has an effect on the emittance spectra. These effects should not influence a PLS calibration as long as they are reproducible. In a set of experiments carried out earlier, we had established that reproducible IR reflectance spectra can be collected from these samples if the beam striking the wafer was slightly defocused [115]. Presumably the defocused beam included enough of the underlying structure that any differences averaged out. We used the same general approach to collect the IR emittance data from the product wafer set, collecting radiation from a spot, approximately 8 mm in diameter, on the wafer large compared to the feature size, 0.65 μ, to average out any localized feature differences. This approach yielded very reproducible spectral data that were insensitive to position as long as the area viewed consisted totally of product structures. In addition to product structures, each wafer has three regions that are covered with test structures. The spectra obtained from the test structure regions were noticeably

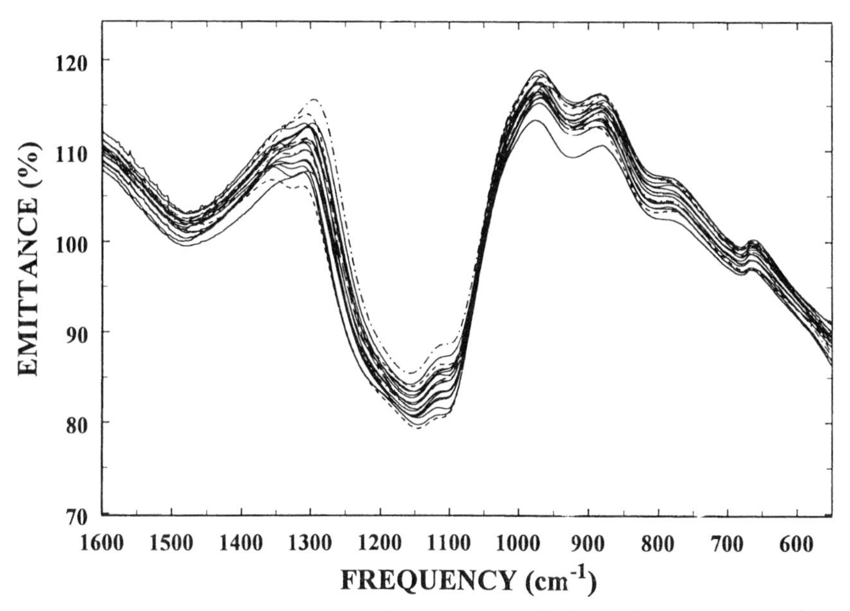

Fig. 9.23. Infrared emittance spectra collected from 21 BPSG samples on product wafers at 400°C.

different than those obtained from product regions and thus significantly affected the calibration.

The IR emittance spectra for the product wafer set obtained at 400°C are shown in Fig. 9.23. The spectra are nearly identical to the emittance spectra measured for the films deposited on monitor wafers (see Fig. 9.20). The similarity of the data suggests that the structure below the film has very little effect on the emission from the films. The spectra shown in Fig. 9.23 were combined with spectra collected at 20°C intervals between 300°C and 380°C, and the resulting data set subjected to a PLS calibration. The calibration results for spectral resolution of 4 cm^{-1} and 32 cm^{-1} are summarized in the first two columns of Table 9.2. The SEPs for all four parameters predicted are statistically equivalent (using a two-tailed F-test at the 10% significance level) at the two resolution levels, indicating no compromise is made for operation at low resolution. Calibration plots for all four parameters look exactly like the calibration plots shown for the monitor wafers (Figs. 9.21 and 9.22).

There is some degradation in the SEPs obtained from the product wafer data when compared to the SEPs obtained from the monitor wafers. The differences in the SEPs for the B content, temperature, and thickness are not statistically different, but those for P content are. These differences might be

TABLE 9.2 Calibration Results for BPSG Thin Films on Product Wafers

	SEP (4 cm^{-1}) (21 Wafers)	SEP (32 cm^{-1}) (21 Wafers)	Repeatability (15 Repeats)	Reproducibility (10 Repeats)
Phosphorus (wt%)	0.14	0.15	0.005	0.041
Boron (wt%)	0.10	0.10	0.006	0.049
Thickness (Å)	57	53	15	38
Temperature (°C)	2.5	2.8	0.8	3.2

due to the underlying structure on the product wafers. The increase in the phosphorus content SEP, and the smaller increases in the other calibration SEPs, are more likely due to the fact that the reference measurements were obtained from "sister" wafers. The CVD apparatus was set to produce a BPSG film at one of the 21 design points, and a monitor wafer and a product wafer ran back-to-back through the CVD apparatus. The monitor wafer was divided in half, and the reference measurements made on one half and the IRES measurements made on the other half. Thus for the monitor wafer set the reference measurements were made on each wafer, while for the product wafers the presumption was that the films exactly match those produced on the sister monitor wafer. This may not be the case and might be the reason for the increase in the SEPs obtained from the product wafers.

The repeatability and reproducibility of the measurements on the product wafers were also determined. The repeatability was determined by placing a sample in the emission apparatus, collecting 15 spectra from the sample without moving it or changing the temperature, and performing predictions of the 4 film parameters from each of the 15 spectra. The standard deviations for these 15 determinations are reported in Table 9.2. The standard deviations for the 4 determinations should be a measure of how noise in the spectral data affects the PLS predictions. The standard deviations for all predictions are smaller than the corresponding SEPs. This is an indication that the noise in the spectral data is not a limitation. The consequence is that the spectral data can be collected faster (less signal averaging), or collected with instrumentation of lesser quality than that used here, without hurting the predictions.

The reproducibility was determined by repeatedly (ten times) placing a sample in the emission apparatus, bringing the sample to temperature, collecting a spectrum, letting the apparatus cool to room temperature, and finally removing the sample. The standard deviations of the ten repeat determinations are shown in the last column of Table 9.2. There are two additional sources of variation in these measurements when compared to the repeatability measurements, and the increases in the standard deviations are indications that these variations are significant. First, the sample has to be repositioned in the apparatus. The increases in the SEPs might be caused by the film parameters not being constant across the area sampled. It is unlikely that the film parameters change enough across the distances represented by the repositioning error to affect the predictions. Differences in the product structure beneath the films due to repositioning error could affect the spectral data. We do not believe this to be an important factor. The repositioning error is small, and IR reflection spectra collected from these same wafers are unaffected by the repositioning error. One might expect the reflection data to be more sensitive to changes in the underlying structure than the IR emission data. The second source of variation is more likely the cause for the increase in the SEPs. This source of variation is due to errors in reproducing the temperature at which the 10 repeat emittance spectra were collected. An indication of this being the cause is that the standard deviation of the temperature determination is greater than the SEP determined from the entire sample set, while standard deviations for B content, P content, and film thickness remain below the corresponding SEPs.

9.5. CONCLUSION

The development of Fourier transform infrared spectrometers has made the collection of infrared emission spectral data a relatively straightforward process. IRES has been employed in a wide variety of applications. None of the applications has become routine because of the multitude of complexities that affect IR emission data. One of the key problems is the IR emission spectra become featureless (i.e. blackbody-like), for thick, opaque samples. For this reason IRES has been found to be most useful in the characterization of thin films. The TIRES approach discussed above solves the problem associated with thick samples, effectively creating a thin sample by heating (or cooling) only a very localized region of the sample. This approach has great potential for application in a wide variety of areas but especially in process monitoring.

Until recently the complexity of IRES data had limited the use of the technique to nonquantitative applications. Only recently have reports of quantitative IRES appeared in the literature. All of the quantitative applications have employed some form of multivariate calibration. The use of multivariate calibration to interpret IRES data is probably necessary if quantitative results are desired.

We believe the most significant future applications for IRES will be in the area of process monitoring. There are some features of IRES that give it significant advantages for this application when compared to other techniques. The sample is the source. Thus no conventional source is required and no allowance has to be made to get a probe beam to the sample. IRES measurements are nondestructive and can be made rapidly. IR emission spectra are high-information-content data; that is to say, the data contain information about a wide variety of sample properties. The application of IRES to the determination of thin-film coatings, both organic (polymer) and inorganic (dielectrics), show that the technique has especially good potential as an *in situ* monitor for the production of thin films.

REFERENCES

1. J. D. Ingle, Jr., and S. R. Crouch, *Spectrochemical Analysis*. Prentice Hall, New Jersey, 1988, pp. 87–89.
2. P. W. Atkins, *Molecular Quantum Mechanics*, 2d ed. Oxford University Press, Oxford, 1983, pp. 3–9.
3. P. R. Griffiths and J. A. deHaseth, *Fourier Transform Infrared Spectrometry*, vol. 83, Wiley, New York, 1986, pp. 202–203.
4. P. Baraldi, *Chim. Oggi.*, **8**, 53–57 (1990).
5. P. V. Huong, In *Advances in Infrared and Raman Spectroscopy*, vol. 4, R. J. H. Clark and R. E. Hester, eds. Heyden, London, 1978, pp. 85–107.
6. J. Hvistendahl, E. Rytter, and H. A. Øye, *Appl. Spectrosc.*, **37**, 182–187 (1983).
7. F. J. DeBlase and S. Compton, *Appl. Spectrosc.*, **45**, 611–618 (1991).
8. J. R. Aronson, in *Progress in Nuclear Energy. Series IX: Analytical Chemistry*, vol. 11, H. A. Elion and D. C. Stewart, eds. Pergamon Press, Oxford, 1972, pp. 137–156.
9. H. O. McMahon, *J. Opt. Soc. Am.*, **40**, 376–380 (1950).
10. R. Gardon, *J. Am. Cer. Soc.*, **39**, 278–287 (1956).
11. J. B. Bates, In *Fourier Transform Infrared Spectroscopy: Applications to Chemical Systems*, vol. 1, J. R. Ferraro and L. J. Basile, eds. Academic, New York, 1978, pp. 99–142.
12. P. Baraldi and G. Fabbri, *Spectrochim. Acta*, **38A**, 1237–1244 (1982).
13. P. Baraldi and G. Fabbri, *Spectrochim. Acta*, **39A**, 669–675 (1983).
14. E. Rytter, *Spectrochim. Acta*, **43A**, 523–529 (1987).
15. J. A. McGuire, B. Wangmaneerat, T. M. Niemczyk, and D. M. Haaland, *Appl. Spectrosc.*, **46**, 178–180 (1992).
16. B. Wangmaneerat, J. A. McGuire, T. M. Niemczyk, D. M. Haaland, and J. H. Linn, *Appl. Spectrosc.*, **46**, 340–343 (1992).
17. B. Wangmaneerat, T. M. Niemczyk, and D. M. Haaland, *Appl. Spectrosc.*, **46**, 1447–1453 (1992).
18. T. M. Niemczyk, J. E. Franke, B. Wangmaneerat, C. S. Chen, and D. M. Haaland, in *Computer-Enhanced Analytical Spectroscopy*, vol. 4, C. L. Wilkins, ed. Plenum Press, New York, 1993, pp. 165–185.
19. J. E. Franke, L. Zhang, T. M. Niemczyk, D. M. Haaland, and J. H. Linn, *J. Electrochem. Soc.*, **140**, 1425–1429 (1993).
20. J. E. Franke, L. Zhang, T. M. Niemczyk, D. M. Haaland, and K. J. Radigan, *J. Vac. Sci. Technol. A*, **13**, 1959–1966 (1995).

21. T. M. Niemczyk, J. E. Franke, L. Zhang, D. M. Haaland, and K. J. Radigan, In *Diagnostic Techniques for Semiconductor Materials Processing, Mat. Res. Soc. Symp. Proc.*, vol. 324, O. J. Glembocki et al., eds. Boston, 1994, pp. 93–98.
22. D. M. Haaland, *Anal. Chem.*, **60**, 1208–1217 (1988).
23. I. S. Adhihetty, J. A. McGuire, B. Wangmaneerat, T. M. Niemczyk, and D. M. Haaland, *Anal. Chem.*, **63**, 2329–2338 (1991).
24. J. E. Franke, T. M. Niemczyk, and D. M. Haaland, *Spectrochim. Acta*, **50A**, 1687–1723 (1994).
25. T. R. Kozlowski, *Appl. Opt.*, **7**, 795–800 (1968).
26. J. K. Wilmshurst, *J. Chem. Phys.*, **39**, 2545–2548 (1963).
27. D. B. Chase, *Appl. Spectrosc.*, **35**, 77–81 (1981).
28. S. V. Compton, D. A. C. Compton, and R. G. Messerschmidt, *Spectrosc.*, **6**(6), 35–39 (1991).
29. P. R. Griffiths, *Appl. Spectrosc.*, **26**, 73–76 (1972).
30. M. J. D. Low and I. Coleman, *Appl. Opt.*, **5**, 1453–1455 (1966).
31. G. Keresztury and J. Mink, *Appl. Spectrosc.*, **46**, 1747–1749 (1992).
32. S. F. Kapff, *J. Chem. Phys.*, **16**, 446–453 (1948).
33. P. R. Griffiths and J. A. deHaseth, *Fourier Transform Infrared Spectrometry*, vol. 83. Wiley, New York, 1986.
34. M. Handke and N. J. Harrick, *Appl. Spectrosc.*, **40**, 401–405 (1986).
35. R. T. Rewick and R. G. Messerschmidt, *Appl. Spectrosc.*, **45**, 297–301 (1991).
36. R. J. Karpowicz, *Rev. Sci. Instrum.*, **61**, 188–190 (1990).
37. D. C. Tilotta, K. W. Busch, and M. A. Busch, *Appl. Spectros.*, **43**, 704–709 (1989).
38. L. T. Lin, D. D. Archibald, and D. E. Honigs, *Appl. Spectrosc.*, **42**, 477–483 (1988).
39. O. Koga, T. Onishi, and K. Tamaru, *J.C.S. Chem. Comm.*, 464 (1974).
40. R. W. Jones and J. F. McClelland, *Anal. Chem.*, **61**, 650–656 (1989).
41. R. W. Jones and J. F. McClelland, *Anal. Chem.*, **61**, 1810–1815 (1989).
42. R. W. Jones and J. F. McClelland, *Anal. Chem.*, **62**, 2074–2079 (1990).
43. R. W. Jones and J. F. McClelland, *Anal. Chem.*, **62**, 2247–2251 (1990).
44. R. J. Pell, C. E. Miller, B. R. Kowalski, and J. B. Callis, *Appl. Spectrosc.*, **47**, 2064–2070 (1993).
45. T. Hirschfeld, *Appl. Opt.*, **17**, 1400–1412 (1978).
46. D. M. Haaland and E. V. Thomas, *Anal. Chem.*, **60**, 1193–1202 (1988).
47. H. Martens and T. Naes, *Multivariate Calibration*. Wiley, New York, 1989.
48. S. Wold, *Technomet.*, **20**, 397–405 (1978).
49. R. K. Koopman and A. R. Saunders, *J. Quant. Spectrosc. Radiat. Transfer*, **10**, 389–401 (1970).
50. F. Roux, D. Cerny, and J. Verges, *J. Mol. Spectrosc.*, **94**, 302–308 (1982).
51. M. Elhanine, R. Farrenq, and G. Guelachvili, *Proc. SPIE-Int. Soc. Opt. Eng.*, **1575**, Int. Conf. Fourier Transform Spectrosc., 8th, 1991, 1992, pp. 314–315.
52. W. T. Rawlins. R. R. Foutter, and T. E. Parker, *J. Quant. Spectrosc. Radiat. Transfer*, **49**, 423–431 (1993).
53. X. Wang, H. Li, Q. Ju, Q. Zhu, and F. Kong, *Chem. Phys. Lett.*, **208**, 290–294 (1993).
54. G. V. Hartland, D. Qin, and H. L. Dai, *J. Chem. Phys.*, **100**, 7832–7835 (1994).

55. W. Schatz, K. F. Renk, L. Fusina, and J. R. Izatt, *Appl. Phys. B: Lasers Opt.*, **B59**, 453–465 (1994).
56. H. W. Hermann and S. R. Leone, *J. Chem. Phys.*, **76**, 4759–4765 (1982).
57. G. Henderson, C. S. Ko, and T. C. Huang, *J. Chem. Educ.*, **59**, 683–686 (1982).
58. J. Vandooren and M. Vanpee, *Bull. Soc. Chim. Belg.*, **97**, 797–808 (1988).
59. V. Braun and P. F. Bernath, *J. Mol. Spectrosc.*, **167**, 282–287 (1994).
60. R. S. Ram and P. F. Bernath, *J. Mol. Spectrosc.*, **169**, 268–285 (1995).
61. C. I. Frum, R. Engleman, Jr., and P. F. Bernath, *J. Mol. Spectrosc.*, **150**, 566–575 (1991).
62. R. S. Ram and P. F. Bernath, *J. Mol. Spectrosc.*, **155**, 315–325 (1992).
63. R. S. Ram, C. N. Jarman, and P. F. Bernath, *J. Mol. Spectrosc.*, **160**, 574–584 (1992).
64. K. Q. Zhang, B. Guo, V. Braun, M. Dulick, and P. F. Bernath, *J. Mol. Spectrosc.*, **170**, 82–93 (1995).
65. R. S. Ram, P. F. Bernath, R. Engleman, Jr., and J. W. Brault, *J. Mol. Spectrosc.*, **172**, 34–42 (1995).
66. K. Kunimori, T. Uchijima, and G. L. Haller, *Shokubai*, **29**, 430–433 (1987).
67. S. Chiang, R. G. Tobin, and P. L. Richards, *J. Electron Spectrosc. Relat. Phenom.*, **29**, 113–118 (1983).
68. S. Chiang, *Energy Res. Abstr.*, **8**, Abstr. No. 47694 (1983).
69. S. Chiang, R. G. Tobin, P. L. Richards, and P. A. Thiel, *Phys. Rev. Lett.*, **52**, 648–651 (1984).
70. S. Chiang, R. G. Tobin, and P. L. Richards, *J. Vac. Sci. Technol. A*, **2**(2, Pt. 2), 1069–1074 (1984).
71. R. G. Tobin, S. Chiang, P. A. Thiel, and P. L. Richards, *Surf. Sci.*, **140**, 393–399 (1984).
72. R. G. Tobin, *Energy Res. Abstr.*, **11**, Abstr. No. 13650 (1986).
73. G. Fabbri and P. Baraldi, *Ann. Chim. (Rome)*, **62**, 740–748 (1972).
74. G. Fabbri and P. Baraldi, *Atti Soc. Natur. Mat. Modena*, **103**, 255–260 (1972).
75. L. Aina, H. Hier, M. Mattingly, J. O'Connor, and A. Iliadis, *Inst. Phys. Conf. Ser.*, **91** (Gallium Arsenide Relat. Compd. 1987), 601–604 (1988).
76. G. Fabbri and P. Baraldi, *Atti Soc. Natur. Mat. Modena*, **103**, 241–245 (1972).
77. L. M. Gratton, S. Paglia, F. Scattaglia, and M. Cavallini, *Appl. Spectrosc.*, **32**, 310–316 (1978).
78. A. Ueno and C. O. Bennett, *Bull. Chem. Soc. Jpn.*, **52**, 2551–2555 (1979).
79. A. Ueno, Y. Kotera, S. Okuda, and C. O. Bennett, *Chem. Uses Molybdenum, Proc. Int. Conf. 4th*, H. F. Barry and P. C. H. Mitchell, eds. Climax Molybdenum Co., Michigan, 1982, pp. 250–255.
80. S. Bleux, F. Seronde, P. Echegut, and F. Gervais, *J. High Temp. Chem. Processes*, **3**, 213–219 (1994).
81. M. Primet, P. Fouilloux, and B. Imelik, *Surf. Sci.*, **85**, 457–470 (1979).
82. B. Jonson, B. Rebenstorf, R. Larsson, and M. Primet, *Appl. Spectrosc.*, **40**, 798–803 (1986).
83. H. M. Barnes, Jr., W. F. Herget, and R. Rollins, *Analytical Methods Applied to Air Pollution Measurements*, R. K. Stevens and W. F. Herget, eds. Ann Arbor Sci., Ann Arbor, MI, 1974, pp. 245–266.
84. W. F. Herget and D. D. Powell, *Joint Conference on Sensors for Environmental Pollutants 4th.* ACS, Washington, DC, 1978, pp. 541–544.
85. W. F. Herget and J. D. Brasher, *Top. Meet. Atmos. Spectrosc., Dig. Tech. Pap.* Optical Soc. Am., Washington, DC, 1978, ThA5/1–ThA5/4.

86. D. J. W. Kendall and T. A. Clark, *Int. J. Infrared Millimeter Waves*, **2**, 783–808 (1981).
87. M. J. Bangram, A. Bonetti, R. H. Bradsell, B. Carli, J. E. Harries, F. Mencaraglia, D. G. Moss, S. Pollitt, E. Rossi, and N. R. Swann, *Proc. Quadrenn. Int. Ozone Symp.*, vol. 2, J. London, ed. Natl. Cent. Atmos. Res., Boulder, CO, 1981, pp. 759–764.
88. B. Carli and J. H. Park, *J. Geophys. Res. D: Atmos.*, **93**(D4), 3851–3865 (1988).
89. K. V. Chance, D. G. Johnson, W. A. Traub, and K. W. Jucks, *Geophys. Res. Lett.*, **18**, 1003–1006 (1991).
90. C. E. Blom, H. Fischer, N. Glatthor, T. Gulde, M. Hoepfner, and C. Piesch, *J. Geophys. Res.*, **100**(D5), 9101–9114 (1995).
91. K. V. Chance, *Proc. SPIE-Int. Soc. Opt. Eng.*, vol. 1576. Conf. Dig. — Int. Conf. Infrared Millimeter Waves, 1991), pp. 34–37.
92. A. F. H. Goetz, *J. Geophys. Res.*, **73**, 1455–1465 (1968).
93. F. H. Murcary, D. G. Murcray, and W. J. Williams, *J. Geophys. Res.*, **75**, 2662–2269 (1970).
94. Y. Andrillat and J. P. Swings, *Bull. Soc. R. Sci. Liege*, **47**, 229–236 (1978).
95. C. J. Chandler, J. E. Carlstrom, N. Z. Scoville, and W. R. F. Dent, *Astrophys. J.*, **412** (2, Pt. 2.), L71–L74 (1993).
96. P. R. Christensen, H. Kieffer, and S. Chase, *Lunar Planet. Sci.*, **16**, 125–126 (1985).
97. P. R. Christensen, D. L. Anderson, S. C. Chase, R. N. Clark, H. H. Kieffer, M. C. Malin, J. C. Pearl, J. Carpenter, N. Bandiera, F. G. Brown, and S. Silverman, *J. Geophys. Res.*, **97**(E5), 7719–7734 (1992).
98. A. M. Vassallo, P. A. Cole-Clarke, L. S. K. Pang, and A. J. Palmisano, *Appl. Spectrosc.*, **46**, 73–78 (1992).
99. P. A. Cole-Clarke and A. M. Vassallo, *Fuel*, **71**, 469–470 (1992).
100. A. M. Vassallo and K. S. Finnie, *Appl. Spectrosc.*, **46**, 1477–1482 (1992).
101. L. L. Baxter, G. H. Richards, D. K. Ottesen, and J. N. Harb, *Energy Fuels*, **7**, 755–760 (1993).
102. M. J. D. Low and H. Inoue, *Anal. Chem.*, **36**, 2397–2399 (1964).
103. M. J. D. Low, J. C. McManus, and L. Abrams, *Appl. Spectrosc. Rev.*, **5**, 171–210 (1971).
104. J. L. Lauer and M. E. Peterkin, *J. Lubr. Technol.*, **97**, 145–150 (1975).
105. J. L. Lauer and M. E. Peterkin, *Am. Lab.*, **7**, 27–33 (1975).
106. J. L. Lauer and M. E. Peterkin, *J. Lubr. Technol.*, **98**, 230–235 (1976).
107. J. L. Lauer, *J. Lubr. Technol.*, **101**, 67–73 (1979).
108. J. L. Lauer and V. W. King, *Infrared Phys.*, **19**, 395–412 (1979).
109. J. F. Durana, *Polymer*, **20**, 1306–1307 (1979).
110. Y. Nagasawa and A. Ishitani, *Appl. Spectrosc.*, **38**, 168–173 (1984).
111. K. Tochigi, H. Momose, Y. Misawa, and T. Suzuki, *Appl. Spectrosc.*, **46**, 156–158 (1992).
112. R. J. Pell, B. C. Erickson, R. W. Hannah, J. B. Callis, and B. R. Kowalski, *Anal. Chem.*, **60**, 2824–2827 (1988).
113. R. J. Pell, J. B. Callis, and B. R. Kowalski, *Appl. Spectrosc.*, **45**, 808–818 (1991).
114. J. S. Wong and Y. S. Yen, *Appl. Spectrosc.*, **42**, 598–604 (1988).
115. L. Zhang, Determination of dielectric thin film components and thickness using FTIR reflection-absorption spectroscopy. Ph.D. dissertation, University of New Mexico, Albuquerque, 1994, pp. 113–122.

10 Fiber Optics in Molecular Spectroscopy

Chris W. Brown

10.1. INTRODUCTION

Description of Optical Fibers

Optical fibers are a type of wave guide in which light can be transferred from one point to another without going in a straight line. The fibers are made of materials transparent in the wavelength range of interest. They are cylindrical in shape and range in diameter from a few micrometers ($\sim 4\ \mu m$) up to a few millimeters (~ 2 mm). Light launched into a fiber is confined to the fiber core by internal reflections along the walls. Internal reflections depend on the angle of incidence at the walls of the core, the refractive index of the core, and the surrounding medium. Light can be attenuated by bending the fiber or by changing the refractive index of the surrounding medium. Both of these effects have led to the use of fibers as the intrinsic part of optical sensors.

Modern Techniques in Applied Molecular Spectroscopy, Edited by Francis M. Mirabella.
Techniques in Analytical Chemistry Series.
ISBN 0-471-12359-5

Fibers have been manufactured from various materials including glass, plastic, and fused salts. The type of material depends on the spectral region of interest and the particular application. There has been considerable interest in making fibers from materials that are transparent in the longer wavelength infrared region, since loses due to Rayleigh and Raman scattering are reduced at the longer wavelengths. Various fibers are now available that cover the electromagnetic spectrum from less than 200 nm to approximately 15 μm; however, the throughput and specific ranges vary with the type of fiber.

Internal Reflection Considerations

All light rays incident on the walls of a fiber at an angle greater than the critical angle, θ_c, will be totally reflected, as shown in Fig. 10.1. The critical angle depends on the refractive index of the core material, η_1, and that of the surrounding material, η_2. The critical angle is equal to the reciprocal sine of the ratio of the refractive index:

$$\theta_c = \sin^{-1}\left(\frac{\eta_1}{\eta_2}\right), \tag{10.1}$$

which is Snell's law. It can be shown that all rays launched into a fiber from

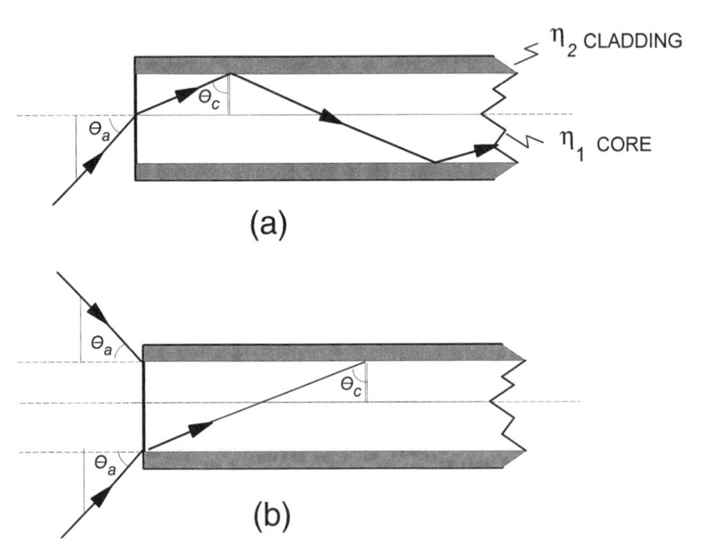

Fig. 10.1. (*a*) Schematic of internal reflections inside of a fiber optic showing the critical angle, θ_c, the angle of acceptance, θ_a, and the refractive indexes, η_i. The light cone accepted by a fiber also depends on the diameter of the fiber as shown in (*b*).

air at an angle less than the acceptance angle,

$$\theta_a = \sin^{-1}(\eta_1^2 - \eta_2^2)^{1/2}, \tag{10.2}$$

will undergo internal reflection at the core walls. Often fiber optic manufacturers refer to the numerical aperture

$$\text{NA} = \sin\theta_a = (\eta_1^2 - \eta_2^2)^{1/2} \tag{10.3}$$

or the $F\#$

$$F\# = \frac{1}{\sin\theta_a} \tag{10.4}$$

The larger the acceptance angle, the smaller the $F\#$.

Core-cladding

Generally, the core of a fiber is surrounded by a cladding material as shown in Fig. 10.1. The cladding may be formed from material similar to that of the core or it may be an entirely different material. For example, a silica core may be surrounded by a doped silica cladding or by a silicone polymeric cladding. This type of cladding causes an abrupt refractive index change at the core-cladding interface, and this is referred to as a step-index fiber. In addition to providing a step-index change, the polymeric cladding provides mechanical support and cladded fibers are more robust than the all silica fibers. In another type of fiber, the core may be doped at increasing concentrations from the center outward to the cladding at which point there is a constant concentration of the dopant in the cladding. This type is referred to as a graded-index fiber. The best type fiber for a specific application depends on the spectral region of interest and the purpose of the application; these requirements will be discussed later.

Evanescent Wave

Interferences between incident and reflected rays at the core-cladding (or other surrounding material) interface creates a standing wave in the cladding, which is perpendicular to the reflecting surface [1, 2]. The energy in this wave falls off exponentially from the interface outward, but it accounts for the fact that the cladding or surrounding material can attenuate the intensity of the light at resonant wavelengths of the surrounding material. The depth of penetration into the surrounding material can be calculated, but as a good rule-of-thumb, the depth is about one-tenth of a wavelength.

For example, a ray with a 1000-nm wavelength will penetrate the cladding or surrounding material about 100 nm. This effect becomes a more serious problem (or advantage) in the longer wavelength infrared region. Interferences caused by absorptions of polymeric claddings are a problem; however, the effect can also be used to produce evanescent wave spectra for monitoring surrounding materials [1]. This latter feature will be demonstrated at the end of this chapter by using infrared fibers to monitor the curing process that takes place during lamination.

Fiber Throughput

The propagation of waves through fibers depends on the core size and numerical apertue [2, 3]. Step-index fibers with very small cores generally transmit only a single ray (or mode) and are referred to as single-mode fibers; these fibers transmit a limited ranges of wavelengths. Larger step-index or graded-index fibers transmit multiple rays and multiple modes over a large range of wavelengths; these are referred to as multimode fibers.

The wavelength range and the throughput as a function of wavelength for multimode fibers is dependent on the composition of the core and cladding. The transmission through a fiber is characterized by the attenuation of the light, which is given in terms of decibels (dB) or absorbance multiplied by 10. The attenuation depends on the length and it is given for unit lengths as dB/m or dB/km.

10.2. SELECTING FIBERS

Optical Regions

The first consideration is to select an appropriate fiber for the optical region of interest. The throughput for each optical region from the UV to the mid-IR will depend upon the core and cladding material. The amount of throughput will then dictate the diameter of the fiber or the need to use a bundle of fibers to obtain the needed throughput.

Types of fibers for various optical regions are listed in Table 10.1. Currently the regions with restricted throughputs are at each end of the optical spectral range, namely in the short wavelength UV and the long wavelength mid-IR regions. The main difficulty in both regions is caused by optical absorptions of the core or cladding material. Fused silica is used for the UV, but it has to be very pure since impurities (primarily aromatic compounds) absorb at the short wavelengths below about 250 nm. Absorption spectra for two different fused silica fibers in the UV-visible region are shown in Fig. 10.2. These spectra were measured on a Beckman DU-7000 diode array spectrometer using an empty cell compartment as background; both fibers were 2 m long and 500 μm in diameter. The upper, fused silica

TABLE 10.1 Fiber Types and Typical Optical Wavelength Range

Fiber Type	Approximate Optical Range
High-purity fused silica	200–2000 nm
Plastics	350–1100 nm
Glass	350–1800 nm
Low OH fused silica	250–2700 nm
Zirconium fluoride	2–5.5 μm
Chalcogenide	3–11 μm
Silver halides	5–20 μm

spectrum was that for an earlier UV-transmitting fiber. The new fused silica spectrum was obtained with a fiber produced in about 1992. Both fibers have flat absorbances in the visible region from 700 down to 400 nm, with the newer fiber having about half the absorbance of the earlier type fiber. The newer type fiber remains flat down to 200 nm, whereas the absorbance

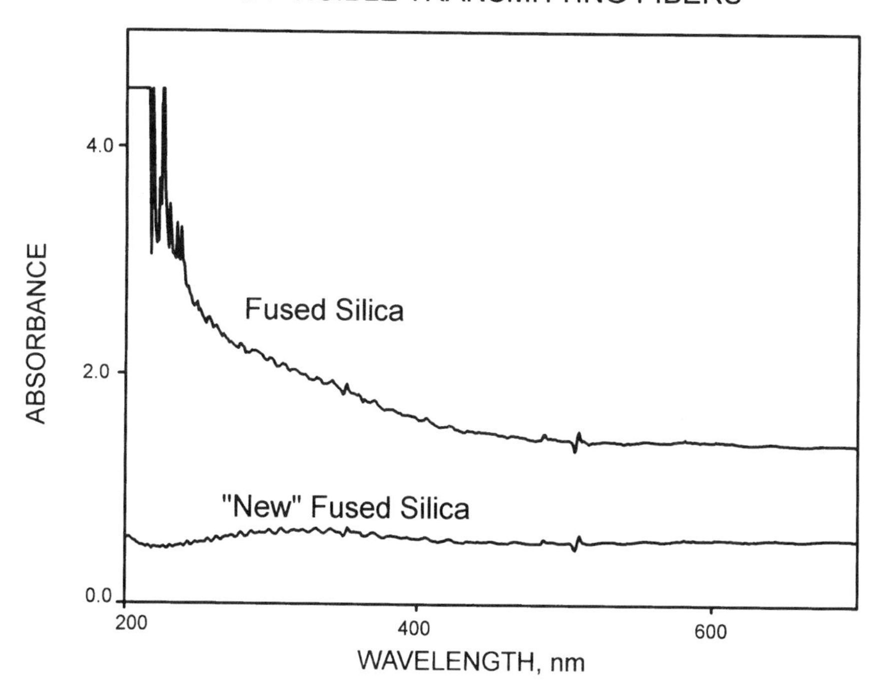

Fig. 10.2. UV-Visible absorption spectra of two fused silica fibers; both fibers are 2 m long with 500 μm diameters.

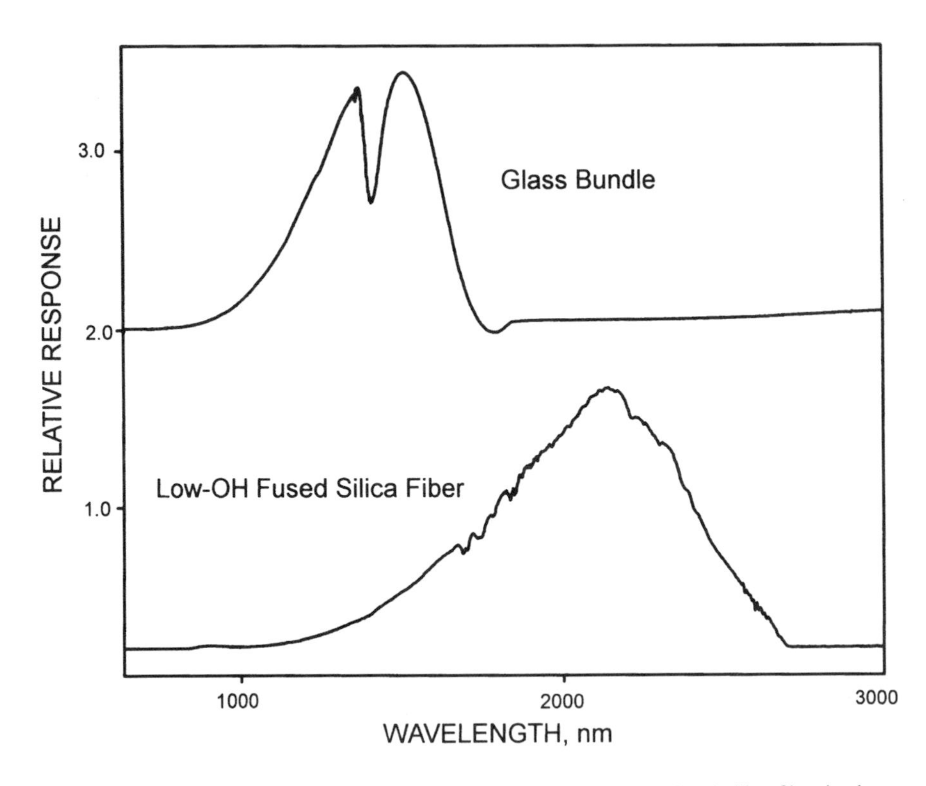

Fig. 10.3. Relative responses of a glass bundle fibers and a low-OH fused silica fiber in the near-IR region. Both spectra were measured on an FTIR and shown the throughput of the fiber superlimposed on the instrument response.

of the earlier type increases gradually down to about 250 nm at which point there is a dramatic increase in the absorbance. This comparison indicates that the processes for producing "cleaner" fibers are improving and that the UV region is accessible with fiber optics.

Fibers for the visible and near-IR regions are robust and inexpensive. Plastic, glass, and fused silica can be used in the visible and short wavelength near-IR regions. Low hydroxy fused silica fibers are required for the longer wavelength near-IR region. These low hydroxy fibers are being used in the communication industry and presently have high throughputs at low cost. The availability of these fibers has had a major impact on the increase use of near-IR for process monitoring. The response spectrum for a 1-m length of a low hydroxy silica fiber with a 400-μm core is compared to that for a glass bundle in Fig. 10.3. These transmission curves were measured on a Bio-Rad FTS-40 FTIR using InSb detector; thus the transmission spectrum

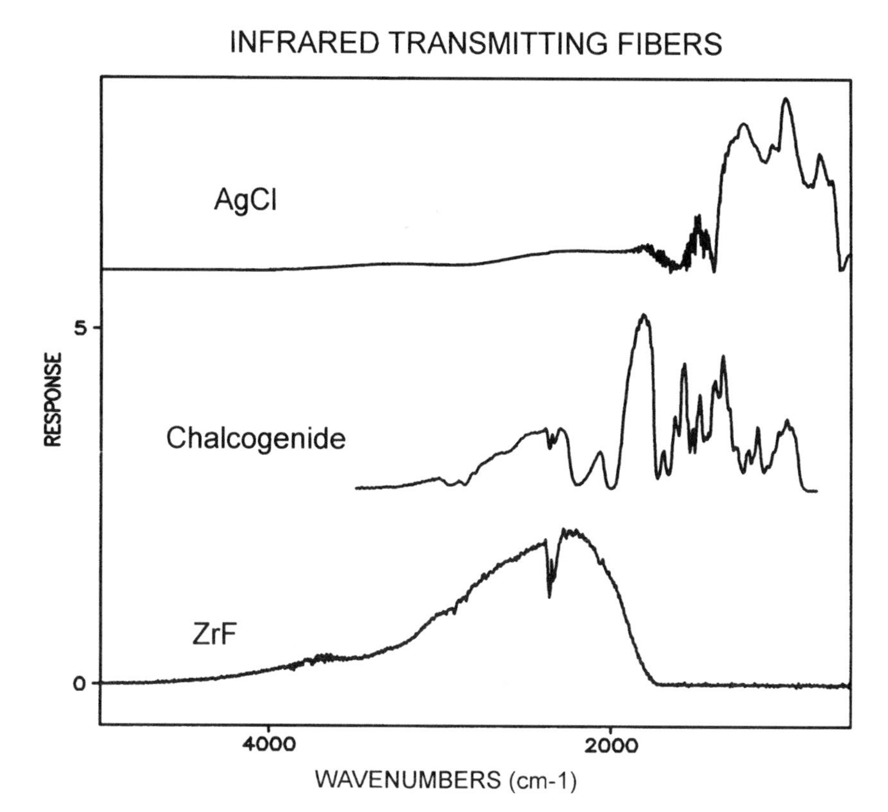

Fig. 10.4. Relative response of three mid-IR fibers with the throughputs superimposed on the instrument response.

of the fibers are superimposed on the instrument response curve of the interferometer which is dependent on the source, detector, and throughput of the interferometer. There are two important features to note in the transmission curves; the maximum throughput for the silica fiber is at a much longer wavelength, and there are only a few minimal absorbances (due to the polysilicone cladding) at about 1600 nm. The glass bundle has a strong OH absorption at 1400 nm.

In the mid-IR region, chalcogenides (As and Ge salts of S, Se, and Te), zirconium fluoride glass, and silver halides (AgBr + AgCl) crystalline materials have been used for making fibers. Response curves for these three types of fibers are shown in Fig. 10.4. These spectra were all measured on a Bio-Rad FTS-40 with a MCT detector so that the response curves are really a superposition of the transmission spectrum of each fiber onto the instrumental response curve. The chalcogenide and halide fibers have rather large attenuations. All three fibers are still reasonably expensive, although there have been significant price decreases in the last five years. Finally both the chalcogenide and fluoride fibers are brittle and have to be handled with care. Both have polymeric claddings, which improve the mechanical stability; however, it is difficult to remove the claddings so that the fibers can be used as an evanescent wave.

Single Fibers or Bundles as Probes

A bifurcated fiber is one of the most commonly used probes for spectroscopic measurements. These probes can be made from single fibers or from bundles and are often used for absorption and/or absorption-reflection measurements, as shown in Fig. 10.5. In the single-fiber configuration, light from the source is launched into the input fiber which directs the light onto or through the sample; the reflected or transmitted light is collected by the output fiber and returned to the detector.

Bundles of fibers [4] are ideal for a bifurcated probe, which consists of an input leg, output leg and a common leg. Each leg of the bifurcated bundle may contain a few sizable fibers or many (several hundred) very fine fibers. As shown in Fig. 10.6, the fibers in the common leg can be randomly distributed between input and output, or they can have an ordered distribution such as a concentric configuration with input fibers in the center surrounded by output fibers, and vice versa. The size, type, and configuration depend on the particular spectroscopic measurement.

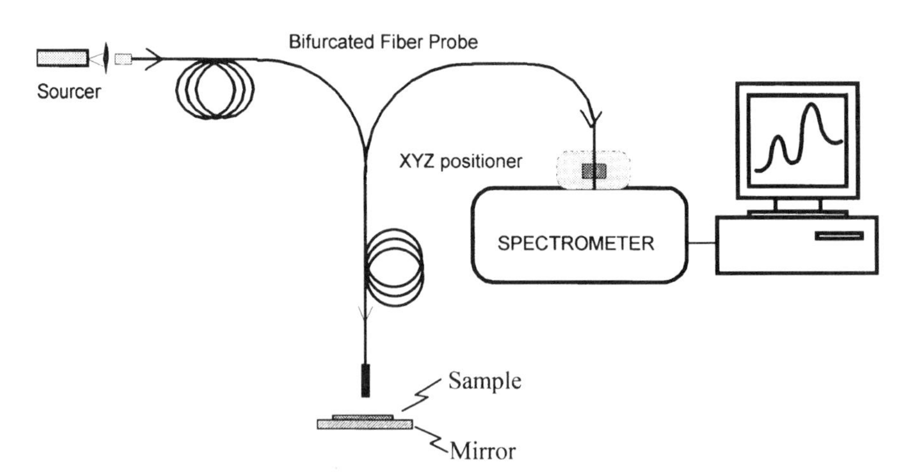

FIG. 10.5. A typical bifurcated fiber probe being used to measure absorption/reflection spectra.

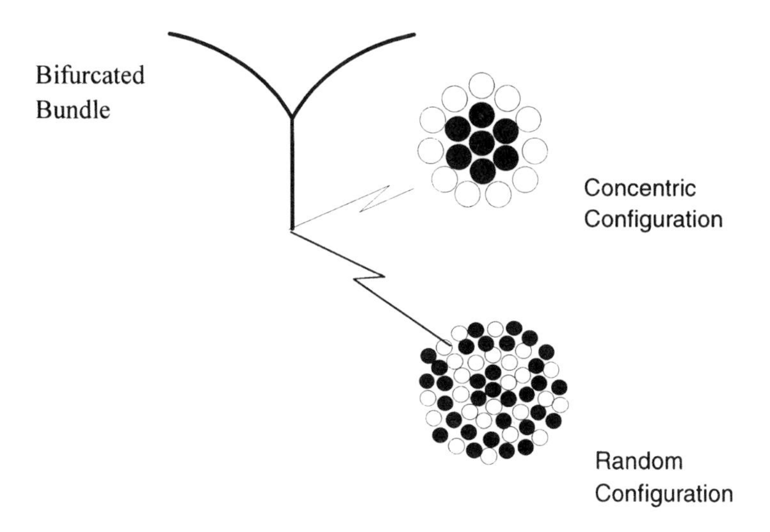

Fig. 10.6. Two possible fiber combinations in the common leg of a bifurcated probe.

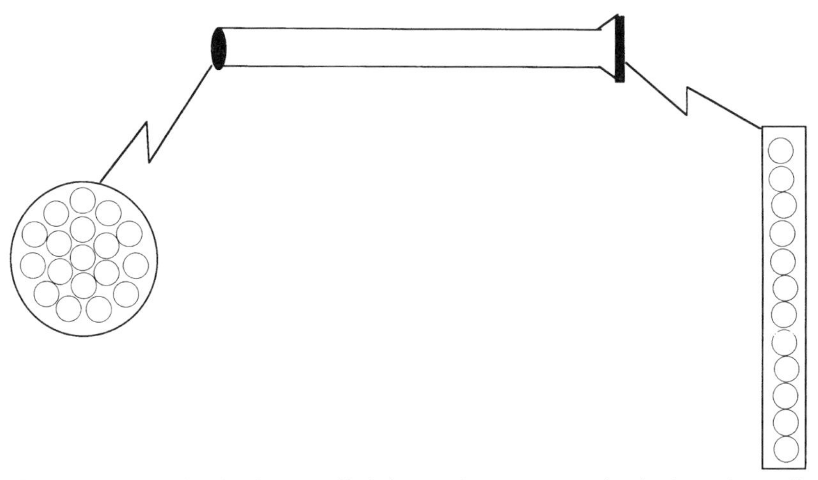

Fig. 10.7. Schematic of a fiber bundle being used to convert a circular beam into a linear beam.

Fiber optic bundles can also be used to change the shape of an optical image. For example, a circular beam of light can be converted into a slit image simply by using a bundle of fibers arranged in a circular pattern at one end and a slit image at the other end, as shown in Fig. 10.7. Basically the fibers can be used to distribute the light in any desired spatial configuration.

10.3. INTERFACING FIBERS

Spectrometer Accessories

Fiber optic attachments are available for many spectrometers, and generally, they fit easily into the sample compartment. These accessories can be obtained from the instrument manufacturer or from accessory suppliers. In addition it is reasonably easy to build an optical interface in-house. The basic design we have used in the UV-visible, near-IR, and mid-IR is shown in Fig. 10.8. The device sits in the sample compartment; the light is directed out of the compartment by a plane mirror through a lens that focuses the light onto the input end of the fiber optic. The reverse sequence of lens and mirror can be used to return the light from the output fiber to the original light path of the spectrometer. A design similar to the one in Fig. 10.8 was used to build an interface for a Beckman DU-7000; fused silica lenses were used for focusing the input and output light. A similar attachment is provided by Bio-Rad for their FTIR instruments; the lenses are made from ZnSe which is transparent in the infrared region. The attachment shown in Fig. 10.8 can also be used for interfacing fiber optic bundles to spec-

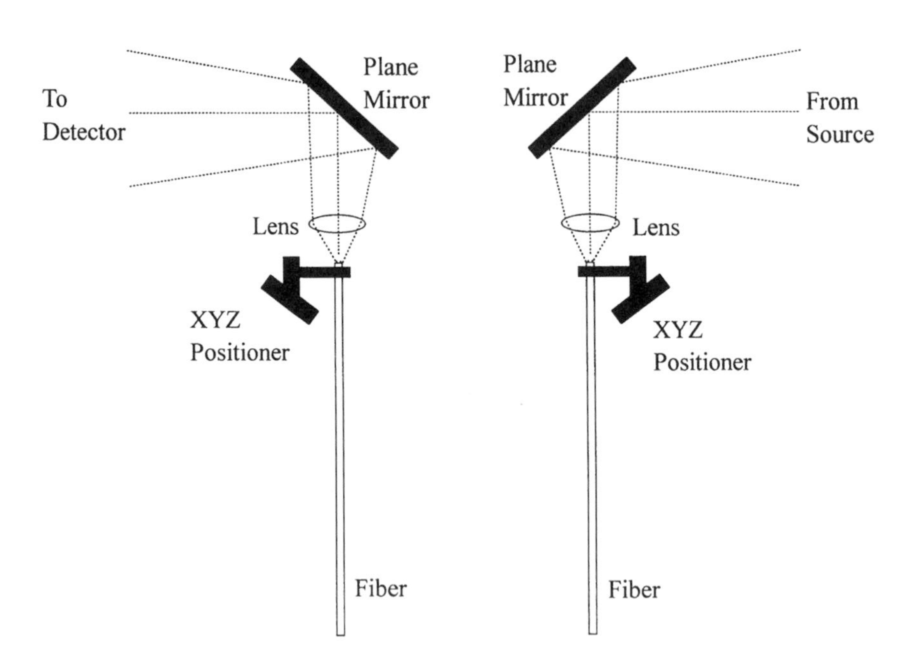

Fig. 10.8. Optical schematic of a fiber optic attachment for a typical spectrometer sample compartment.

trometers. The two unique legs of a bifurcated bundle can be easily attached to the interface.

Coupling Fibers to Spectrometer Accessories

One of the first concerns of a novice to fiber optic probes and sensors is how to handle the fibers and how to couple them to input and output devices. Many fiber probes come with some type of standard coupling and can be mated to the receptacle in the input or output device. This is very similar to interfacing a VCR to a TV; that is to say, it is only necessary to match the male and female connectors. There are a number of standard connectors, but matching pairs have to be used.

Often times one wants to use single-strand fiber or fiber bundles that do not come with a standard connector. One simple way to couple single-strand fibers to a device is to pass the fiber through a small piece of rubber or synthetic tubing and place the tubing/fiber into a Swagelok fitting. The tubing acts as a cushion for the fiber and makes it possible to lightly tighten the Swagelok to hold the fiber firmly in place. It is convenient to have the Swagelok terminal mounted on an XYZ-translator as shown in Fig. 10.8. The translator is used to position the end of the fiber at the focus of the lens. Fiber bundles can also be coupled in a similar manner by using larger Swagelok fittings. In either case the fibers are held firmly by the Swagelok and can be maneuvered into focus by the translator. The alignment of the fiber, especially a single-strand fiber, at the focus takes some practice, but it is much easier using a XYZ-translator with the fiber held in position.

10.4. SAMPLING CONFIGURATION

Straight-through Transmission Probes

Possibly the simplest fiber probe is made using single-strand fibers for the input and output with the sample held in the gap between the distal ends of the two fibers [5]. A diagram of a simple liquid measuring device is shown in Fig. 10.9. The fibers are attached with epoxy to a small piece of plexiglass, which has a small well (dimple) drilled in the center to hold a couple of drops of a liquid. This type of probe eliminates the need for liquid sample cells. It requires a minimum amount of sample, and it ensures that all of the light reaching the detection system passes through the sample. Moreover the device can be easily modified to allow variable path determinations; however, light will be lost at the sample unless lenses are used on each side of the sample (these problems will be addressed later). A 2-mm pathlength is reasonable for the UV-visible and near-IR regions.

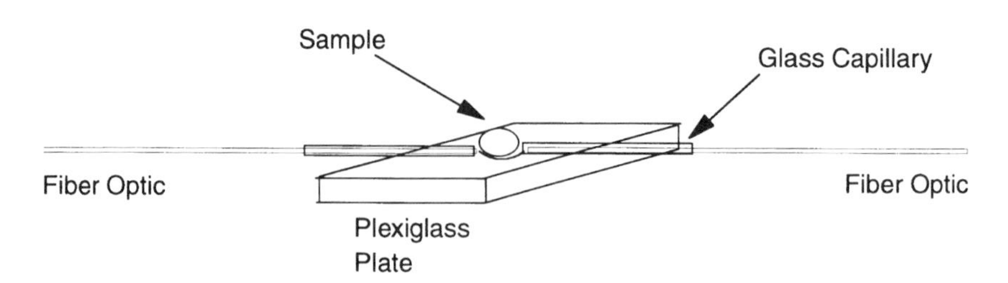

Fig. 10.9. Fiber optic "sample cell" for a drop of liquid.

Reflection-Absorption Probes

As mentioned earlier bifurcated bundles are ideal as reflection-absorption probes. The sample can be a completely reflecting material, or it can be partially transparent and backed with a mirror. The latter probes are perfect for monitoring solutions. The distance between the distal end of the fiber and the mirror can be adjusted to vary the pathlength, and the mirror can be easily removed for cleaning. The distal (common) end of a bifurcated probed can be fitted with a matching lens for collimating or focusing the light. (This type of probe can be used effectively to monitor gases as shown later in Fig. 10.19.) We [6] found that best combination was to use two lenses; the first collimates the light exiting the fiber, and the second lens, a short distance away, focuses the light on a distant mirror. The light reflected by the mirror returns over the same optical path.

Evanescent Wave Probes

As was mentioned earlier, the evanescent wave penetrates, on average one-tenth of the wavelength. In the mid-IR the average depth is enough to use the bare core of a fiber as the probe. The maximum number of reflections along the length of a fiber can be calculated from the diameter and the critical angle θ_c. The distance between reflections b is given by

$$b = d \tan \theta_c \tag{10.5}$$

where d is the diameter of the fiber. For a 500-μm fiber with $\theta_c = 65^\circ$, the distance between reflections is slightly over 1 mm; thus there would be about 10 reflections per centimeter. The cladding can be removed from a section of fiber to produce an evanescent wave probe. In the mid-infrared region, say, at 10 μm, an exposed 10-cm section of 500-μm fiber provides an optical pathlength of approximately 100 μm. A good spectrum of a typical organic solvent can be obtained with a 25-μm pathlength liquid cell; thus the 100-μm pathlength obtainable with a fiber optic probe is ideal for monitoring small spectral changes that might be observed in spectra of compounds dissolved in organic solvents.

Fiber Optic Lenses

Graded-index (GRIN) lens are commonly used for fiber optic couplers in the communication industry [7]. GRIN lens are rod shaped with flat ends. A light ray is curved inside of the lens due to the gradient change in the refractive index of the material from the axis of the lens outward. Actually these lens work the same as the graded-index fibers, except that they are cut at such lengths to produce a collimated beam or a focused beam of light as shown in Fig. 10.10. For fiber optic couplers, one lens might be used to form a collimate beam of light from one fiber and direct the beam to a lens that will focus the light into second fiber. Couplers like this can also be used to mix two or more beams of light.

Currently GRIN lenses are made from borosilicate glass and are primarily manufactured for the red to near-IR regions for the communications industry. Thus their spectroscopic applications have been limited, since they are only usable in the visible and short wavelength near-IR regions of the electromagnetic spectrum which contain very little spectral information. It would certainly be desirable to have lens made of fused silica, since these would open up the UV and the longer wavelength near-IR regions, which contain a lot more spectral information.

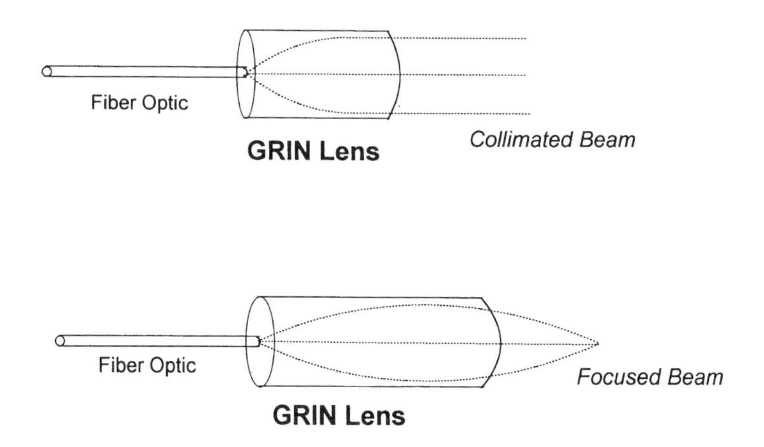

Fig. 10.10. Two examples of GRIN lens, for collimating a beam and focusing a beam exiting from a fiber optic.

10.5. APPLICATIONS

Drug Dissolution

Immediate and timed-released pharmaceutical formulations have to undergo extensive dissolution studies. Generally, six tablets from each manufactured lot are tested concurrently. Each table is placed in a flask containing the diluent and aliquots removed at set times for analysis. For example, the dissolution of a 12-hour release capsule might be sampled every half hour producing 25 aliquot for each of the six tablets. Currently the accepted method of analysis is HPLC, but analyzing this many samples requires considerable effort and time. Recent methods of chemometrics makes it

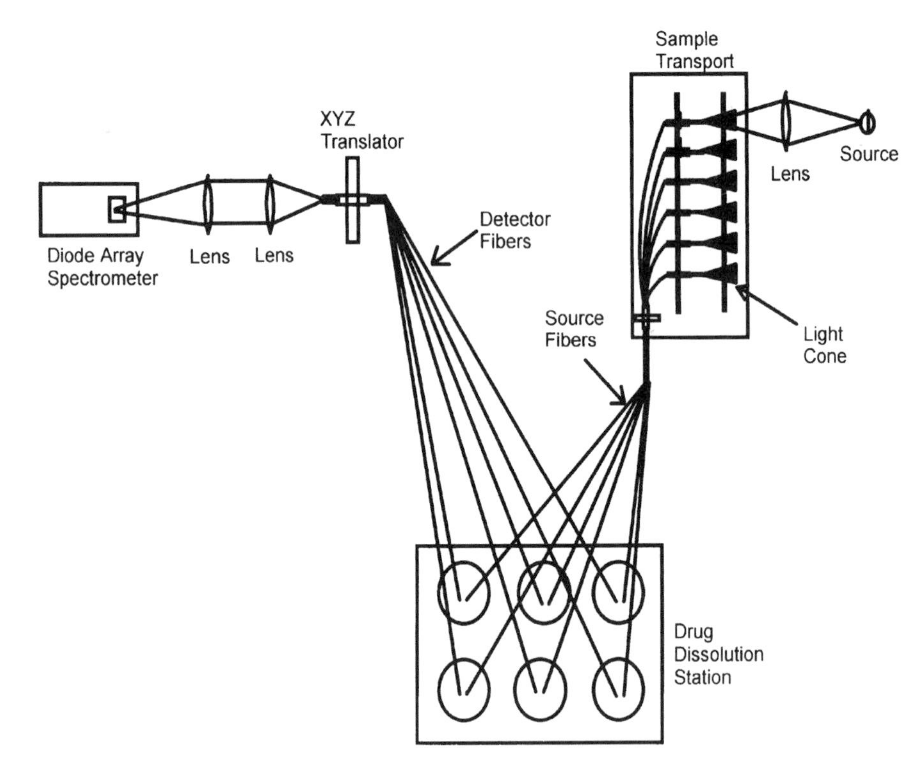

Fig. 10.11. Diagram of interfacing a six-sample drug dissolution station with fiber optic probes from a diode array spectrum using a six-cell automated sampler launching light into the fibers.

possible to perform the same analyses with much less effort and in short time periods.

Fiber optic methods in both the UV [8, 9] and the near-IR [10] have been proposed as alternatives to the HPLC method. A network of six fibers was used to monitor six dissolution baths in the UV region. A schematic of the system is shown in Fig. 10.11. Using this scheme, it was possible to monitor the two active ingredients of a commercial product in near-real time without collecting a single sample. Typical UV spectra measured on-line through one of the fiber optic probes during the dissolution are shown in Fig. 10.12, and the concentrations of the two active ingredients as a function of time are shown in Fig. 10.13. After completing one experiment with six tablets, it was only necessary to clean the fibers and the vessels before starting a new experiment. Thus, the entire investigation requires very little operator time, since the entire system runs without operator intervention.

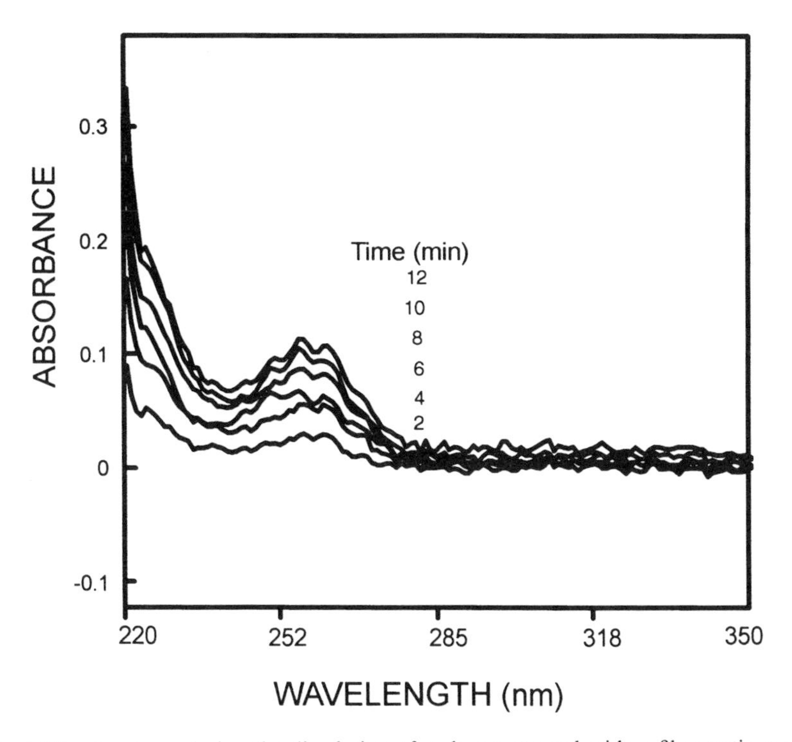

Fig. 10.12. Spectrum during the dissolution of a drug measured with a fiber optic probe shown in Fig. 10.11.

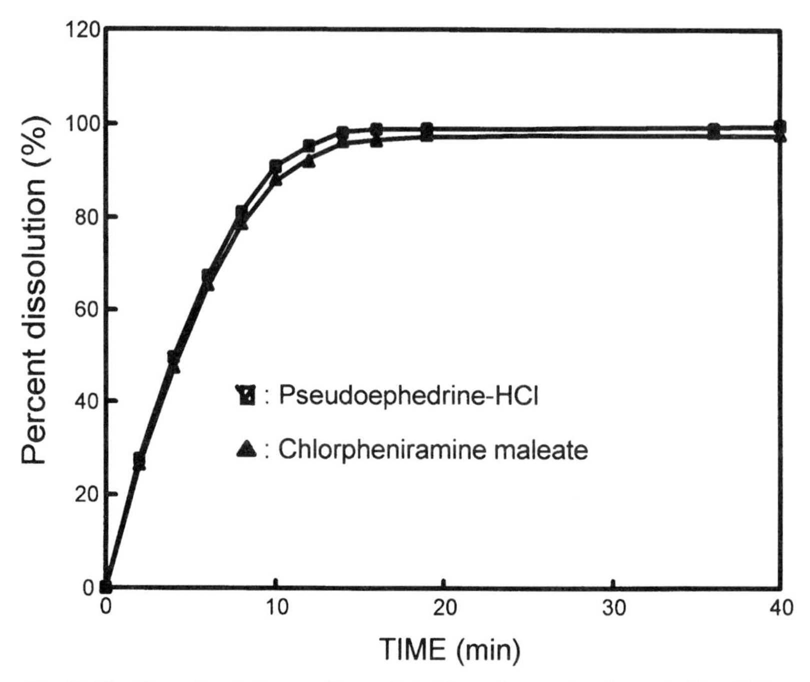

Fig. 10.13. Drug dissolution profile predicted from the spectra shown in Fig. 10.11.

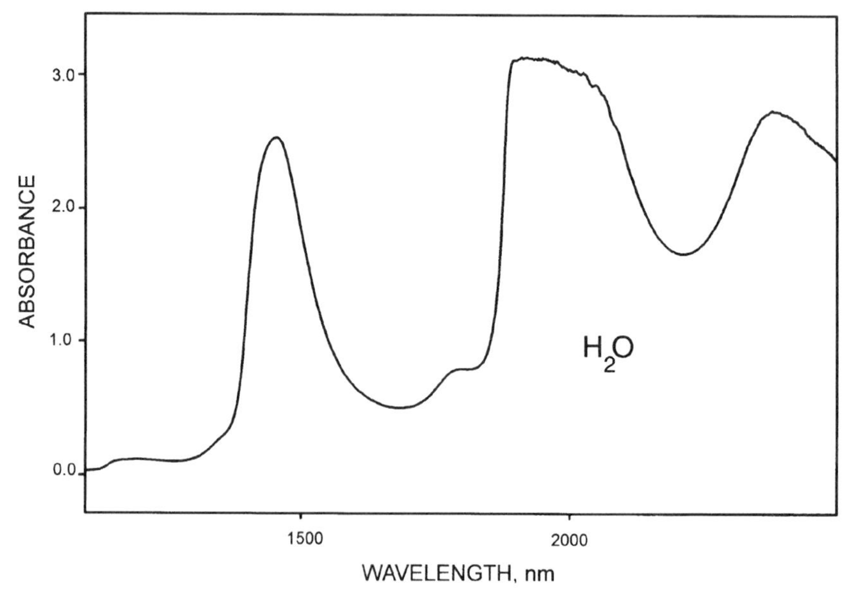

Fig. 10.14. Near IR-spectrum of water.

Water Analysis in the Near-IR

At the present time the near-IR is the optimum spectral region for using fiber optics for remote monitoring. The combination of high-throughput fibers and high-information content makes the near-IR the best region. This will undoubtedly change as fibers with higher throughput becomes available in the mid-infrared region which has a higher-information content. To demonstrate the potential of near-IR for remote monitoring, we will consider the effects of temperature and electrolytes on the spectrum of water. Water is strongly hydrogen bonded, which is very evident in the spectrum. Changes in temperature [11] and electrolytes [12] can alter the water spectrum; both affect hydrogen bonding, and electrolytes also interact with the water molecules, which causes addition changes in the spectrum.

The near-IR spectrum of water is shown in Fig. 10.14. The difference spectra of water at various temperatures minus the spectrum at 5°C are shown in Fig. 10.15. The water band centered at 1404 nm shifts to shorter

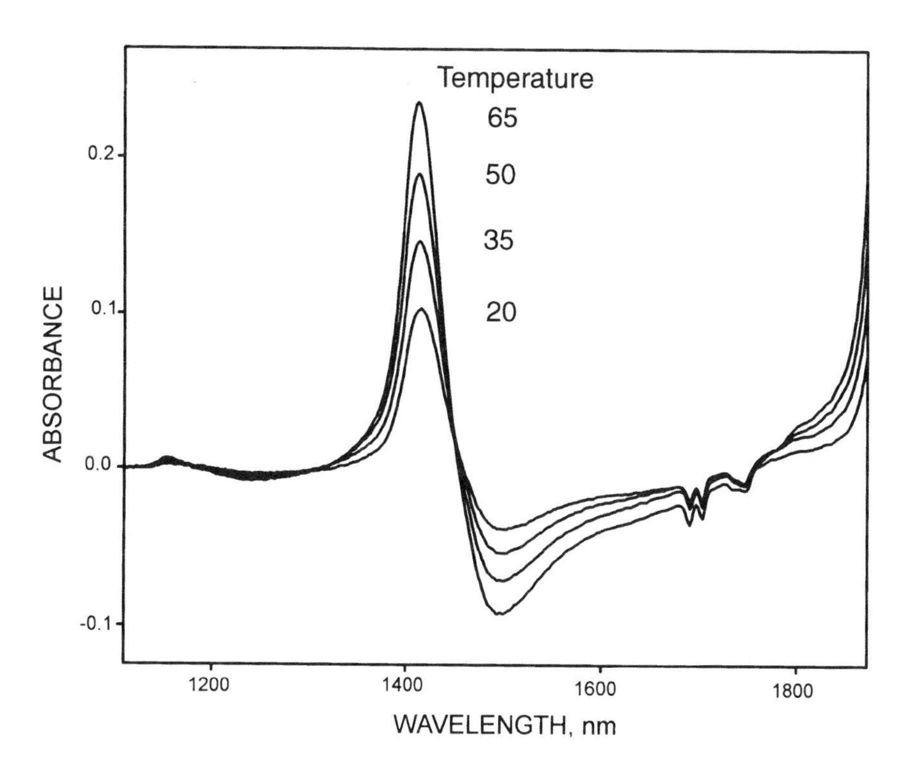

Fig. 10.15. Near-IR difference spectra of water as a function of temperature. The background spectrum was measured at 5°C.

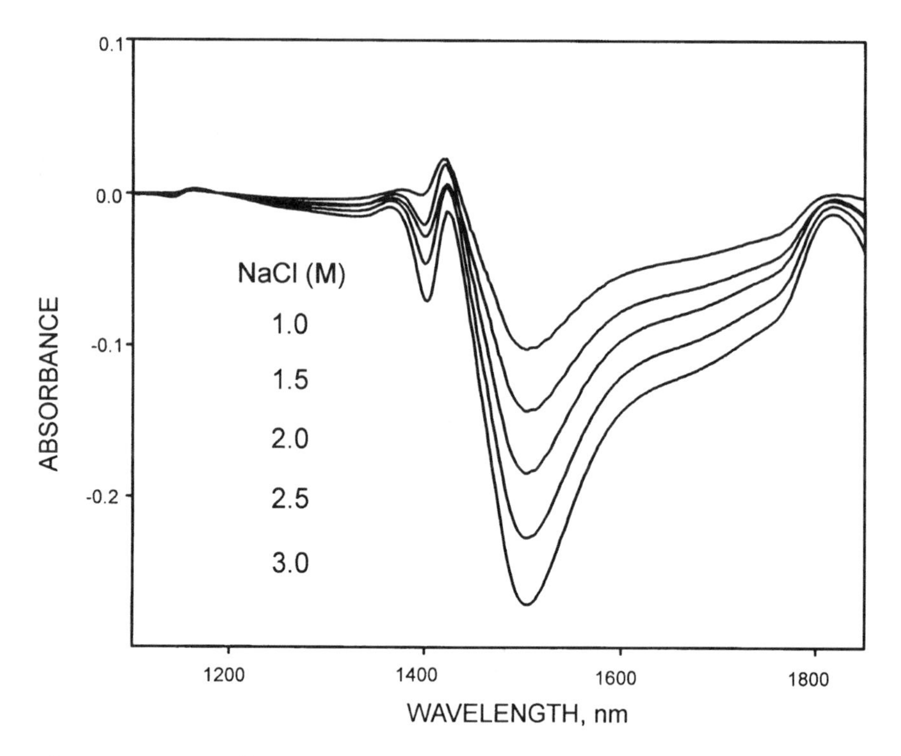

Fig. 10.16. Near-IR difference spectra of aqueous NaCl solutions using the spectrum of water at the same temperature as the background.

wavelengths as the temperature increases, and the spectral changes can be used to predict the temperature to better than ±0.25 over a temperature range of 5 to 85°C.

Addition of electrolytes to water can change the wavelength [12], the intensity and the shape of the absorption bands due to water. The overall changes in the water spectrum are relatively small, but these differences as shown for NaCl in Fig. 10.16 can be used to quantitate the amount of electrolyte. Over a concentration range of 0 to 3.5 M the concentration of NaCl can be predicted to ±0.015 in aqueous solution containing only NaCl. In solutions containing 0 to 1.0 M NaCl and 0 to 1.0 M HCl, the concentration of NaCl can be predicted to ±0.010 and HCl to ±0.004. In solutions containing 0 to 0.7 M of NaCl–$NaHCO_3$–Na_2CO_3 in various combinations of concentrations, the three components can be predicted to ±0.008, ±0.012, and ±0.005, respectively.

Various electrolytes affect the near-IR spectrum of water in different ways, and these differences can be used to identify the electrolytes present in

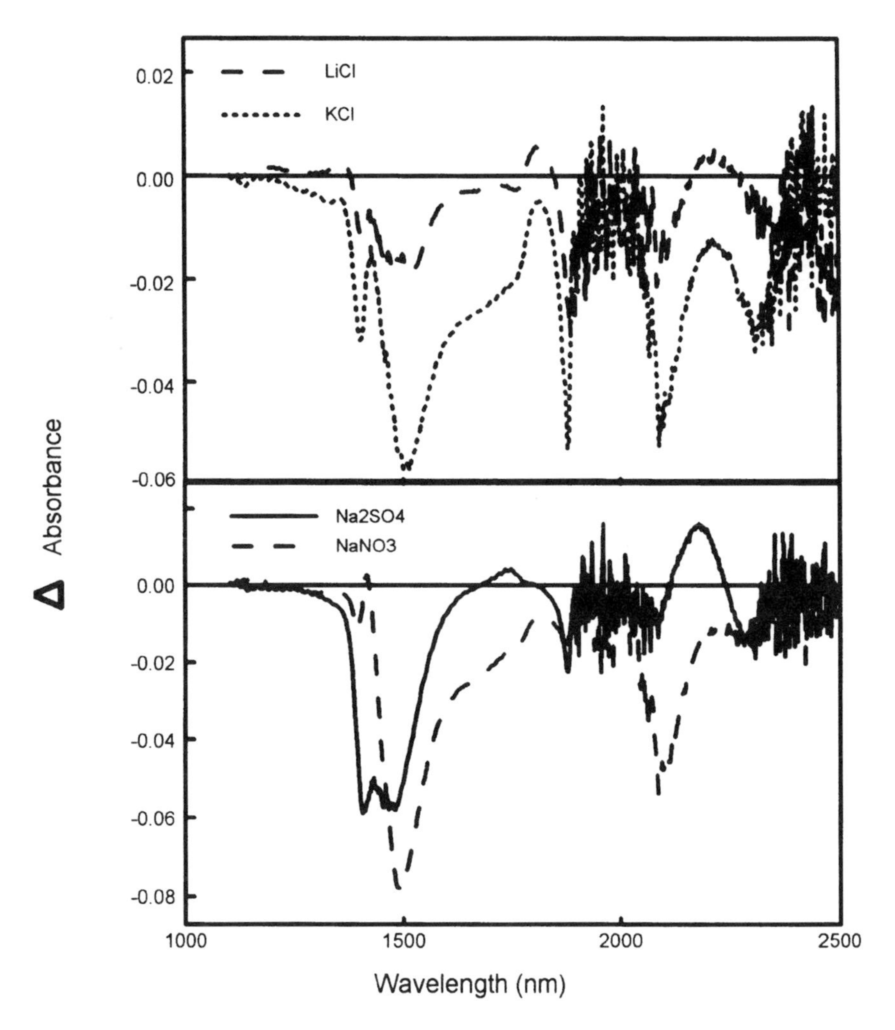

Fig. 10.17. Near-IR spectrum of LiCl, KCl, Na_2SO_4, $NaNO_3$ in water at 0.5 M concentrations.

the solutions [13]. The difference spectra of four randomly selected electrolytes are shown in Fig. 10.17. A near-IR spectral library of 71 solutions of single electrolytes was generated. Sixteen unknown solutions of single electrolytes were prepared. Their spectra were measured and the library was searched using a novel algorithm for mixtures. The algorithm identified 10 of the 16 unknowns correctly, and one of the ions (cation or anion) was correctly identified for the remaining 6 samples. For eight unknown mixtures of two electrolytes, both electrolytes were correctly identified in four of the mixtures and one electrolyte was correctly identified in three of the other four mixtures, while three of the four ions were correctly identified in the fourth mixture. For the total of 16 single component and 8 two component mixtures, 55 of the ions out of a total of a possible 64 were correctly identified.

Gas Analysis in the Near-IR

The hydrocarbons present in natural gas have reasonably strong absorptions in the near-IR region. Since natural gas transmission lines often operate at pressures up to 1000 psia in rather hostile locations, fiber optics are ideal for measuring spectra remotely. It is possible to have a spectrometer in a laboratory and measure spectra on-line using fibers to carry the analytical signals. Preliminary work on using fiber optics for remote monitoring of natural gas were performed in the laboratory on a research grade FTIR [6]. A typical spectrum of natural gas at 100 psia is shown in Fig. 10.18. From these high-resolution spectra it was determined that similar quantitative results could be obtained using much lower resolution [14].

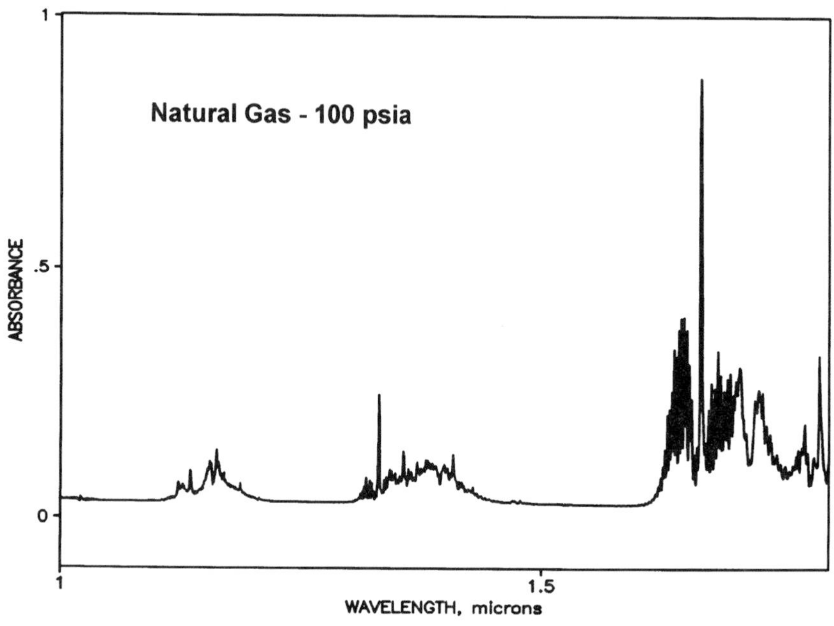

Fig. 10.18. Near-IR spectrum of a natural gas sample at 100 psia.

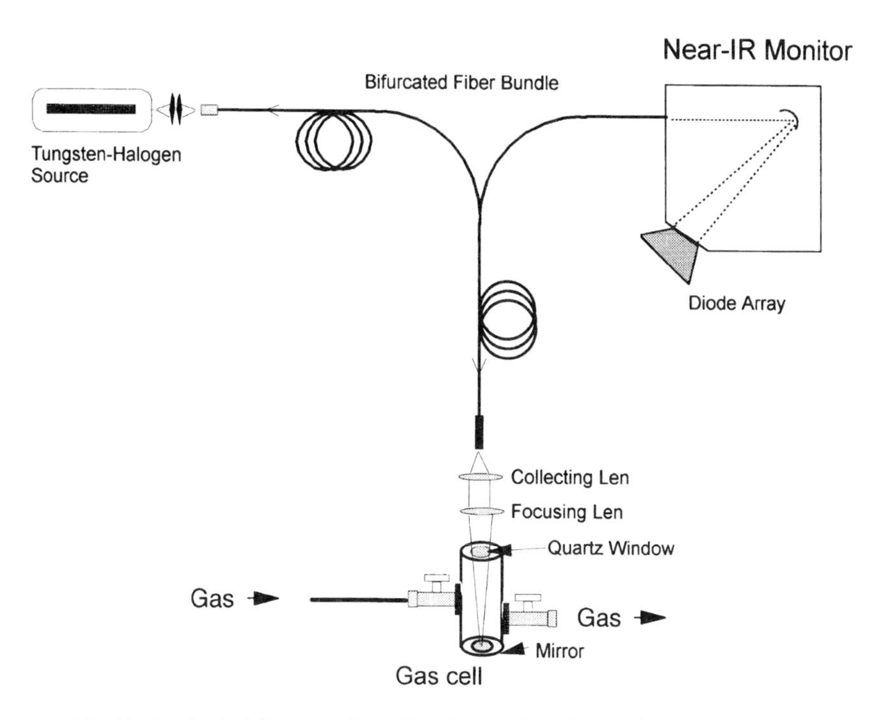

Fig. 10.19. Optical diagram of near-IR BTU monitor showing interface to gas cell.

Thus a minimum resolution instrument with a 32 element array detector was assembled. A schematic of this instrument is shown in Fig. 10.19. The instrument was tested first under laboratory conditions and later in the field at a gas transmission pumping station [15]. The results in terms of predicted BTU from monitoring every 20 minutes over a 12-hour period are compared to gas chromatographic results on the same gas line as shown in Fig. 10.20. The optical BTU monitor was able to operate without an operator for two months and measure BTU and composition values in real time; the measurement and calculation cycle required one minute. This same system is currently being considered as a permanent monitoring system.

Lamination Monitoring in the Mid-IR

Often times lamination of polymeric materials are performed in an autoclave or similar device in which it is difficult to monitor the changes taking place in the material. As a consequence it is very difficult to determine the optimum conditions for lamination. Typically procedures are developed by

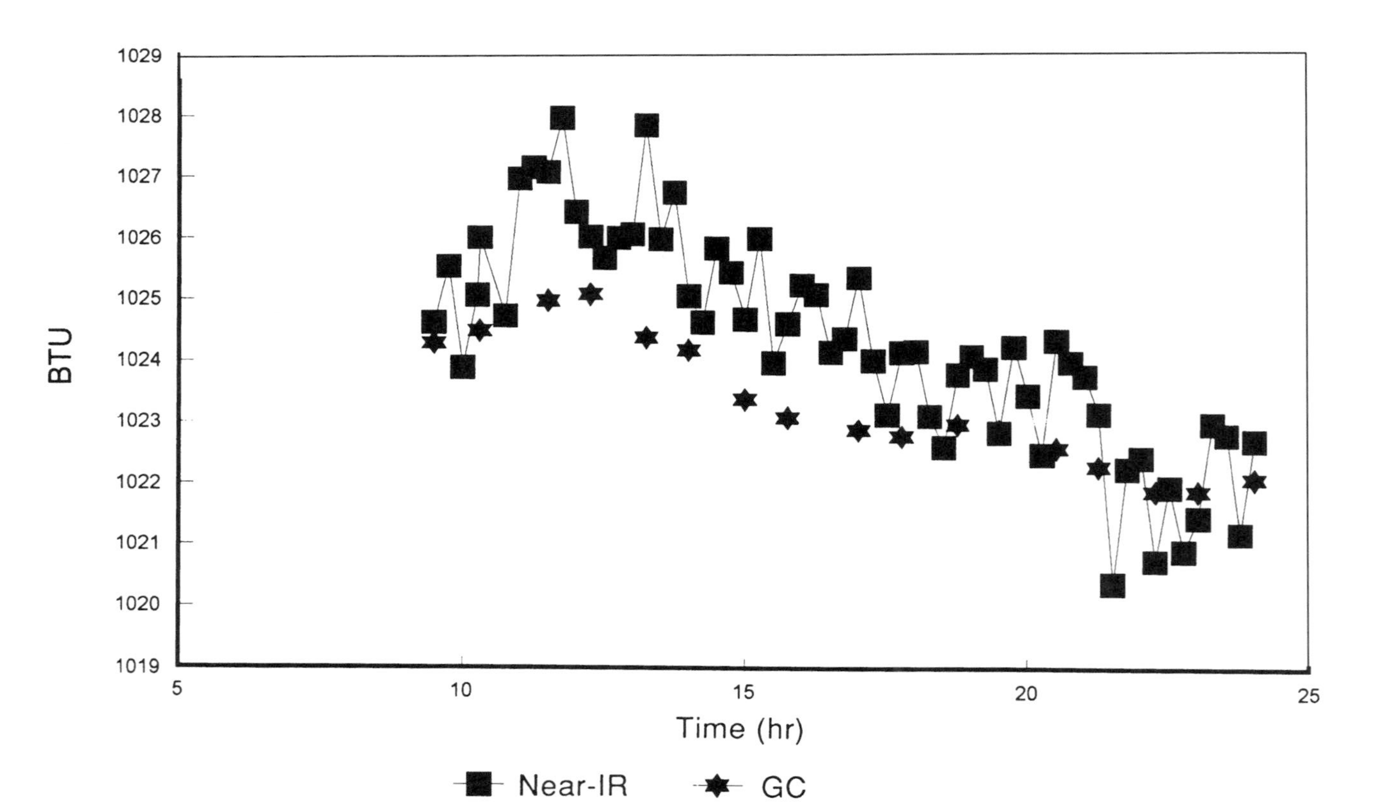

Fig. 10.20. Near-IR BTU values compared with gas chromatographic values during on-line experiments.

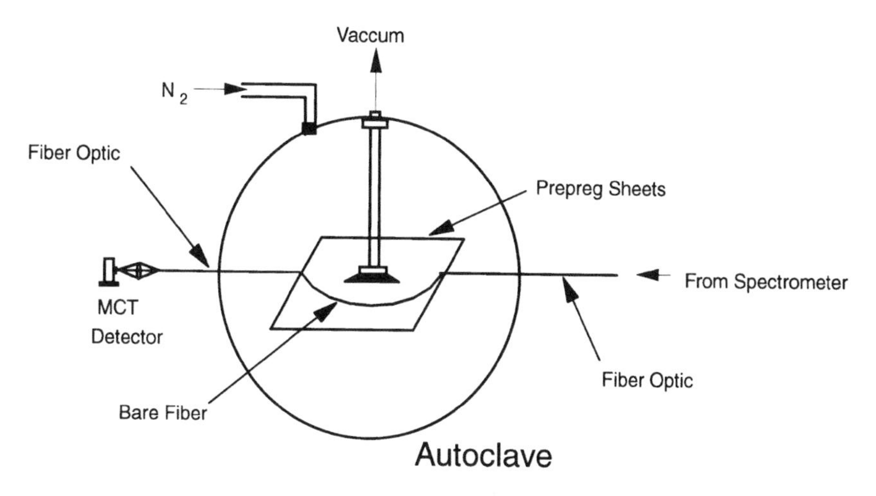

Fig. 10.21. Diagram of autoclave with fiber optic passing through prepreg lamination material.

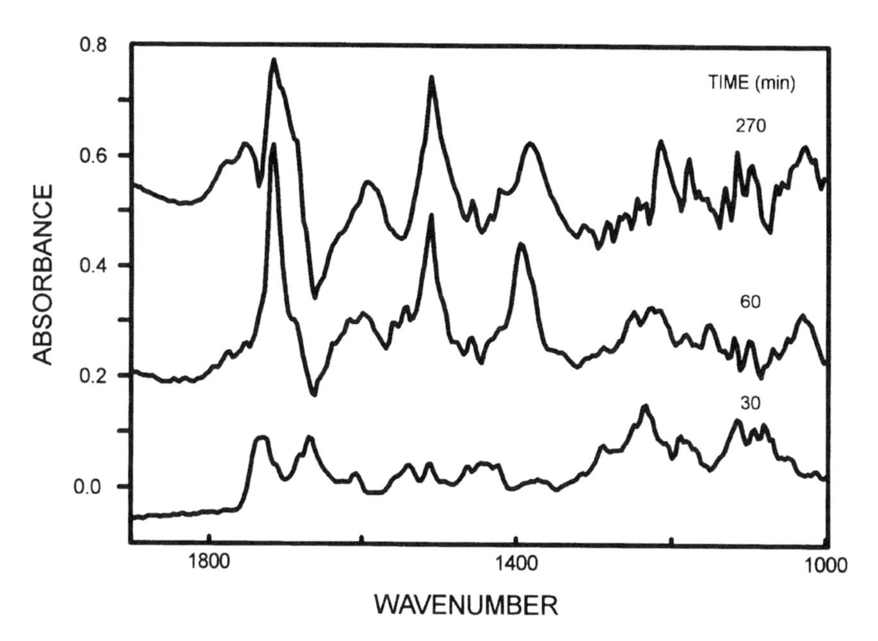

FIG. 10.22. Mid-IR spectra of polyimide prepreg during the lamination process at 30, 60, and 270 minutes.

trial and error, and there is no real scientific basis for increasing temperatures and/or pressures. In an effort to understand the changes taking place during the lamination of prepreg materials (partially polymerized materials), a bare chalcogenide fiber was placed between sheets of the material inside of a lamination package [16]. The package was placed inside of a laboratory autoclave in such a way that the input end of the fiber was fitted through a Swagelok fitting on one side and interfaced to an interferometer; the output of the fiber was passed through a similar fitting on the other side of the autoclave, and the light was focused onto the detector. A schematic of the system is shown in Fig. 10.21. The evanescent wave in the bare fiber passing through the prepreg material was attenuated at wavenumbers corresponding to the material. Example spectra of a polyamide prepreg during the lamination process at 30, 60, and 270 minutes are shown in Fig. 10.22. For this particular prepreg it was shown that the spectra did not change extensively after about 150 minutes. This would seem to indicate that the lamination under the prescribed conditions of temperature and pressure was constant after that time period, and that the process could be concluded. In reality it would be necessary to test the other physical parameter of that particular specimen and to correlate these parameters with the spectra. However, it is obvious that the evanescent wave spectra can be used to monitor such a system and to provide a scientific basis for the process.

10.6. CONCLUSION

Numerous applications for fiber optics have already been found in the near-IR. Robust, transparent fibers are available for this region, and many materials have characteristic absorptions in this spectral region. Fibers are still rather questionable for the UV region, although they have been improved in the last couple of years. Fibers for the visible are very robust and have high throughput, but the region lacks information content. Ideally the mid-IR region offers the greatest information content. However, fibers for this region can be brittle such as ZrF and chalcogenide, and all are expensive and have limited throughputs. Nevertheless, these latter problems are quickly improving and less-expensive fibers with greater throughput are being developed. Eventually there will be inexpensive fibers with high throughputs available for all spectral regions.

REFERENCES

1. C. W. Brown, S. M. Donahue, and S.-C. Lo, In *Advances in Near-Infrared Measurements*, G. Patonay, ed. JAI Press, Greenwich, CT, 1993.
2. D. A. Krohn, *Fiber Optic Sensors—Fundamentals and Applications*. Instrument Society of America, Research Triangle Park, 1988.

3. M. J. Webb, *Spectrosc.*, **4**(6), 26 (1988).
4. G. C. Burke, *Adv. Instrument.*, **43**(2), 841 (1988).
5. J. Lin and C. W. Brown, *Anal. Chem.*, **65**, 287–292 (1993).
6. C. W. Brown and S.-C. Lo, *Appl. Spectrosc.*, **47**, 812 (1993).
7. D. A. Landis and C. J. Seliskar, *Appl. Spectrosc.*, **49**, 547 (1995).
8. C. W. Brown and J. Lin, *Appl. Spectrosc.*, **47**, 615 (1993).
9. C.-S. Chen and C. W. Brown, *Pharmaceutical Res.*, **11**, 979 (1994).
10. P. K. Aldridge, D. W. Melvin, and S. S. Sekulic, *J. Pharm. Sci.*, **84**, 909 (1995).
11. J. Lin and C. W. Brown, *Appl. Spectrosc.*, **47**, 62 (1993).
12. J. Lin and C. W. Brown, *Anal. Chem.*, **65**, 287 (1993).
13. J. Lin, J. Zhow, and C. W. Brown, *Appl. Spectrosc.*, **50**, 444 (1996).
14. C. W. Brown, *Appl. Spectrosc.*, **47**, 619 (1993).
15. C. W. Brown, *Optical BTU Sensor Development*. Final report, Gas Research Institute, GRI-93/0083, 1993.
16. Z. Ge, C. W. Brown, and M. Brown, *J. Appl. Poly. Sci.*, **56**, 667 (1995).

INDEX